# Inuit, Whalers, and Cultural Persistence

# Inuit, Whalers, and Cultural Persistence

Structure in Cumberland Sound
and Central Inuit Social Organization

Marc G. Stevenson

OXFORD UNIVERSITY PRESS
Toronto   New York   Oxford
1997

Oxford University Press
70 Wynford Drive, Don Mills, Ontario M3C 1J9

*Oxford New York*
*Athens Auckland Bangkok Bombay*
*Calcutta Cape Town Dar es Salaam Delhi*
*Florence Hong Kong Istanbul Karachi*
*Kuala Lumpur Madras Madrid Melbourne*
*Mexico City Nairobi Paris Singapore*
*Taipei Tokyo Toronto*

and associated companies in
*Berlin Ibadan*

*Oxford* is a trademark of Oxford University Press

**Canadian Cataloguing in Publication Data**

Stevenson, Marc
   Inuit, whalers, and cultural persistence : structure
in Cumberland Sound and Inuit social organization

Includes bibliographical references and index.
ISBN 0-19-541217-6

1. Inuit – Northwest Territories – Cumberberland Sound – Social life
and customs.   2. Social structure – Northwest Territories – Cumberland
Sound.   3. Inuit – Northwest Territories – Cumberland Sound – Foreign
influences.   I. Title

B99.E7S838 1996      306'.089'97107195      C96-932006-X

Design: Max Gabriel Izod
Compositor: Indelible Ink

Frontispiece: *Whale hunting*, stone cut by the late Nowyook Nicketimoosie,
stencil print by the late Solomon Karpik, Pangnirtung Print Collection, 1975
[Author's Note] The hunters pulling together symbolize *ungayuk*, the
co-operative egalitarian nature of Inuit social organization. The helmsman
symbolizes its structure, order, hierarchy: *naluktuk*. Traditional clothing and
the whaleboat speak of the issues of cultural persistence after the coming of
the whalers.

Copyright © Oxford Unversity Press Canada 1997

1 2 3 4 - 00 99 98 97

This book is printed on permanent (acid-free) paper ∞.

Printed in Canada

# Contents

List of Tables, x

List of Figures, xi

Acknowledgements, xiii

Orthography, xv

Introduction, xvi
    Cumberland Sound, Baffin Island   xvii
    A Whale Hunt   xviii
    The Approach   xxi

## PART ONE
### MODELS AND THEORIES OF INUIT SOCIAL ORGANIZATION

1. 'Eskimo' Type Kinship and Social Organization   2
    The Construction of 'Eskimo' Type Kinship and Organization   2
    Unravelling 'Eskimo' Type Kinship and Social Organization   4
    Kinship: the Foundation of Inuit Social Organization?   8
    Negotiation: the Foundation of Inuit Social Organization?   13
    Composite or Patrilocal Band?   14
    Central Inuit Comparisons: the Search for Structure   16
    Recent Research   23
    Anarchy or Structure: Kinship or Locality?   27

## PART TWO
### THE CUMBERLAND SOUND INUIT IN HISTORY AND PREHISTORY

2. Cumberland Sound Before Qallunaat   32
    The Environment of the Cumberland Sound Inuit   33
        Tides, Rock, Ice, and Climate   33
        Animals Hunted   39
    Reconstructions of Cumberland Sound Inuit Prehistory   45
        An Alternative View   47
    Aboriginal Social Groups and Interaction   50
    Cumberland Sound Inuit Social Organization: Traditional Features   58

        Local Group Size   58
        Leadership   63
        Marriage, Residence, and Descent   64
   A Cumberland Sound Type of Social Organization?   70

## 3. Cumberland Sound Inuit-Qallunaat Interaction to 1970   72
   Precontact, 1820–1840   73
   Early Contact, 1840–1857   74
   Cumberland Sound Whaling, 1857–1870   77
   Economic Diversification: Whaling and Sealing, 1870–1894   82
   A Clash of Ideologies: A New Way of Believing, 1895–1906   88
   Decreasing Expectations, Continuing Adjustments, and General
        Trading: 1906–1921   90
   Pangnirtung and the Hudson's Bay Company, 1921–1962   93
        Economic Concessions and the Fur Trade   93
        Law, Order, and 'No Loitering'   100
   The Approach of Modern Times, 1962–1970   102

## 4. Culture Change and Continuity, 1840–1970   105
   Forces of Production   106
        Changes in Technology and Subsistence   108
        Changes in Organization of Production   110
   Relations of Production   115
        Interregional Group Relations   115
        Leaders and Followers   117
        Epidemics and Social Change?   123
   Christianity and Leadership: a New Order?   125
   Enduring Features of Cumberland Sound Inuit Social Organization   130
        Social Stratification: the Emergence of Class?   130
   Conclusion   139

## PART THREE
### CUMBERLAND SOUND INUIT SOCIAL STRUCTURE

## 5. Cumberland Sound Inuit Kin and Local Groups, 1920–1970   142
   Methodology and Presentation   143
   Qikirtarmiut Settlements   145
        Nunaata   145
        Idlungajung   155
        Avatuktoo   167
        Tuapait   172

　　　　Sauniqtuajuq   174
　　　　Naujeakviq   185
　　　　Other Qikirtarmiut Camps   189
　　Umanaqjuarmiut Settlements   190
　　　　Ussualung   190
　　　　Iqalulik   196
　　　　Kingmiksoo   201
　　　　Opinivik   210
　　　　Kipisa   214
　　　　Illutalik   220
　　　　Other Umanaqjuarmiut Camps   223

6. **The Structure of Cumberland Sound Inuit Social Organization   225**
　　Structural Tendencies   226
　　　　Leadership   233
　　　　Group Size, Residential Stability, and Individual Mobility   235
　　　　Territoriality   238
　　　　Marriage Patterns   240
　　　　　Marrying Out   240
　　　　　Marrying Up   243
　　　　　Marital Residence   243
　　　　Adoption   244
　　　　Caching and Sharing   245
　　　　Kinship   246
　　*Naalaqtuq* and *Ungayuq* Social Structure   247

7. **Cumberland Sound Inuit Prehistory Revisited   250**
　　The Talirpingmiut and Kinguamiut   250
　　　　Leadership   250
　　　　Local Group Size   251
　　　　Core Group Structure   251
　　　　Productive Activity and Relationships   253
　　The Kingnaimiut: Big Groups, Big Men, Big Problems?   255
　　　　Local Group Size   255
　　　　Conflict and Contradiction in Kingnaimiut Society?   255
　　　　Productive Forces and Relationships   256
　　Tasaiju: 'the Odd Site Out'   257
　　　　Communal Houses: a Unique Occurrence?   257
　　　　History Repeats Itself?   258
　　Environment and Society in Cumberland Sound   259

# PART FOUR
## Central Inuit Social Structure

**8. Iglulingmiut and Netsilingmiut Social Organization** 266
   The Iglulingmiut: Coping with *Naalaqtuq* 266
      History of Contact 266
      Culture Change and Continuity 270
      Enduring Features of Iglulingmiut Sociopolitical Organization 272
         Hierarchy 273
         Solidarity 274
      Structural Problems and Solutions in Iglulingmiut
        Socioeconomy 276
      Iglulingmiut/Qikirtarmiut Comparisons 278
   The Netsilingmiut: *Ungayuq* Intensified 279
      History of Contact 279
      Culture Change and Continuity 280
      Traditional Features of Netsilingmiut Social Organization 282
         Closed Groups 282
         Cousin Marriage 284
         Female Infanticide 285
         Closed Groups and Ecological Necessity 287
      Contradiction and Integration in Netsilingmiut Society 288

**9. The Copper Inuit: Antithesis of Central Inuit Social Structure?** 289
   History of Contact 289
      Nineteenth Century Contact and Cultural Change 289
      The Contact-Traditional Period and Cultural Change 291
   Features of Traditional Copper Inuit Society 295
      Economy 295
      Kinship and the Nuclear Family 296
      Voluntary Alliances 297
      Individual Mobility and Group Membership 298
      Mythology, Ideology, and Diametric Dualism 299
      Female Infanticide and Seal Sharing 300
   Breaking the Rules: a Rejection of Central Inuit Social Structure
      and Ideology 301
      Reconciling Theories 302
      Environment and Society in Copper Country 302
   Conclusion 304

## PART FIVE
### IN CONSIDERATION OF CENTRAL INUIT SOCIAL STRUCTURE

**10. Canadian Arctic Prehistory Reconsidered** 308
  Thule Inuit Out of Alaska 309
    Speculations on the Initial Thule Expansion 310
    A Second Thule Wave 315
    A Third Expansion 316
  Copper, Netsilik, and Caribou Inuit Origins 318
    Late Precontact Central Arctic Population Movements and
      Their Causes 321
  Copper Inuit and Paleoeskimo Affinities 325
  Conclusion 326

**11. Central Inuit Social Structure and Kinship Theory** 327
  Complex Structures 328
    Variations on a Theme 329
    Consanguineal Solidarity and Organization 331
  Conclusion 334

**12. The Politics of Survival: Central Inuit Social Structure and Nunavut** 335
  Barriers to Survial 336
  A Proposal for a Nunavut Goverment 336
  A Missing Link 337
  Other Solutions 338
  Conclusion 340

**Bibliography** 341

**Glossary of Common Inuktitut Terms Used in Text** 359

**Glossary of Common Anthropological Terms Used in Text** 362

**Notes** 365

**Index** 383

# List of Tables

1. Comparisons of three Central Inuit regional groups.  19
2. Occurrence of same generation and intergenerational relationships between nuclear family heads in 'multi-family' households at Kingmiksoo, fall 1846.  69
3. Number of ships wintering in Cumberland Sound, 1851 to 1880.  78
4. Prices of Arctic whale oil and 'whalebone' on American markets, 1868–1880.  83
5. Cumberland Sound and Davis Strait census by Boas, December 1883.  84
6. Returns of whales and seals from Noble's station at Kekerten and Umanaqjuaq, 1883–1903.  87
7. White whale and ringed seal returns, Pangnirtung Post, 1923–1940.  87
8. Populations of Cumberland Sound settlements, 1923–1936.  99
9. Populations of Cumberland Sound settlements, 1944–1966.  102
10. Pangnirtung seal, whale, fox, and other returns.  103
11. Inventory of hunting equipment, Cumberland Sound, summer 1966.  109
12. Heads of whaleboat crews participating in HBC white whale drive prior to 1928 and between 1930 and 1932.  114
13. Structure and strength of kinship ties among Qikirtarmiut and Umanaqjuarmiut central group cores.  228
14. Umanaqjuarmiut and Qikirtarmiut differences in structure of primary kinship ties among central core adults/families.  229
15. Umanaqjuarmiut and Qikirtarmiut differences in strength and structure of primary kinship ties among central core adults/families.  230
16. Umanaqjuarmiut and Qikirtarmiut differences in strength of primary kinship ties among parent-child cores with resident adult sons and/or children of both sexes present.  231
17. Marriage arrangements between Umanaqjuarmiut and Qikirtarmiut.  241
18. Local group endogamy, marriages resulting in multiple affinal ties among co-resident consanguines, and hypergamy within Qikirtarmiut and Umanaqjuarmiut camps.  242
19. Adoptions among Umanaqjuarmiut and Qikirtarmiut.  245
20. Summary of Qikirtarmiut and Umanaqjuarmiut structural tendencies.  247
21. Occurrence of single, double, and triple platform/room dwellings at Anarnitung and Kingmiksoo.  252
22. Occurrence of triple platform dwellings at Tasaiju and three other major Kinguamiut sites.  258
23. Iglulingmiut, Netsilingmiut, and Copper kinship terminology for Ego's and first ascending and first descending generations, male speaking.  275

# List of Figures

1. Leader or *angajuqqaq* of a recent whale hunt in Pangnirtung Fiord.   xix
2. Cumberland Sound cousin terminology for male Ego, early 1860s (from Morgan 1870).   3
3. Iglulingmiut cousin terminology, male Ego.   9
4. Naalaqtuq and Ungayuq directives for Iglulingmiut male.   11
5. Map of eastern Canadian Arctic and Baffin Island.   34
6. Map of Cumberland Sound with selected geographical place names.   35
7. Early and late winter ice conditions in Cumberland Sound.   38
8. Map of Cumberland Sound. Drawn by Eenoolooapik and associates for W. Penny in 1839, showing locations of settlements and principal whaling and hunting grounds.   41
9. Map of Cumberland Sound with selected historical and archaeological sites.   51
10. Site plans of five late prehistoric winter villages in Cumberland Sound.   57
11. Modern and 1839 maps of Kingmiksoo.   61
12. Boas' 'tribal' divisions on sub-regional groups superimposed over Eenoolooapik's 1839 map of Cumberland Sound.   62
13. Cumberland Sound cousin terminology for female Ego, *c.* 1860.   67
14. Graphic depiction of household census of Kingmiksoo, in the fall of 1846.   68
15. Constituent components of a social formation and their systemic relationships.   107
16. Areas and animals exploited by Inuit at Kekerten around 1918.   111
17. Present-day Cumberland Sound Inuit kinship terminology, Ego's and first ascending generations (female Ego).   136
18. Locations of contact-traditional period settlements.   146
19. Social composition of Nunaata camp during the mid-1920s.   148
20. Map of Nunaata, mid-1920s.   148
21. Social composition of Nunaata during the early 1940s.   151
22. Plan of Nunaata, 1940–1942.   151
23. Social composition of Nunaata during the early 1960s.   153
24. Angmarlik and Jim Kilabuk standing in front of the last bowhead taken by Angmarlik, Kingua Fiord, August 1945.   155
25. Social composition of Idlungajung during the early 1920s.   157
26. Plan of Idlungajung, 1922–1924.   157
27. Social composition of Idlungajung during the mid-1930s.   161
28. Plan of Idlungajung during the mid-1930s.   161
29. Social composition of Idlungajung during the mid-1940s.   164
30. Plan of Idlungajung during the mid-1940s.   165
31. Social composition of Avatuktoo during early 1940s and late 1950s.   170

xii    List of Figures

32. Plan of Avatuktoo during the early 1940s.   171
33. Plan of Avatuktoo during the late 1950s.   171
34. Social composition of Tuapait during the early 1960s.   173
35. Plan of Tuapait during the early 1960s.   173
36. Social composition of Sauniqtuajuq in 1923–1925.   175
37. Plan of Sauniqtuajuq, 1923–1925.   175
38. Social composition of Sauniqtuajuq during the late 1930s.   180
39. Plan of Sauniqtuajuq in the late 1930s.   180
40. Social composition of Sauniqtuajuq around 1950.   182
41. Plan of Sauniqtuajuq around 1950.   182
42. Social composition of Naujeakviq during the mid-1930s and mid-1950s.   186
43. Plan of Naujeakviq during the mid-1950s.   186
44. Social composition of Ussualung during the early 1920s.   192
45. Plan of Ussualung during the early 1920s.   192
46. Social composition of Iqalulik during the early 1930s.   197
47. Social composition of Iqalulik during the late 1950s.   200
48. Plan of Iqalulik, late 1950s.   200
49. Social composition of Kingmiksoo around 1926–1928.   204
50. Plan of Kingmiksoo, c. 1927–1929.   204
51. Social composition of Kingmiksoo during the mid- to late 1930s.   207
52. Plan of Kingmiksoo, c. 1936–1938.   208
53. Social composition of Opinivik around 1926–1928 and 1937–1939.   212
54. Plan of Opinivik during the late 1930s.   212
55. Social composition of Opinivik during the mid-1950s.   213
56. Social composition of Kipisa during the late 1930s.   215
57. Plan of Kipisa during the late 1930s.   215
58. Social composition of Kipisa during the late 1940s.   217
59. Plan of Kipisa during the late 1940s.   217
60. Social composition of Kipisa during the late 1950s.   219
61. Plan of Kipisa, late 1950s.   219
62. Social composition of Illutalik during the early and late 1930s.   221
63. Social composition of Illutalik during the early 1950s.   222
64. Plan of Illutalik during the early 1950s.   222
65. Populations of Idlungajung and Kingmiksoo between 1923 and 1944.   237
66. Winter hunting areas in Cumberland Sound in 1965–1966.   254
67. Whaling scene on ivory bow drill recovered from LlDj-1, near Imigen Island.   254
68. Distributions of Iglulik, Netsilik, and Copper Inuit.   267
69. Schematic of interrelationships among Netsilingmiut socioeconomic features.   288
70. Four fundamental social structures.   328

# Acknowledgements

This book is a revised version of my doctoral dissertation, *Central Inuit Social Structure: The View from Cumberland Sound, Baffin Island, Northwest Territories* (Stevenson 1993). To all those individuals and institutions who made this work possible, I wish to express my sincere gratitude. If I have missed someone, it is because of oversight, not intention.

My 1989 field research in Cumberland Sound was funded by the Canadian Circumpolar Institute (then, the Boreal Institute for Northern Studies), University of Alberta, and the Department of Culture and Communications, Government of the Northwest Territories (GNWT). For this support, I am truly grateful. The Social Sciences and Humanities Research Council of Canada through their Doctoral Fellowship Award Program provided financial support during the course of my doctoral program, as did the University of Alberta in the form of the Andrew Stewart Prize, the Graduate Fellowship Award, Graduate Faculty Scholarships, and the University of Alberta Doctoral Dissertation Award. To each of these institutions I extend my appreciation.

My research in Cumberland Sound began in 1983 while employed with the Prince of Wales Northern Heritage Centre, and I wish to thank Drs Robert Janes and Chuck Arnold, for their support of and interest in my work. This research continued under contract to the Department of Economic Development and Tourism, GNWT, in the context of developing Kekerten Historic Park and the Angmarlik Cultural Centre. Katherine Trumper, Gary Magee, and Dave Monteith deserve considerable credit for believing in these projects and 'going to bat' for both them and me on various occasions. Dave took over from Gary, and his ongoing interest and understanding of the issues, problems, and solutions as well as willingness to share ideas made my time in Cumberland Sound especially memorable. Research in the south Amundsen Gulf region was carried out under the auspices of Environment Canada, and I want to thank Gordon Hamre, Tony Green, and Aimé Ahegoona for making this experience so rewarding. I am also grateful to those individuals in Paulatuk and Coppermine who allowed me into their homes and shared their time and knowledge.

Over the half dozen years I was actively involved with the people of Pangnirtung, many individuals contributed to my research endeavours and assisted me in a variety of ways. As my main guides, Joavee Alivaktuk and Kaneea Etuangat were without parallel, as were my major field assistants and interpreters Margret Karpik, Meeka Kilabuk, Meeka Mike, Koni Alivaktuk, Sara Tautuajuk, July Papatsie, Ami Papatsie, Simionee Akpalialuk, and Moe Keenainak. Most of all I would like to thank those Inuit elders who shared their personal experiences and histories with me over the years. I am especially grateful to the late Etuangat Aksayuk, who was more than a friend, teacher, and mentor to me, and the late Qatsu Eevic for making my research so unforgettable and personally rewarding, *qujannamik!*

For their assistance during the archival phase of my research, I would like to thank the staffs of the Hudson's Bay Archives in Winnipeg and the Anglican Archives in Toronto, as well as Anne Keenleyside and Dr Philip Goldring. Phil deserves special recognition, for, in addition to sharing ideas and information during the formative period of my work in Cumberland Sound, he greatly facilitated archival investigations by allowing me access to his collection of RCMP and medical records. I also want to thank Elaine Maloney for her patience and assistance in the reproduction of this manuscript.

The ideas, approaches, and perspectives advanced and developed throughout this study owe much to my dissertation committee members, Drs Michael Asch, Milton Freeman, and Cliff Hickey, though convention dictates that I alone must assume responsibility for the former. Courses from and discussions with Michael Asch sharpened my thinking about northern hunter-gatherer socioeconomic organization, particularly notions regarding mode of production and dialectical materialism. At the same time, Milton Freeman freely shared information, materials from his extensive library, and most of all his personal experiences and perspectives on Inuit society with me. I do not really know how or where to begin to thank Cliff Hickey. For several years we worked closely together, and I would especially like to thank him for his sage advice and direction as well as his support of and contribution to the ideas expressed throughout this thesis. The open and stimulating discussions in which we frequently engaged will not soon be forgotten. Discussions with Dr D. Cole, Dr R. Darnell, Dr J. Ives, Dr M. Magne, K. Morris, and Dr D. Stenton have also contributed to this work. For pointing out inconsistencies in Inuktitut orthography and for his time and patience as my external examiner, I would like to thank Dr N. Graburn. Finally, I owe an intellectual debt of gratitude to Dr D. Damas for his pioneering research into Central Inuit social structure.

Very special recognition is reserved for my wife Kathie and my children Saara and Ben, to whom this book is dedicated. While Kathie drew the maps contained in this paper, Saara and Ben demonstrated patience beyond their years. For their love and understanding, while persevering through what must have seemed an eternity of misdirected goals and questionable behaviour, *qujannamik!*

# Orthography

The Inuit words used throughout this book endeavour to follow the standardized orthography developed by Gagné (1961) and adopted by Spalding (1979), among others. At the same time, however, out of respect for local preferences for the alphabetic spellings of various personal and place names, some words in this study are not consistent with this orthography.

Prior to attempts at standardization, names, places, objects, actions, etc. often were spelled inconsistently, even by the same writer. One of the greatest sources of this confusion was and is that Inuktitut contains only three major vowels, *i, a, u*, while English (Qallunaatitut) possesses five: *a, e, i, o, u*. Another and perhaps the greatest phonemic distinction is that, while Inuktitut distinguishes between the velar k and uvular q sounds, Qallunaatitut does not have the latter. Consequently, many spellings, while questionable, if not technically incorrect, have been 'captured' into local and official use. For example, Tooloogakjuaq, an historic leader and lay preacher, is the popular spelling of the more grammatically correct Tulugaqjuaq. Similarly, whereas Pannirtuuq is the standardized spelling of the settlement where most Cumberland Sound Inuit now reside, bilingual locals prefer to use the historical spelling, Pangnirtung, which originated with Boas (1964).

Having noted why some personal and place names deviate from Gagné's orthography, it is important to point out that these are the exceptions—the spellings of most words conform with accepted orthography. For easy reference glossaries of common Inuktitut terms and of anthropological terms used in this book are provided at the end of this study.

# Introduction

In most descriptions of Canadian and Alaskan Inuit the environment is seen as the major, and sometimes only, factor conditioning social organization. The more productive the environment, the more complex the social organization, which also encompasses economic and political dimensions. However, as Fienup-Riordan (1983: xi) observed, the 'ability to survive in a frigid and inhospitable environment has often been emphasized . . . to the exclusion of a comprehensive account of the value system that makes such survival meaningful'.

It is easy to see why Arctic anthropologists have idealized Inuit survival ability; it is the one aspect of their way of life that is 'most comprehensible in terms of our own cultural system' (ibid.). And I must confess that it was this feature, particularly their elegantly efficient technology, that initially stimulated my interest in the Inuit. Our agricultural traditions and typically western modes of thinking have taught us that the Arctic represents a startling, even terrifying, challenge to the human condition. Cold is to be feared, and it is this apparent atavism that has made such powerful appeals to our imaginations (Brody 1987). The Inuit fascinate us for we wonder how, and rejoice in the fact that, they can eke out an existence under such intolerably harsh conditions.

In conjunction with our view of the Inuit as the quintessential example of 'culture as adaptive response' we have romanticized their relationship with nature. It comforts us to know that somewhere in this increasingly alienating, complex, and fast-paced world there live a people who exist in harmony with nature. Unencumbered by the problems and vexations of modern industrial civilization, life is simpler, more fulfilling, perhaps more meaningful. Yet, when we learn that Inuit buy snowmobiles, rent videos, and eat 'fast food' we become disillusioned, even disparaging, relegating Inuit culture to the past, to dusty museum drawers, to a society that once was but is no more. Consequently, we dismiss modern Inuit as having any efficacy in shaping a cultural identity and destiny distinct from our own. While archaeologists actively avoid sites contaminated by contact with foreigners, anthropologists attempt to reconstruct the 'aboriginal condition'. Recent and modern changes in Inuit society are of little interest.

However, under the acculturated exterior of many contemporary Canadian Inuit communities there still lies a remarkable cultural vitality. This fact did not become apparent to me at first. Rather, it came only gradually after several seasons of research among the Pangnirtarmiut of Cumberland Sound, Baffin Island, Northwest Territories. To me, this vitality was a clue that there was (and is) some underlying structure to Inuit social organization. As my conclusions regarding the structure of Inuit social organization in the Canadian Arctic began to take shape in Cumberland Sound, I would like to relate some of the experiences that served as 'benchmarks' and directed me along the way.

## Cumberland Sound, Baffin Island

As an anthropologist with an interest in hunting and gathering societies, I have always been intrigued by the Inuit. Yet, the opportunity to experience Inuit culture and society first hand presented itself only in 1983 when I was offered a two-year position as the historic archaeologist of the Northwest Territories at the Prince of Wales Northern Heritage Centre in Yellowknife. One of my two primary functions was to assist the hamlet of Pangnirtung in developing the historic Inuit whaling station of Kekerten as a tourist attraction. The seal skin market had just recently collapsed owing to the success of the anti-sealing campaign, and the community was looking for alternative sources of income to support its hunters and hunting economy. After working for Parks Canada in Winnipeg for several years, it was time for a change, and I accepted.

In August of 1983 I arrived at Kekerten (Qikirtan) 100 years almost to the day after Franz Boas, the father of American anthropology, landed at the same spot to conduct his pioneering study of the Central Inuit. The term Cental Inuit is used for the same regional groups that Boas (1964), Damas (1975b), and other anthropologists call the 'Central Eskimo'. These include the Caribou, Copper, Netsilik, Iglulik, and Baffin Island Inuit, as well as the Labrador and east Hudson Bay Inuit—in effect, all historic Inuit populations in Canada, with the exception of the Mackenzie Inuit who are directly related to north Alaskan Inupiat. Over the next two years I was to draw extensively on the knowledge and experiences of several Inuit elders in my efforts to reconstruct the social and occupational history of Kekerten and other historic Inuit settlements in Cumberland Sound. To interview directly the people who actually inhabited the areas and features that were being investigated and excavated was a rare opportunity for an archaeologist, and I seized the moment.

Through many long hours of discussion with elderly Qikirtarmiut, I gradually became aware of what life must have been like at Kekerten during the first few decades of this century. For most informants, especially Etuangat Aksayuk, these were the years they remembered best for there was an integral vitality to life in those days. 'Everybody had things to do' and 'nobody questioned their responsibilities', Etuangat told me. The influence and authority of men of substance, and the respect and reverence given them, particularly the leader of the Qikirtarmiut, Angmarlik, was remarkable for a society where social (and political and economic) relationships were supposed to be egalitarian, or at least so I thought. While most Qikirtarmiut were well enough off, there was also marked material inequity and social inequality. These findings led me to propose in an extremely perfunctory paper (Stevenson 1986), that this social differentiation was formalized into a class system and the result of Inuit participation in the commercial whaling industry of the last century.

That I could have been so naïve amazes me now. Over the next several years, I had the opportunity to continue my research into the cultural history of the Cumberland Sound Inuit. After I left the Prince of Wales Northern Heritage Centre, I was contracted by Economic Development and Tourism, Government of the Northwest Territories (GNWT), to conduct research and assist in the development of a number of cultural tourism attractions in Pangnirtung. Pangnirtung had been especially hard hit by the anti-sealing campaign, and both local Inuit and GNWT officials felt that the development of cultural and historical attractions would (a) provide a viable alternative to wage labour employment for native hunters/guides in order to keep up the hunt, and (b) instill a sense of pride within the community in its unique cultural heritage. Combined with Pangnirtung's natural beauty, these initiatives had a good chance of succeeding.

Two attractions, Kekerten Historic Park and the Angmarlik Cultural Centre in Pangnirtung, were the main focus of my efforts as I continued to carry out archaeological investigations, conduct archival research, and, most importantly, collect oral histories. During the course of my interviews with Pangnirtung's elders two things became apparent: (1) not all Pangnirtarmiut emphasized leadership and 'followership' to the extent that the Qikirtarmiut did—they were simply not that important among the other regional subdivision to have occupied the Sound during the historic period, the Umanaqjuarmiut, and (2) hierarchy continued to characterize social relations at most Qikirtarmiut camps long after commercial whaling ended around 1920. Indeed, as I grew more familiar with the language and the people, I became increasingly aware of the fact that hierarchical relations were still an essential part of social and political reality in Pangnirtung. In particular, elders continued to play important leadership roles and were still shown considerable respect and deference.

## A Whale Hunt

Nowhere did this reveal itself more vividly than during an unplanned whale hunt in Pangnirtung Fiord in August of 1989. As a follow-up to my work in cultural tourism, I was conducting a heritage inventory of Pangnirtung's historic buildings, when the town came alive in a flurry of excited activity. In the midst of their work, whether in construction, municipal service, or arts and crafts, scores of men dropped what they were doing and sprinted to the beach where their boats were secured. A pod of narwhal had just swum by the town towards the head of the fiord, and not wanting to waste this gift of *maqtaak*, the most prized of all 'country foods', all available males in the community (myself included) jumped into boats in hot pursuit.

What immediately followed was something that, in retrospect, I was fortunate to escape unscathed. Nearly a hundred men, half of them with rifles, in 25 or more boats trapped the pod, which I was told numbered around 35, near the head of the fiord. Shots rang out from every direction as boats sped

FIGURE 1. Leader or *angajuqqaq* of a recent whale hunt in Pangnirtung Fiord.

past our bow and stern. After a period of about 40 minutes or so, we spotted several boats on a beach. Here, two whales had been hauled ashore and a dozen or so Inuit were busily cutting *maqtaak* off the whales, helping themselves to small pieces in the process. But this scene was in striking contrast to the apparent chaos that had just transpired. There was order, structure to the events unfolding before my eyes. This was 'anthropology in action'. On the periphery of this activity, passive and observant, teenage boys watched the flensing operations, while gorging on *maqtaak*. Meanwhile their fathers, older brothers, and/or uncles cut off slabs of this delicacy and transported them to the boats. The job of flensing the whales, however, was reserved only for the boat owners. Yet, in the midst of this flensing, a hunter with a captain's hat, the eldest man on the hunt, freely helped himself to the choicest parts of both whales, i.e, the flukes, flippers, and cheeks (Figure 1). This man was probably not the individual who shot the whales, although who could tell. Nor was he the richest or most influential; there were younger men on the hunt who held 'better jobs' and had more political acumen on the hamlet council. He was simply the eldest and no hunter had more life experience, or *isuma*. Here, in the increasingly contradictory and changing world in which the Pangnirtarmiut found themselves, traditional productive relationships and social order were being acted out and reaffirmed through the hunt.

I did not conduct a follow-up on what happened to the *maqtaak* after it arrived in the community, though I wanted to. I was scheduled to take several Inuit elders back to their original camps in order to reconstruct the social composition of these settlements for my doctoral dissertation research. Nevertheless, perhaps more than any other incident, this whale hunt provided a glimpse into how Inuit hunters in Pangnirtung still thought about and acted out their concept of what it means to be allied in productive enterprise. It also demonstrated to me the inherent, if latent, cultural integrity of modern Inuit populations in the face of pervasive acculturative influences of the last few decades—the many external forces to which the Cumberland Sound Inuit were subject during the previous 120 years seem to pale in comparison. Yet, after a quarter century of forced assimilation, traditional productive relationships still found expression.

In the few weeks just prior to undertaking my field research I had finished re-reading David Damas' (1963) seminal work *Iglulingmiut Kinship and Local Groupings: a Structural Approach*. I was convinced that the two dimensions of interpersonal social behaviour which he described, *ungayuq* (closeness-affection) and *naalaqtuq* (respect-obedience), possessed considerable explanatory power, and possibly even held the key to understanding Inuit social organization. In particular, these concepts seemed to provide a viable alternative to the heavy-handed environmental determinism that has increasingly characterized anthropological inquiry and explanation across the Arctic.

It is absurd to think that the environment plays no role in shaping Inuit social organization. People 'gotta eat'. Nevertheless, the structure of productive social, political, and economic relationships—who hunts with whom, how the product of the hunt is distributed, who marries whom, etc.—and how these are given value and meaning is culturally determined, and not as preordained by environmental factors as many might suppose. And while the diversity of Central Inuit cultures may reveal 'less common sense environmental determinism than cultural imagination' (Fienup-Riordan 1983: xi), this imagination

Introduction   xxi

is not boundless. Rather, it operates within parameters, within existing systems of social reproduction. It is these structures which this study seeks to address.

## THE APPROACH

The search for structure in Inuit social organization has been a frustrating quest for Arctic anthropologists. The inability to explain variability in social organization within and between regional groups has led to accommodative arguments, which hold that Inuit social organization is somehow less structured than other preliterate societies, or that the environment, often in combination with poorly understood historical processes, is the ultimate architect of Inuit society. This study, in exploring the structural basis of Inuit social organization, directly challenges the validity of both assumptions.

The search begins in Cumberland Sound, Baffin Island, one of the first regions in Arctic Canada to be studied by ethnographers, but least understood in terms of prehistory and social organization. This study offers a new interpretation of the late prehistory of the Sound, and provides a history of contact period relations. Few other Central Inuit groups experienced as long or as intense an association with Qallunaat as did the Cumberland Sound Inuit. Qallunaat is the Inuktitut word for people of European descent. The word does not refer to skin colour, but is a reference to either eyebrows (*qallut*), e.g., 'those who pamper or have beautiful eyebrows', or to the unnatural or man-made cloth (*qallunataq*) that Europeans brought with them (Minnie Aodla Freeman, personal communication, 1995). Thus, Qallunaat could be derived from the word *qallunaraluit*, 'people who make unnatural things or tamper with nature'. The Cumberland Sound Inuit appear not to have undergone a significant change in social organization as a consequence of contact with commercial whalers, other Qallunaat, and foreign diseases. An analysis of local groups' composition elucidates the structural basis of historic Cumberland Sound Inuit social organization.

Differences between the two major regional subdivisions to have occupied the Sound during the contact-traditional period, the Qikirtarmiut and Umanaqjuarmiut, manifest two structural tendencies inherent within all Central Inuit social relationships. Whereas the former were governed largely by hierarchical directives *(naalaqtuq)*, productive relationships among the latter were constituted more on egalitarian behaviours *(ungayuq)*. This model permits detailed re-analysis of the late prehistory of the Sound and a closer examination of structural variability in Iglulingmiut, Netsilingmiut, and Copper Inuit social organization. Specifically, the former two regional populations are found to be embellishments, respectively, of *naalaqtuq* and *ungayuq*. Alternatively, the Copper Inuit are seen to be the antithesis of Central Inuit social structure and ideology.

The archaeological and anthropological implications of these findings advance alternative models of Canadian Arctic prehistory and the origins of 'complex' social systems, such as those exemplified by the Central Inuit and European society. Finally, we consider the implications of this study for political development in the new Canadian territory of Nunavut.

# Part One

# Models and Theories of Inuit Social Organization

# Chapter 1

## 'Eskimo' Type Kinship and Social Organization

THE CONSTRUCTION OF 'ESKIMO' TYPE KINSHIP
AND SOCIAL ORGANIZATION

The 'Eskimo' have long fascinated western civilization. The fact that a preliterate people, with nothing more than their ingenuity and what little they could wrest from the land, could eke out an existence under such intolerably harsh conditions has captured the imagination of explorers, scientists, and the public alike. This fascination, in turn, has generated a body of literature that distinguishes the Inuit as one of the most thoroughly studied peoples in the world. As one student of Inuit culture put it, 'rarely has so much been written by so many about so few' (Hughes 1963: 452). Indeed, interest among anthropologists in Inuit culture has produced what another scholar has termed the 'Arctic small paper tradition' (Adams 1972: 9). Yet, despite all this interest and attention, examinations of Inuit social organization have taken a back seat to studies of material culture, folklore, social customs, history, prehistory, archaeology, cultural ecology and, more recently, cultural change. After all, was not the lack of formal organization found among most Inuit groups a direct function of the cold, desolate environment in which they lived?

Although studies of Inuit social organization have remained subordinate to more tangible areas of interest, they have a long if somewhat undistinguished history. Most analyses of social systems in preliterate societies begin with a consideration of kinship structure. Experience has led most anthropologists to expect a correlation between terminologically prescribed statuses and the actual behaviour that pertains among individuals (Damas 1963: 34). In this regard, L.H. Morgan's (1870) presentation of three Inuit kinship terminologies from the Canadian Arctic and Greenland represents an important beginning in the study of Inuit social organization. Morgan made no attempt to relate these schedules to specific features of Inuit social life; he was more concerned with tracing the distribution of kinship systems world-wide. His terminologies, however, became the basis for what later anthropologists came to regard as 'Eskimo' type kinship. One of Morgan's schedules, in particular, warrants presentation here, at least in part (Figure 2). Not only did it constitute the

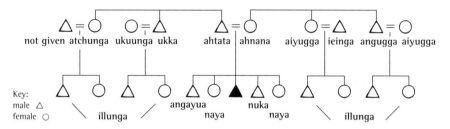

FIGURE 2. Cumberland Sound Inuit cousin terminology for male Ego, early 1860s (from Morgan 1870). Spellings of kin terms differ somewhat from those of conventional usages. However, pronunciations remain roughly the same, with the possessive ending '*ga*' attached in some cases, e.g., *angugga* = my *angak*.[1]

cornerstone of the 'Eskimo' type kinship system, but it is the terminology of the regional group that is the focus of this study, the Cumberland Sound Inuit. The key feature of this terminology, and 'Eskimo' type kinship in general, is the lumping of cross and parallel cousins under the same term. If cousins were differentiated at all, it was done so only on the basis of gender.

During his pioneering ethnographic fieldwork among the Inuit in 1883–84, Franz Boas, the father of American anthropology, recorded many details about Cumberland Sound Inuit material culture, folklore, and economic behaviour, but very little about traditional Inuit social organization. Although Boas (1964: 170–1) observed that the Inuit had no marriage rules beyond one not being able to marry his/her closest relatives (e.g., cousin, aunt and uncle, nephew and niece, etc.), and that bride-service[2] and child betrothal were common, he made no specific effort to explore the area of kinship and social organization. Perhaps Boas' most significant contribution to our understanding of Inuit social organization was his observation that 'the social order of the Eskimo is entirely founded on ties of consanguinity (blood) and affinity (marriage) between . . . individual families' (1964: 170).

Beginning with Holm's (1914) fieldwork among the Angmasalik of east Greenland in the mid-1880s, ethnographers began to collect data on group composition and social life that were to shape thinking about Inuit social organization for the next half century. Foremost among these studies were the accounts of Stefansson (1919), Jenness (1922), Mathiassen (1928), Rasmussen (1931, 1932), and Birket-Smith (1924, 1929). Although these reports presented information of use to social anthropologists, even to the extent of listing terminologies, they contained very little thinking on the structure of Inuit social organization. What little there was consisted largely of rebuttals of Mauss (Mauss and Beuchat 1904–5), who evolved a general theory of Inuit social life relating seasonal fluctuations in the density of human aggregations to the existence of two distinct social configurations, the summer society and the winter society. Mauss was perhaps the first anthropologist to explore the structural foundations of Inuit social organization. However, his theory was not well received in the climate of Boas-influenced historical-particularism—a rejection of focused inquiry and theorizing in favour of documenting all aspects of society—that dominated anthropology at the time.

Interest in Inuit kinship was rekindled in 1925 when Spier (1925: 79) defined, on the bases of several ethnographies, the 'Eskimo' type kinship system. This system differed from that of the Iroquois in that cross and parallel cousins were merged,[3] while siblings were differentiated according to relative age. In addition, there were four terms for parent's siblings and separate terms for grandfather and grandmother, but only one term for grandchild. This model laid the groundwork for further theorizing on Inuit kinship and social organization. Most notably, parallels were soon drawn between Inuit and European society; both employed 'Eskimo' cousin terminology and were strong practitioners of conjugal organization (Linton 1936). As Damas (1963: 7) noted, Linton and Spier's stereotypes of the conjugal family and a common cousin term separate from siblings began to play important roles in thinking about Inuit social life, especially since this particular cousin system has been associated with an emphasis on the nuclear family.

On the basis of Jenness' Copper Inuit and Holm's Angmasalik (east Greenland) ethnographies, Murdock (1949) went further to propose an 'Eskimo' type of social organization that had broader cross-cultural applicability. He defined 'Eskimo' type social organization as one that includes all societies characterized by 'Eskimo' cousin terminology, an absence of exogamous kin groups, monogamy, independent nuclear families, lineal terms for aunts and nieces, and such bilateral kin groups as demes and kindreds (1949: 226–7)—the latter being defined, respectively, as 'exogamous local groups in the absence of unilinear descent' and 'kin groups of a typically bilateral type' (1949: 63, 45). A survey of the social organization of 13 Inuit groups by Valentine supported Murdock's 'Eskimo' type of social organization insofar as he found 'no unifying unilinear . . . group which unequivocally associates each individual with a single clearly defined series of relatives and segments the community or the tribe into discrete social entities' (Valentine 1952: 162–3). However, where Valentine saw homogeneity, others began to see diversity.

## Unravelling 'Eskimo' Type Kinship and Social Organization

Sperry (1952), for example, found not one but three distinct types of cousin terminologies represented in the literature, each corresponding to a separate ecological zone. At Nunivak and the Fox Islands in the Aleutian chain he reported a system that distinguished cross-cousins, while equating parallel cousins with siblings. In west Alaska and northeast Siberia, Sperry noted the use of three distinct cousin terms whereby father's brother's (FB) children, mother's sister's (MZ) children, and cross-cousins were differentiated from one another. Interestingly, this three-cousin system, which was found in both Inupik and Yupik-speaking groups, was considered by Sperry to represent the ancestral form of Eskimo kinship structure. In central Canada, Sperry observed that the Caribou Eskimo distinguished cousins depending on whether they were on the father's or mother's side, a system he regarded as transitional between the Alaskan three-cousin and the eastern single-cousin systems (1952: 13). Sperry also found divergences from Murdock's 'Eskimo' type in terms of residence patterns and marriage regulations.

'Eskimo' as a system of kinship reckoning and a form of social organization came under increasing attack throughout the 1950s with the works of Giddings (1952), Hughes (1958), and Heinrich (1960), among others. Giddings (1952: 5) noted that an emphasis on one or the other side of descent occurred frequently in western Alaska. For example, while the Malemiut separated the children of father's sister (FZ) from all other first cousins, the Unalit distinguished children of the mother's brother (MB). Like Sperry, Giddings found that the merging of parallel cousins with siblings among the Nunivagmiut was more characteristic of Iroquois kinship reckoning than Spier's and Murdock's 'Eskimo' type. Moreover, a close examination of Lantis' (1946) ethnography indicated that the Nunivagmiut displayed patrilineal inheritance, but not descent, similarly oriented marriage rules, and matrilocal residence (Giddings 1952: 8). In short, Giddings (1952: 9) found no reason to include the Nunivagmiut within either Spier's or Murdock's types of 'Eskimo' kinship or social organization.

Hughes' (1958) study of the Yupik-speaking St Lawrence Islanders similarly discovered marked discrepancies from the 'Eskimo' type. For example, there was one common term for both cross-cousins, but two separate terms for paternal and maternal parallel cousins. The children of two brothers were particularly close as sibling terms often replaced the use of the cousin term between them (Hughes 1958: 1141). In this connection, Ego maintained father- and mother-like relationships, respectively, with FB and his wife, and there was a greater development of terms for one's paternal relatives than one's mother's kinsmen (1958: 1143). Most importantly, Hughes demonstrated the existence of patrilineal clans among the St Lawrence Islanders; there was an explicit unilinear rule of descent that united its central core of members (Murdock 1949: 68), although no explicit rules of exogamy or endogamy prevailed. Hughes (1958: 1146) ultimately attributed the development of unilineal tendencies and clan organization on St Lawrence Island to the recent merging of several patrilocal bands into larger villages whereby 'relatively greater emphasis was placed upon descent than locality in defining a person's social identification'. As Hughes believed that these clans evolved from social units which were anciently similar to those found in the central and eastern Arctic regions, he advanced a theory for how such territorially defined groups could, under the right circumstances, evolve eventually into patri-clans.

In northwest Alaska Heinrich (1960) found terminological differences between kinship systems of groups living on the coast and in the interior. Specifically, three-cousin systems and affinal-excluding structures were associated with the more permanent coastal dwelling groups. Conversely, two-cousin terminologies (similar to that described above for the Caribou Inuit) and affinal-incorporating structures, wherein most in-marrying individuals were merged with corresponding blood relatives, were associated with the more nomadic inland dwellers (1960: 113–14). Spencer (1959), however, in analysing data collected from the same areas as Heinrich, concluded that there were only minor differences between coastal and inland Alaskan groups in these respects. Moreover, Spencer found many resemblances between north Alaskan societies and the perceived structure of central and eastern Inuit groups, while pointing out the importance of quasi-kinship or voluntary alliances in regulating interpersonal behaviour.

Befu (1964) found both diversity and uniformity in the distribution of Inuit kinship terms in his analysis of 14 consanguineal schedules from Alaska, Canada, and Greenland. Specifically, he distinguished three regional systems: west Alaska, north Alaska, and central/eastern Arctic. Kinship schedules in west Alaska and the central/eastern Arctic resembled each other in that both (a) possessed a bifurcate collateral pattern for parent's siblings, (b) used speaker's sex as a component for sibling terms, (c) utilized the component of relative age, while having alternate terms which ignored it, (d) bifurcated nepotics on the basis of sibling's sex, (e) used speaker's sex as a component for nepotics, and (f) lacked a reciprocal term for the third ascending and descending generations. Conversely, north Alaskan schedules manifested opposite patterns in all these respects. However, schedules from north Alaska and west Alaska were more similar to each other than either of them were to central/eastern schedules insofar as parallel and cross-cousins were differentiated.

Fainberg (1967) in a paper first written in 1955 examined various lines of evidence across the Arctic for vestigial remnants of matrilineal clan organization. In west Alaska, Fainberg found support for his theory in (a) the existence of totems, though their inheritance was in the patri-line among some groups, (b) the extension of *ujohuk* (sister's child, or *uyuruk*) to other consanguines, (c) a corresponding use of MB, or *angakok* (*angaquk*) for 'chief', (d) the existence of men's houses, or *kazhim*, and (e) the division of cross and parallel cousins. In regard to the latter, Fainberg (1967: 250) argued that 'without an exogamous-clanship system there was no basis whatever for the division of cousins into parallel and cross-cousins'.

In the central/eastern Arctic, the survival of a former separation of society into two exogamous sections/clans was recognized in, among other things, (a) the 'Sedna' (or sea goddess) ceremony, where people were divided into two totemic groups, the ptarmigans and ducks, for ritual activity (Boas 1907, 1964), and (b) the designation of single terms for wife's sister and brother's wife, and for husband's brother and sister's husband. Spousal exchange, the predominance of female deities in religious ideology, the levirate and sororate where they existed, inheritance of names in either the matri- or patri-line, the avoidance relationships between consanguineal and affinal relatives, as well as the preference for male adoption were all reasoned to be survivals of former dual clan organization. The disappearance of dual exogamy, and matrilineal clans in particular, was attributed to the movement and intermingling of clans in the 'context of the vast unpopulated stretches of the Arctic' (Fainberg 1967: 255). As the process of disintegration of the matrilineal clan was completed relatively recently, the patrilineal clan did not have time to develop.

During the late 1950s, renewed interest in the welfare of northern native people by the Canadian government led to a number of federally sponsored field studies among Inuit in the central and eastern Canadian Arctic. In the Port Harrison region of western Quebec Willmott (1961) described Inuit social customs, group composition, and leadership patterns, which he compared with Murdock's model of 'Eskimo' type social organization. Willmott noted both convergences and discrepancies with Murdock's type, e.g., although eight cousin terms were found to be in use, most of which were a derivation of FB's children (male speaking), parallel and cross-cousins were not differentiated, while all first cousins were separated from siblings. In addition, there also appeared to be an indiscriminate extension of the term *akka* (FB) to most males in the first ascending generation.

In the Sugluk region of northern Quebec, Graburn (1964, 1969) undertook a structural analysis of Tak(q)amiut kinship and group composition. He identified several structural principles of social life which related directly to Takamiut kinship terminology. Specifically, he found that Takamiut social structure was based on a model of social reality that included, among other ideals, (a) patrilocal residence following bride-service, (b) local and kin group exogamy, (c) dominant-submissive behaviours between consanguines and affines of adjacent generations and siblings, (d) co-operative behaviours among Ego's generation and adjacent generations, and (e) separation of consanguines from affines, and original camp members from site visitors (1964: 477–8). An analysis of group composition indicated that over 80 per cent of the households studied conformed to the ideal model of social organization expressed by these structural principles (1964: 108). Group composition showed a virilocal bias superimposed on a bilateral base (1964: 192), with male sibling bonds being strengthened by such means as the use of affectional terms between female Ego and husband's brother's children (HBS=*irniajuk* or 'little son', HBD=*paniakjuk* or 'little daughter').

Steenhoven (1959) did not attempt any cross-cultural comparisons or intensive analyses of kinship structure, but his work among the Netsilingmiut of Pelly Bay did provide important new information on Netsilik group composition and leadership patterns. Specifically, he noted the presence of preferential first-cousin marriage and weakly developed leadership in local bands. Of greater theoretical interest was the work of Balikci (1960, 1964), who sought generalizations concerning differing directions of culture change brought about by the introduction of the rifle and the institution of trader-trapper relationships among the Netsilingmiut and Inuit of east Hudson Bay. Most notably, Balikci felt that the rifle produced a marked individualization of hunting practices in Pelly Bay, which led to increased band isolation, local group endogamy, and ultimately to preferential first-cousin marriage (1960: 144, 151). At Eskimo Point on the west coast of Hudson Bay, Van Stone and Oswalt (1960) undertook fieldwork, the results of which they compared with two acculturated villages in Alaska, Napaskiak and Point Hope. While each of the three communities differed markedly in social structure and economic activities (e.g., cousin marriage was preferred among the whale-hunting Point Hope residents, but not among the caribou-hunting Eskimo Point Inuit or the fishing/trapping Napaskiamiut), contact undermined leadership roles and positions of authority in all three communities.

The most influential work of this genre, and arguably the most important study of Inuit kinship and social organization ever undertaken in the Canadian Arctic, was conducted by Damas (1963, 1964, 1968a) amongst the Iglulingmiut of Foxe Basin. As a model of Inuit social organization, Murdock's 'Eskimo' type was perhaps doomed to extinction from the start. Not only were its two essential features, nuclear family orientation and single cousin terminology, abstracted from different groups, but this arrangement appeared to exist more in theory than in reality. If interregional comparisons of Inuit societies cast into doubt the universality and validity of Murdock's 'Eskimo' type of social organization, Damas' work laid to rest once and for all any utility this construct may have had for the study of Inuit social organization.

## KINSHIP: THE FOUNDATION OF INUIT SOCIAL ORGANIZATION?

During the early 1960s Damas (1963, 1964, 1968a) undertook analyses of Iglulingmiut group composition, networks of co-operation in economic activity, as well as kinship statuses and their behavioural content. Aside from a few attempts to interpret terminological patterns in wider social contexts, until Damas' study most investigations of Inuit kinship had been concerned with tracing typological and distributional patterns—Hughes' (1958) and Graburn's (1964) interest in the behavioural content of kinship terms and statuses alone being noteworthy in this respect. Like Graburn's research, Damas' work demonstrated that kinship factors provided the most pervasive means of aligning personnel amongst the Iglulingmiut, and thus were the most important factors in the formation of groups (1963: 11). Moreover, his study was not merely a 'snapshot' of Iglulingmiut social structure, but provided analyses of

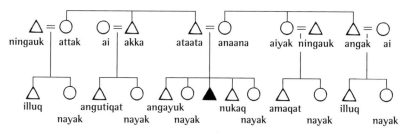

FIGURE 3. Iglulingmiut cousin terminology, male Ego (after Damas 1963, 1964, 1968a).

changing kinship relations, group composition, and authority patterns over several decades.

Iglulingmiut kinship terminology conformed to Spier's model in the use of four terms for parent's siblings, two terms for grandparents, one for grandchild, and a four-phase nepotic system. However, lineal terms for aunts and nieces were missing and, most importantly, sharp distinctions were made between maternal and paternal parallel cousins (Figure 3). The fact that emphasis was placed on the first ascending and Ego's generations is evident in the fact that pairs of paternal and maternal parallel cousins were referred to, respectively, as *angutikattigiit* (children of two brothers) and *arngnakattigiit* (children of two sisters).

Damas recognized that the social context of kinship terminology, rather the terminology itself, determined and organized social relations. Yet his appraisal of the Iglulingmiut system as a conceptually consistent arrangement of behavioural norms was not forthcoming until he realized (a) the frequency with which the terms *ungayuq* or 'affection-closeness' and *naalaqtuq* or 'respect-obedience', were used in informants' descriptions of status relationships, and (b) that the extent to which these directives were observed was based exclusively on the principles of age and generation, solidarity of the sexes, and consanguineal-affinal boundaries (1963: 48–51). Viewed from the perspective of *ungayuq* the three-cousin system of the Iglulingmiut attained a kind of logic whereby the three types of cousins of the same sex could be ranked hierarchically with respect to emotional ties or affectional closeness according to the principle of solidarity of the sexes:

> For male Ego, paternal parallel cousins as the sons of two brothers form the closest sort of brother-like bond outside the actual sibling group. Most distant of the three cousins for male Ego is the maternal parallel cousin. The logic of that arrangement is that the maternal parallel cousins are related to one another through parents who are both opposite in sex to either of the cousins. Intermediate in affectional closeness are the cross-cousins. This is still consistent with the principle of the solidarity of the sexes, for in that

case one of the connecting relatives is of the same sex as the cousins in question and one is of opposite sex. A complementary picture obtains for female Ego. Since the maternal parallel cousins are related through two females, this is the strongest sort of cousin bond, the most sister-like outside the actual sibling group. (1963: 48)

The principle of solidarity of the sexes also allowed collateral relatives in the first ascending generation to be conceived of as surrogate parents. For males, FB (*akka*) is more father-like than MB (*angak*) because he is related through the father (F), while FZ (*attak*) is a closer mother-surrogate than MZ (*aiyak*). Again, the situation is reversed for females, whereby MB and MZ are closer than FB and FZ. So pronounced was the principle of gender solidarity that the terms for sisters and female cousins for male Ego were merged (*nayak*), as were the terms for brothers (B) and male cousins for female Ego (*anik*), although siblings were recognized as being closer.

Father-son (F-S) and mother-daughter (M-D) relationships formed the closest bonds of any in the actual social life of the people as same-sexed offspring were gradually introduced to, and encouraged to assume, the roles of adulthood (1963: 49–50). But such relationships were also the most respectful, or *naalaqtuq*-directed, of any in Iglulingmiut society. While bonds of great closeness and co-operation existed between parents and their same-sex children, these relationships were also strongly oriented towards leadership and 'followership'. Within the sibling group, age as well as sex determined the subordinate-dominance hierarchy:

> For male Ego, the older brother is terminologically distinguished from the younger, and obedience is along the lines of age, as is the case with females who show a complementary terminology. Between males and females, however, the female should obey the male sibling regardless of age differences ... (though) in actuality, this operates only when both the brother and sister are mature. (1963: 50)

*Naalaqtuq* directives also structured relationships for in-marrying people, and the terminology reflects the dominance and solidarity of the consanguineal group. Thus, *ningauk* for male Ego (e.g., MZH, FZH, ZH, and DH) and *ukuaq* for female Ego (e.g., FBW, MBW, BW, and SW) were subordinate irrespective of relative age and generational differences between affines and consanguines (1963: 51). With respect to the first ascending generation, in-marrying males (*ningauk*) and females (*ukuaq*) must obey their parents-in-law or *sakkiik*, as certain respect relationships prevail. Although bonds of affection and closeness could, and often did, develop between sons-in-law and fathers-in-law, this relationship (*ningaugiik*) traditionally was the most *naalaqtuq*-skewed (i.e., asymmetrical along the respect-obedience axis) of any kin relationship in

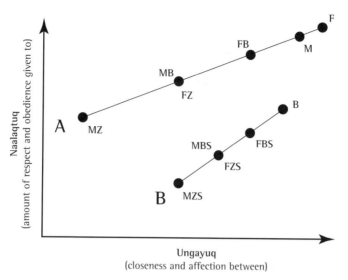

FIGURE 4. *Naalaqtuq* and *ungayuq* directives for Iglulingmiut male. Line A = consanguines in first ascending generation. Line B = male consanguines in Ego's generation, older than self. Relationships, for any given context, are likely more curvilinear than illustrated. Figure based on Damas (1963, 1968a).

Iglulingmiut society.[4] *Ningauk*, however, attained a kind of sibling-like status with his co-affines (wife's consanguine's spouses) in that co-operative associations pertained and subordinate-dominance relationships revolved around age distinctions within the sibling core (e.g., husbands of wife's older and younger sisters were, respectively, *angayuunnguk* and *nukaunnguk*).

*Ungayuq* and *naalaqtuq* directives did not transmit contradictory information to individuals nor function independently of each other. Rather, these axes of social behaviour were mutually influencing and together laid the foundations of interpersonal relationships among kinsmen in Iglulingmiut society. Although only a select group of relationships has been described here,[5] enough information has been provided to illustrate graphically the operation of the principles of age, gender solidarity, and blood ties with respect to the *ungayuq* and *naalaqtuq* axes (Figure 4). In the first ascending generation male Ego generally maintained the closest bonds of affection with, and displayed the most respect-obedience to (beginning with the most structured relationship) (1) F, (2) M, (3) FB, (4) FZ and MB, and (5) MZ. For male relatives older than self in male Ego's generation, the most respectful relations and greatest amount of affection were shown to, in descending order, (1) B (*angayuk*), (2) FBS (*angutiqat*), (3) FZS and MBS (*illuq*), and (4) MZS (*arnaqat*). For female consanguines in male Ego's generation the following hierarchy was observed: (1) Z, (2) FBD, (3) FZD and MBD, and (4) MZD (all *nayak[saq]*). However, after childhood opposite-sex kin relationships were characterized more by avoidance and respect than by affectional closeness (Damas 1963, 1968a). Based on the principle of solidarity of the sexes, a complementary picture obtained for female Ego. The principles of age and gender solidarity clearly indicate that, as male Ego moves away from father, relationships with the first ascending generation become governed more by *naalaqtuq* and less by *ungayuq* directives. Conversely, in Ego's generation relationships are regulated more by *ungayuq* behaviours. Affinal-consanguineal ties, however, were perhaps the most skewed or imbalanced. While they score low on the *ungayuq* axis, Ego's relationships with his affines and co-affines are clearly characterized by either subordination or dominance depending on whether he is an in-marrying male or not.

On the basis of the above discussion it is clear that, for male Ego, the F-S (*irniriik*) dyad constitutes the most stable foundation of group membership. In this regard, it is not surprising that Damas found the majority of Iglulingmiut social groups to be founded on *irniriik* cores. In addition, exogamy prevailed among kinsmen and local group members, residence tended to be patrilocal after a year's bride-service, and leadership was very well-developed, sometimes extending beyond the local group. While theory anticipates that the next most stable groups would be those founded on brother-brother (*nukariik*), brother-sister (*nayagiik*), and paternal uncle-nephew (or *qangiariik*) dyads, *nukariik* and other same-sex, same-generation cores proved less viable than the kinship system, and the cousin terminology in particular, would predict. Ideal behaviour predicted by the Iglulingmiut kinship system did not always correspond with actual behaviour. Indeed, Damas noted numerous divergences between prescribed and actual behaviour. Yet, while a certain amount of electiveness to norms of behaviour existed between people, Damas generally found a rough correspondence between ideal and actual behaviour:

> He (Damas' informant) expressed ... that general feelings of good-naturedness (*sic*) should be maintained in contacts between relatives, and that no matter what dominance-subordinance hierarchy should be called for, the latter should be secondary to the joking and the warmth that should prevail in co-operative enterprise. Actual observations in the course of hunts, boat launchings, and journeys convinced the writer (Damas), however, that the *ningauk-sakkiaq* and *angayungruk-nukaungruk* (*sic*) roles were followed rather closely in the execution of co-operative activity and were constantly reinforced through verbal reference, kin terms being more frequently used than names. (1963: 57)

Damas clearly established the importance of kinship directives in structuring interpersonal relationships among the Iglulingmiut. However, about the same time, others, most notably Guemple (1972a, 1972b, 1972c), began to question the importance of kinship relations *vis-à-vis* non-kinship alliances in regulating social behaviour in Inuit society.

## NEGOTIATION: THE FOUNDATION OF INUIT SOCIAL ORGANIZATION?

Work among the Belcher Islanders in the 1960s led Guemple to suspect that non-kinship alliance mechanisms such as child betrothal (1972c), spouse exchange (1961, 1972c), naming (1965, 1972c), ritual sponsorship (1969, 1972c), and adoption (1972c) played a larger role in structuring Inuit social relationships than kinship. On the assumptions that any social relation constitutes an alliance if it possesses an institutionalized form and its content is based to a very large extent on negotiation, Guemple (1972b: 2) argued that Inuit social organization 'is most appropriately explained as a system of alliance in which kinship is, more than anything else, a rhetoric which obscures the underlying form'. Put another way, various forms of alliance did not function in support of kinship to overcome its deficiencies, but were symptomatic of the negotiable character of all relations in Inuit society:

> Viewed in this light, the alliance system constitutes the heart of the social organization and kinship becomes an unimportant means by which Eskimo rationalize what is essentially a pragmatic social system based on reciprocity. (1972b: 5)

Guemple's analysis of marriage patterns among the Belcher Islanders, for instance, indicated that role congruity within the domestic household took precedence over genealogy in the formation of domestic units. Whether the heads of households were the siblings, fathers, or adopting fathers of the females involved was incidental to 'a very strong pattern which demanded that domestic unit roles be congruent so that if a woman filled some part of the role of female household head, she eventually came to play all of it' (1972c: 70–1). The inherent negotiability of the kinship system was further evidenced in the facts that affines were sometimes turned into consanguines and genealogical relationships sometimes played subordinate roles to generational connections in the use of kin terms. Although Guemple (1972c: 73) reasoned that kinship may be an important idiom of social relatedness, it was sufficiently flexible to permit a reallocation of existing kinsmen into new social roles even when these violated genealogical criteria. The Belcher Islanders apparently equated proximity with kinship in that all social relations had to be validated by personal contact to be recognized as genuine (1972c: 74). This suggested to Guemple that consanguinity and affinity were not the major forces in Belcher Island social organization, and that kinship was not an ascribed status system at all, but a negotiable one:

> Social relatedness is a matter to be worked out by the participants in a social encounter, and . . . the major criteria to be used are first, that people must actively participate in the system to be counted as kinsmen in a meaningful

way, and second, that those who enter the field of face-to-face relations must be treated as kinsmen whether genealogical connections exist to support the claim or not. (1972c: 75)

The issue, Guemple argued, is not one of opposition between kinship and alliance as expressions of the difference between prescribed and negotiable social relations, but rather 'the extent to which particular social systems accommodate themselves to negotiation and maneuver in both the content of social relations and in the recruitment of status'. Kinship in the Belcher Islands, as an ascriptive system forming the basic premise in the organization of social relations, was to a very large degree tempered by a social metaphysics which stressed residence and participation as the basis for social connection. Thus, even though consanguinity and affinity were important, they were:

> not the crucial factors governing social life. To put the matter another way, the underlying symbolism of Eskimo society is one of alliance; and kinship, at least as it is manifested in Island society, is more a vehicle for expressing relations of alliance than a basis for locking people into a relatively inflexible, social network. In this context it is the real kinship system which is fictive. (1972c: 75)

## Composite or Patrilocal Band?

Guemple (1972a) continued his exploration into the negotiability of Inuit social status in his examination of Service's (1962) model of hunter-gatherer band organization. Service distinguished three types of bands: patrilocal, composite, and anomalous. Among other traits (e.g., the levirate, sororate, and cross-cousin marriage), patrilocal bands demonstrated the following characteristics: small unit size, virilocal residence, extended families made up of the association of two brothers, moiety organization defined as dual exogamic divisions whether named or not, and bifurcate-merging kinship terminology. For Service, the combination of virilocal residence and reciprocal band exogamy gave rise to moiety organization, whereas the bifurcate-merging terminology was the result of concepts which separated individuals on the basis of age, generation, sex, and blood ties (Guemple 1972a: 81–2). On the other hand, composite bands generally lacked rules of marriage and locality, and exhibited heterogeneous band membership. An anomalous band was regarded simply as an extremely impoverished form of composite organization. The last two band types, Service argued, were caused by the disintegration of patrilocal bands as a result of contact with Europeans. The latter brought epidemic disease, which decimated groups and altered band structure. Trade led to increased geographical mobility between bands and increasing heterogeneity in band composition. Peace-keeping agencies encouraged less hostile relations between groups, which, in turn, resulted in greater mixing of groups and loss of patrilocal band structure.

The utility of Service's model for understanding Inuit social organization, however, was rejected by Guemple; he found no convincing evidence to suggest that the foundation of Inuit social structure was ever based on anything but composite organization. Guemple (1972a: 85) described Inuit residence patterns as being 'practicolocal', even where patrilocal tendencies were well-developed. And the skewing of bilateral kinship ties in favour of patri-kin among some groups was considered to be more apparent than real; kinship is, after all, negotiable. In addition, marriage rules were described as being agamous, and band exogamy could not be found to be 'definitely associated with Eskimo social organization anywhere in the Arctic' (1972a: 87). Similarly, bifurcate-merging terminologies were restricted only to south Alaska, and the lack of reciprocally exogamous marriage sections indicated that moiety organization was not characteristic of Inuit social organization, even though dual ceremonial structures were wide-spread. Finally, outside of Alaska, residence form, marriage practice, kinship nomenclature, and other features of Service's model rarely corresponded. Although the social structure of some groups approached Service's patrilocal form, Inuit bands throughout the Arctic exhibited either composite or anomalous band organization. Moreover, any resemblances to the patrilocal type were regarded as superficial, and 'the underlying form showed a flexible, composite structure based on negotiability of social status' mediated through such institutions as spouse-exchange, childhood betrothal, namesake relationships, ritual sponsorship, and other partnership alliances (1972a: 80).

Guemple further reasoned that Inuit population dynamics, social structure, and intergroup relations were not as dramatically altered by contact as Service had suggested. While foreign diseases and contact-induced starvation brought about by game depletion may have resulted in as much as a 30 per cent loss in population across the Arctic, Guemple noted that Inuit numbers were also susceptible to cyclical fluctuations prior to contact. Moreover, he suggested that trade in items which improved native productivity may have served not so much to disrupt precontact trading patterns, but to increase population size (1972a: 97). Although contact agents may have been beneficial in reducing the amount of raiding and feuding between bands, Guemple observed that the settling of disputes was directed towards individual families and not to the pacification of raiding bands. Additionally, while contact agents contributed indirectly to the formation of mixed bands by stimulating increased movement and interband contact, mixing of groups was noted to be a common precontact pattern (1972a: 102–3). Perhaps Guemple's most convincing argument that significant social change did not accompany contact with Qallunaat is that the process of creating new social forms to meet new organizational problems is very difficult indeed: 'decreases in group size mean only that some parts of the social apparatus fall into disuse, so that while some boxes in the structure collapse it does not mean that the structure itself collapses'. Thus, populations will always choose to adjust by alternate means 'rather than restructuring their social organization' (1972a: 106). As Sahlins and Service (1960) noted, societies change only so much as is essential for them to remain the same.

Consequently, while 'compositization' of Inuit groups may have been accelerated with the appearance of Qallunaat, this feature was not solely nor even largely the result of depopulation by foreign diseases, pacification from external agencies, or the institution of trade with Europeans. Rather, the composite nature of many bands was a direct reflection of the underlying structure of Inuit social organization, i.e., 'negotiability of kinship ties and alliance through ancillary kinship forms' (Guemple 1972a: 91).

Guemple (1972a: 103) ultimately attributed the 'compositization' of Inuit social organization to environmental factors whereby seasonal scarcity of resources encouraged geographical mobility, and social and economic survival necessitated that small highly mobile units periodically reunite into larger aggregations. Those groups which had developed more in the direction of patrilocal band structure did so only because they occupied areas of the Arctic where the biomass was substantially higher and more predictable than elsewhere (1972a: 105). The environment, then, underlay the composite nature of Eskimo social organization and its most significant feature 'the negotiability of status'.[6]

### CENTRAL INUIT COMPARISONS: THE SEARCH FOR STRUCTURE

As part of his ongoing exploration into the structural foundations of Central Inuit society, Damas undertook a number of comparative studies of Copper, Netsilingmiut, and Iglulingmiut socioeconomic features (1968b, 1969a, 1969b, 1972a, 1972b, 1975a, 1975b, 1975c). In terms of economy, the Netsilik and Copper devoted a far greater portion of the year to fishing and caribou hunting than did the Iglulingmiut, who placed much greater emphasis on the hunting of sea mammals. While breathing-hole sealing (*mauliqtuq*) was conducted in all three areas throughout the winter, in the spring the Iglulingmiut turned to the hunting of basking seals, or *uttuq*. Throughout the summer and fall the Copper Inuit and Netsilingmiut were dispersed in small groups away from the coast where they fished and hunted caribou, changing residences frequently. The Iglulingmiut, on the other hand, remained on the coast hunting sea mammals from kayaks. In autumn, however, younger hunters in the Iglulik area ventured inland for caribou, while older men co-operatively hunted walrus, seals, and small whales.

Band composition among all three regional groups was bilateral with a prominence of male-relevant ties. However, nuclear family organization—a key feature of Murdock's 'Eskimo' type social organization—was found only among the Copper Inuit. Household organization among the Netsilingmiut and Iglulingmiut was based on the extended family. Although large aggregations of people gathered together in the winter for *mauliqtuq* sealing in all three regions, Copper Inuit aggregations represented little more than loose clusters of nuclear families tied together by voluntary associations of fleeting duration (1969b: 56). Conversely, winter aggregations in the Netsilik and Iglulik areas were based on strong kinship ties within the extended family. Commensal units among the Copper usually consisted of the entire village, while among the Netsilingmiut men and women of the extended family ate separately. Alternatively, food sharing and eating patterns in Iglulik society were highly variable with nuclear families to whole villages forming commensal units at different times.

Leadership was ephemeral among the Copper Inuit and whatever authority individuals enjoyed was achieved on personality alone rather than any inherent structural feature or tendency within the social system (1969b). Leadership was better developed among the Netsilingmiut and Iglulingmiut, with *ihumataq* (Netsilik) or *isumataq* (Iglulik), i.e., 'the one who thinks', heading extended family units. Only among the Iglulingmiut, however, did the influence of the *isumataq* extend over the entire band, where he regulated game sharing and food distribution. Voluntary associations such as wife-exchange partners had greater importance than kinship in unifying bands among the Copper. As a basis for social and economic alliance, kinship was minimized in Copper Inuit society. Conversely, kinship factors were ascendant over voluntary associations in the Iglulik area. A combination of kinship and non-kinship alliances (e.g., seal-sharing partnerships), however, appeared to have structured socioeconomy among the Netsilingmiut.

Marriage and residence patterns as well as kinship terminologies in all three regions likewise differed. As noted above, both kin and local-group exogamy were preferred among the Iglulingmiut. Marriages among close relatives, however, were permitted in the Netsilik and Copper Inuit regions. Although marriage practices among the Copper could best be described as agamous, since kinship outside the nuclear family played no role in the selection of marriage partners, the Netsilingmiut preferred to marry first cousins. The latter also tended towards local group endogamy. And residence, though patrilocal in orientation, was in part determined by this practice. In general agreement with the lack of importance placed on kinship, residence among the Copper Inuit tended to be neolocal. Alternatively, with a strong emphasis placed on kinship relationships, residence among the Iglulingmiut was patrilocal. While polygamy was not widespread in all three regions, polygyny was far greater than polyandry in the Iglulik area (1975a: 413). Although polygyny was still more common than polyandry among the Copper and Netsilingmiut, there was a more equal occurrence of these features.

As stated previously, the Iglulingmiut have a three-cousin system whereby cross-cousins, patrilateral parallel cousins, and matrilateral parallel cousins are differentiated from each other. According to accepted theory, the separation of cross-sex cousins and siblings is consistent with exogamous marriage practices, as is the separation of consanguines and affines (1972b, 1975c). The Copper Inuit, on the other hand, had a single-cousin terminology whereby siblings and cousins were differentiated. It is interesting to note, however, that FB's children were classed separately from all other cousins, although no behavioural significance was attached to this relationship in ethnographic times. There is also an overriding of affinal-consanguineal boundaries in nepotic terms, which Damas argues is consistent with marriages between relatives (1975c). The Netsilingmiut have a single term for all cousins of Ego's sex (*illuq*), but sibling terms extend to cross-sex cousins, which conforms more with Murdock's (1949) 'Hawaiian' than 'Eskimo' scheme. Yet marriage with classificatory siblings, i.e, one's first cousins, was not only permitted, but preferred among the Netsilingmiut (1975c: 15). Moreover, there was a greater assimilation of affinal terms than that found amongst the Copper Inuit; both nephew-niece and uncle-aunt terms are extended, respectively, to replace separate terms for spouse's sibling's children and the spouses of collateral relatives in the first ascending generation.

Despite variation in social characteristics in all three regions, Damas found that each group possessed an internally consistent combination of features (Table 1), though some arrangements were more integrated than others (1969b, 1975c). For example, while the broad extension of cousin reckoning among the Copper Inuit appeared to run counter to the narrow scope of behavioural directives, there was consistency in (a) the general weakness of kin ties, (b) the use of names rather than kin terms, (c) a general absence of leadership, (d) the presence of agamous marriage practices, neolocal residence patterns, and affinal-including terms, (e) the absence of kinship obligations in sharing beyond the nuclear family, and (f) the relative independence of the nuclear family with regard to household organization. In fact, as Damas has noted, the Copper demonstrated 'a disregard for kinship factors that is remarkable in a hunting society'. Compensation for the consistent lack of emphasis on kinship, however, was found in the importance placed on voluntary associations in aligning personal, especially spouse-exchange and song/dance partnerships (1969b: 49–50).

Damas suggested that the kinship system and effective social structure of the Iglulingmiut were much more coterminous. While voluntary partnerships were only moderately developed, leadership was strongly expressed, with food distribution being regulated by village headmen. The careful separation of consanguines and affines in the Iglulik area was also consistent with the strict observance of marrying outside one's consanguineal group, as was the separation of cross-sex siblings and cousins. In general, the range of kinship

TABLE 1. Comparisons of three Central Inuit regional groups. Adapted from Damas (1968b, 1969a, 1969b, 1972b, 1975a, 1975c).

| FEATURES | COPPER INUIT | NETSILINGMIUT | IGLULINGMIUT |
|---|---|---|---|
| ECONOMIC CYCLE (primary activities) | Winter/Spring: *Mauliqtuq* Sealing. Summer/Fall: Caribou and fish. | Winter/Spring: *Mauliqtuq* sealing. Summer/Fall: Caribou and fish. | Winter: *Mauliqtuq* sealing Spring: *Uttuq* sealing. Summer: Walrus and seal hunting. Autumn: Caribou and walrus. |
| GROUP COMPOSITION | Winter: Bilateral, male relevant ties. Summer: Variable. | Winter: Bilateral, male relevant ties. Summer: Kin-based, viri-oriented. | Winter: Bilateral, male relevant ties. Summer: Kin-based, viri-oriented. |
| HOUSEHOLD ORGANIZATION | Nuclear. | Extended/viri-oriented. | Extended/viri-oriented. |
| LEADERSHIP | Ephemeral. | Extended family head, *ihumataq*. | Local group, extended family head, *isumataq*. |
| SHARING STRUCTURE | Voluntary partnerships. | Kinship and voluntary partnerships. | Kinship, sharing regulated by *isumataq*. |
| COMMENSAL UNITS | Local group. | Division of men/women. | Variable; individual to local group. |
| KINSHIP DIRECTIVES | Attenuated, narrow in scope. | Restricted to extended family, weak dominance hierarchy. | Broad in scope, strong dominance hierarchy and affectional bonding. |
| MARRIAGE PRACTICE | Agamous. | Kin and local group endogamy. | Kin and local group exogamy. |
| RESIDENCE PATTERN | Neolocal. | Patrilocal. | Patrilocal. |
| UNCLE-AUNT TERMS | Affinal-including. | Affinal-including. | Affinal-excluding. |
| CROSS-SEX COUSIN TERMS | 'Eskimo'. | Hawaiian. | Hawaiian. |
| SAME-SEX COUSIN TERMS | Two-cousin system; FBS separate, but no behavioural distinctions. | One-cousin system; all male cousins = *illuq*. | Three-cousin system; FBS closest to male Ego, MZD closest to female Ego. |

behavioural directives was expanded beyond the extended family and operated on a wider scope as an effective regulator of interpersonal behaviour (1972b). In short, while contradictions occurred to varying degrees within some regions, all three societies were considered by Damas to be more or less 'internally adaptive' or 'integrated' on a gradient from west to east (1972b, 1975b).

Analysis of Copper, Netsilingmiut, and Iglulingmiut social features, however, presented certain challenges for Damas. In order to retain the pre-eminent position of kinship in structuring Inuit social relationships, yet acknowledge the important role played by voluntary associations, Damas (1972b, 1975b) saw all three groups on a continuum of increasing complexity in social organization from west to east. For example, kinship and leadership played increasingly larger roles in structuring interpersonal relationships as one moved eastward. However, contrary to expectations, a corresponding decrease in the elaboration of voluntary associations on the same geographical gradient was not observed (1972b: 52–3). In fact, quite the opposite was found. Whereas the Iglulingmiut demonstrated the greatest emphasis on kinship, they also possessed the most complete listing of non-kinship alliance-forming mechanisms. Alternatively, the Copper Inuit possessed the weakest development of kinship coupled with the fewest voluntary associations (1972b: 53–4). On this basis, Damas concluded that the Iglulingmiut demonstrated the most 'internally consistent' social system of the three groups, and that voluntary relationships in this society served 'to augment and reinforce kinship to produce a social system that was more intricately organized than those of the Netsilik and Copper Inuit' (1975b: 26). Conversely, incongruities between kinship terminology and behaviour among the latter two groups, especially the Copper Inuit, suggested to Damas that they possessed less completely integrated and elaborate social structures.

Damas (1969b, 1975c) spent considerable effort exploring the relationship of environment and society among the Iglulingmiut, Netsilingmiut, and Copper Inuit. However, his search for the ecological determinants of variation in the Central Inuit social organization yielded mixed results. Environmental factors were held responsible for large winter aggregations: numerous hunters were required in order to effectively undertake *mauliqtuq* sealing for, depending on the environment, an individual seal may maintain up to a dozen or more breathing holes. General similarities also occurred throughout all three regions in the practice of adoption, child betrothal, and spouse exchange, which Damas argued were well-adapted to external conditions (1969b: 53). Adoption redistributed individuals throughout society to produce social units that matched environmental conditions. As female infanticide in the Netsilik and Copper regions was directly related to survival (1969b: 54, 1975c), and as this practice created a shortage of women, child betrothal provided assurance that males would secure mates in adulthood (1969b: 53, 54). Child betrothal among the Iglulingmiut also functioned to secure marriage partners for males, but for a different reason. Although the Iglulingmiut did not practise female infanticide, their exogamous tendencies often similarly produced local, albeit culturally defined, shortages of eligible females. While spouse-exchange may have had adaptive value insofar as it extended the kinship network, and thus survival chances, Damas (1969b: 54) also considered that it had functions not directly related to environment or social structure.

Despite some similarities in social features among all three regions, it is obvious that substantial differences did exist. The degree of variation in kinship terminology, particularly the existence of three rather distinct arrangements of terms in Ego's generation, was greater than environmental factors would predict. While consistency existed between marriage patterns, cousin terminology, and affinal-including/excluding aunt-uncle terminologies in the Iglulik area, these began to break down in the other two regions, especially in the Netsilik area (1975c: 23). While the correlation of agamous and endogamous marriages with female infanticide among the Copper and Netsilik Inuit was said to be driven by environmental factors, greater productivity in the Iglulik area apparently alleviated the need for this practice (1969b: 54).

The division of the Central Canadian Arctic into two exploitive patterns and areas, one occupied by the Netsilingmiut and Copper Inuit, the other by the Iglulingmiut, continued to yield unsatisfactory results. Parallels between exploitive zones and features of social organization were noted in the case of seal-sharing partnerships, which existed among the Copper and Netsilingmiut but not the Iglulingmiut. With lower levels of subsistence and exploitive efficiency in the Copper and Netsilik areas, structured systems of sharing were considered to provide special insurance for unsuccessful hunters (1969b: 55). Conversely, the lack of these sharing partnerships in the Iglulik area was interpreted as an expression of less urgent ecological pressures, though Damas rationalized that there is no reason why a system of distribution based on kinship would be any less efficient than one based on voluntary partnerships (1969b: 55). As noted above, the Netsilingmiut and the Copper Inuit shared an affinal-including aunt-uncle terminology as well as a tendency towards kin endogamy/agamy. More favourable economic conditions among the Iglulingmiut made larger dog teams possible, facilitating mobility and the practice of exogamy, while greater ecological pressures in the Copper/Netsilik zone rendered exogamy difficult to maintain (1975a: 414).

Beyond these correlations, however, shared similarities in social features were not forthcoming to the extent predicted by the uniformity of resources and exploitive patterns of these two groups. The Copper Inuit pattern of nuclear family organization, egalitarianism, and neolocality contrasted markedly with the Netsilik pattern of extended family organization, extended family leadership, however weakly developed, and patrilocality. The distribution of sharing and commensal practices among the three groups also could not be correlated with environmental conditions; there were three systems in two exploitive zones. Nor were patterns of leadership associated with environmental factors as three different structures existed in two zones. Similarly, a third feature that was split in its distribution was the classification of same-sex cousins. The Copper have a two-cousin system, the Netsilik a one-cousin system, and the Iglulik a three-cousin system. Again, these systems could not be readily interpreted as correlating with ecological factors because two systems existed within one exploitive zone. In violation of ecological predictions, the Iglulingmiut and Netsilingmiut shared some traits which cross-cut their zones, but which were not found among the Copper Inuit. These include a tendency towards patrilocality and extended family organization and the classification of cross-sex cousins as siblings, the latter contrasting sharply with the practice of separating cousins and siblings in the Copper Inuit area (1969b: 56).

While Damas considered a number of social features among the three groups to be related to facts of ecological exploitation, the relationship was far from complete. Many exploitive patterns and social features demonstrated by the three groups were not congruent, and where correlations did exist, Damas concluded that they 'could not be convincingly demonstrated to be other than spurious' (1969b: 57–8). A more complete explanation of structural variation in Central Inuit societies lay elsewhere, and in this regard, Damas turned to an examination of historical factors such as common heritage, migration, diffusion, innovation, and cultural drift (1969b: 58–61, 1975c). Features shared by all three groups were attributed to a common heritage or broad common adaptation, while those restricted to either exploitive zone were considered to represent adaptive innovations to each zone, e.g., seal-sharing partnerships among the Copper and Netsilik. Similarities between the Netsilingmiut and Iglulingmiut were attributed to diffusion, whereas drift, i.e., 'divergent cultural change operating under the conditions of isolation' (1969b: 60), was considered to be responsible for the main differences between these two groups. In this connection, Damas (1969b: 59) suggested that the Thule Inuit ancestors of the Netsilingmiut once practised kin exogamy, but because of a change to a more marginal subsistence economy in the ethnographic era, endogamous practices were adopted out of necessity with changes in the cousin system lagging behind the affinal-including terminologies. Alternatively, the distinctive arrangement of social features among the Copper Inuit suggested an earlier migration of ancestral Thule people into their region than that represented by the Netsilingmiut and Iglulingmiut.

Damas is to be credited for pointing out the remarkable diversity of social features among these Central Inuit groups as well as for exploring the ecological and historical foundations of this variation. Yet, while he examined the relative roles of adaptive and historical forces and their interrelationships in producing both uniformity and diversity, his explanations, as he acknowledged, were far from complete (1969b). Nevertheless, they represent a significant step towards isolating structure in Central Inuit social organization. Subsequent fieldwork among the Copper, Netsilik, and Iglulingmiut, as well as the findings and theoretical orientations of other anthropologists (e.g., Guemple 1972a, 1972b), forced Damas to caution against attaching too much importance to kinship as the exclusive or even major structuring device in Central Inuit society (1972b). Moreover, Damas appears to have relied increasingly upon ecological arguments to account for variation among the Central Inuit (e.g., 1969b, 1975a). Despite these equivocations and shifts in emphasis, Damas' work in the areas of Central Inuit kinship and social organization remain his most significant contributions.

RECENT RESEARCH

Damas set a grand stage for subsequent explorations into the structural foundations of Central Inuit social organization. He illuminated the diversity of social features among three Central Inuit societies, and the lack of congruity between social and environmental variables; his inability to offer more complete explanations for this variation begged, if not demanded, further examination. Unfortunately, few anthropologists since 1970 have investigated the structural underpinnings of Inuit social organization or attempted broader interregional comparisons of Central Inuit societies. While this trend can be attributed to changes in orientation within the discipline itself, and to the centralization of many Inuit groups, as well as other factors, most recent studies of Inuit culture have focused on issues other than kinship and social structure.[7] Many of these studies are germane to questions of Central Inuit social structure and/or social and cultural change. However, few, if any, researchers in recent years have explored the structural foundations of Central Inuit social organization to the extent that Guemple and Damas did during the 1960s and 1970s. Several of the more notable studies addressing the nature of Inuit social organization over the past two decades are described below.

Adams (1972), in contrast to Guemple but in agreement with Service (1962), considered the flexibility of social organization in Inuit society to be the result of acculturative processes, but for different reasons. Frequent and sustained transaction among local group members during the precontact period produced consistent values, behaviours, and responses, while sponsoring stable and durable social forms. However, transactions with traders, missionaries, whalers, police officers, government officials, etc. were variably structured, restricted, and proscribed by situational influences. These, in turn, encouraged variable behaviour resulting in loosely integrated social forms subject to rapid change and variation between groups (Adams 1972: 11). Thus, the flexibility that many researchers found to be characteristic of Inuit social behaviour and organization was not the result of an ethos emphasizing relaxed attitudes toward the demands of living (Honigmann and Honigmann 1959), or a lack of 'value associated with conventional ways of doing things' (Willmott 1960: 59). Rather, flexibility in behaviour and organization was an acculturative process whereby Inuit were unable to acquire consistent complexes of values from transactions with Qallunaat because the latter were themselves 'diverse, inconsistent, and variable as transactional partners' (Adams 1972: 15).

Burch (1975) undertook an examination of change in kinship and family relations in northwest Alaska. Each of the 20 or so traditional northwest Alaskan groups Burch considered were 'overwhelmingly kinship oriented', and 'apparently much more so than most Canadian Eskimo societies'. Both ideally and actually, kinship ties were emphasized at the expense of all others (1975: 22). However, the gradual assimilation of north Alaskans into American society resulted in a number of changes, including (a) a decrease in the activation of some traditional kinship relationships, (b) substantial reduction in emphasis on kinship in strategies of affiliation, (c) considerable simplification in the structure of both domestic and local family units, but not family size, and (d) reduction and/or alteration of traditional mechanisms for changing personnel that left the structure of kinship organization unaffected (1975: 292). Burch's analysis of these societies led him to propose that most Inuit, and a great many non-Inuit, societies might be better regarded as 'family-oriented kinship systems', characterized by, among other features, an emphasis on bilaterally recognized descent and what Levi-Strauss (1969) called 'complex marriage' structure (Burch 1975: 294).

Burch rejected Guemple's view of 'kinship as negotiation'. Burch found at least 27 different patterns in north Alaska which prescribed how related people ought to interact, each with its own strongly institutionalized value system. He viewed north Alaskan social organization as being founded on 'two basic strategies of affiliation', one in which an individual had to be actively associated with kinsmen to have a chance of survival, and one in which, in order to survive over the long term, an individual had to belong to an organization wherein a marital, intergenerational, and/or a same generational relationship pertained (1975: 197–202). Burch also rejected Murdock's (1949) contention that the nuclear family is universal; it certainly did not exist in northwest Alaskan society either in the minds of the people or in reality. Rather, society was organized into what Burch called domestic and local families, the former representing a group of kinsmen occupying a single dwelling, the latter different dwellings (Burch 1975: 237).

In his analysis of alliance mechanisms in north Alaska, Hennigh (1983) discovered an underlying structure in rules of alliance formation. Specifically, Hennigh found a sharp contrast in the potential for *umialit* (boat owners/rich men) and *angaqut* (shamans) on the one hand, and ordinary people on the other, to combine alliances. While *umialit* had several dozen possible ways to maintain two or three alliances with people (usually other *umialit*), and *angaqut* had 18, most individuals had none except his/her name partner. This suggested that the lack of opportunity for individuals to combine alliances was part of a larger social structure existing in the minds of north Alaskans which 'encouraged concentration of power in the hands of a few, and a scattering of alliances among the rest of the people' (1983: 30). This interpretation lent credence to Burch's observations (1975, 1980) that northwest Alaskan societies were ranked, if not stratified.

Wenzel (1981) examined the position of Clyde River Inuit kinship and its associated behavioural concomitants as they effected the patterning of Central Inuit ecological relations. *Ungayuq* and *naalaqtuq* directives were seen to structure not only kin relations, but ecological relations in the organization of production. Specifically, Wenzel focused on the social and economic roles of task group formation and decision-making networks. Thus, social organizational elements, as directed by the kinship system, were seen to provide a framework for the arrangement of environmental as well as sociological relations.

Fienup-Riordan's (1983) exploration of Nelson Island social structure represents perhaps the most thought-provoking research to be undertaken in recent decades. Underlying west Alaskan social organization on Nelson Island were complementary oppositions which reconstituted society through the ritual exchange or cycling of names, souls, objects, and people. If one accepts Fienup-Riordan's definition of ritual as 'ideology in action', this underlying structure appeared in many forms, including the seal party. The latter paralleled the marriage exchange whereby women gave away the product of the hunt of their husbands and sons—the symbolic proof of their potency—to their first same-sex cross-cousins, i.e., to the women who would eventually give their daughters back to the hosting families as brides (1983: xv). Same-sex sibling solidarity within the consanguineal group was continually opposed to opposite-sex mediation between groups, and appeared in work groups as well as ritual configurations (1983: 301). The symbolic cycling of people—e.g., female cross-cousins opposing each other in ritual distribution at one time (the seal party), joined together in ritual (the exchange) dance at another—gave substance to social organization through ritual, ultimately leaving us with a view of social structure as:

> embodying a fundamental complementarity and cycling between descendants of same sex and opposite sex siblings, the former procuring for and distributing in the name of the latter. In the seal party, women join with their same sex siblings to celebrate opposite sex affinal relation, thereby solving the problem of a system by which sisters are closely allied in the natal group, but eventually removed, literally, into the family of their spouse. It is only in celebration of affinal success in the descending generations, for their

sister's children and grandchildren, that the rift is mended and those that affinity pulled asunder to create the unit of cultural generation, reunite outside the village in pursuit of seal and in the village for wild display and distribution. (1983: 302)

The possibility that many Inuit may conceive of their world as a host of complementary oppositions forever alternating from one state to another in a larger process of ritual exchange and the cycling of life forces is evident in Boas' (1907: 139–41, 1964: 196–201) descriptions of the 'Sedna' ceremony among the Oqomiut of Cumberland Sound, wherein many parallels can be drawn with Nelson Island ritual distribution and symbolism. Fienup-Riordan's structural analysis is all the more remarkable when one considers that she made her observations after the Nelson Islanders had been subject to decades of acculturative processes.

In addition to these more case-specific studies of Inuit social structure, a number of detailed cross-cultural analyses of pan-Inuit social customs have been undertaken. Oosten (1976), for example, in his analysis of Netsilingmiut and Iglulingmiut religious behaviour and belief, sought to delineate the theoretical structure of Central Inuit religion. Specifically, Oosten viewed the spiritual world of the Central Inuit as organized into oppositions, each with its own logically consistent structure and gender referent content, with the living world occupying a neutral position in this scheme. Contrary to Fienup-Riordan's thinking, whereby women were cultural mediators between men and animals, in Oosten's view women as givers of life were associated with nature, while men as takers of life were associated with culture. Oosten speculated that the opposition of land and sea was of crucial importance to religious ideology, and that 'when the sea disappeared out of focus (as it did among some interior adapted groups) . . . the whole religious structure disintegrated, leaving only the shamanistic complex intact' (1976: 97).

Kjellstrom (1973) provided an impressive listing of Inuit marriage practices across the Arctic and identified three broad regional marriage regions: (1) the Aleut and Yupik of southwest Alaska, (2) the Inupiat of northwest Alaska, and (3) groups in the central/eastern Arctic. However, he failed to relate similarities and differences in marriage practices in these regions to other social features, and no comprehensive investigation of this potentially important systemic relationship was forthcoming.

In his analysis of Inuit adoption across the Arctic, Guemple (1979) rejected the 'demographic hypothesis' of adoption whereby it served to provide 'an efficient means of co-ordinating population with production'. Clearly, this explanation failed to account for the high frequency of adoptions of grandchildren by grandparents in several areas of the eastern Arctic. Nor was adoption a marginal institution, which made up for the shortcomings of the kinship system. Rather, it was 'a typical manifestation of the underlying structure, having the same flexible, negotiable character as the rest of the social universe'. Thus, Guemple paints adoption with the same brush as other voluntary alliances in Inuit society, in which social structure:

> is best understood as an array of institutionalized forms, including adoption, which can give substance to the claims of the members of the local group to be related to one another as kinsmen. Kinship in this context is not the underlying skeleton of the social system, but rather a kind of rhetoric of social relatedness in terms of which crucial social and economic connections based on *locality* (emphasis added) are expressed. (1972a: 92)

## ANARCHY OR STRUCTURE: KINSHIP OR LOCALITY?

The diversity of social features among Inuit societies has led to an equally diverse number of approaches to the study of Inuit social organization. While a coherent theory of pan-Inuit social structure has not been, or will likely ever be, forthcoming, theoretical perspectives on the foundations of Central Inuit social organization appear to fall into two main camps: (1) those that advocate the primary role of kinship in regulating interpersonal behaviour (e.g., Burch 1975; Damas 1963, 1964, 1968a), and (2) those that emphasize locality and negotiation as the principles upon which Inuit social relationships are forged (e.g., Guemple 1966, 1972c, 1979). For Guemple (1979: 93), 'social relatedness begins in the local group, not in the kinship tie', and its substance is mutually beneficial co-operation, which is partly economic and partly social in character; i.e., kinship is inherently sociocultural, not primarily biological.

While Guemple stressed the importance of geographical propinquity in forming social relationships, his persistent assault on kinship has in no small measure contributed to the view that Inuit social organization is far more flexible or 'formless' than that of other cultures (Adams 1972, Guemple 1972a, Honigmann and Honigmann 1959, Willmott 1960):

> What [flexibility] means is that Eskimo social conventions do not allocate people to social membership in any very unambiguous way; and it also means that there are very few prescriptions, either conscious or unconscious, which state how people ought to treat each other once allocated. (Guemple, cited in Burch 1975: 61)

Yet, as Burch (1975: 61) has pointed out, the research of many anthropologists (most notably, Damas) suggests that 'the stereotypical notion of Inuit anarchy is greatly overdrawn'. Among most Inuit societies there existed definite, strongly institutionalized prescriptions about how kinsmen were supposed to treat each other. If there was any flexibility in Inuit kin relationships it lay in the allocation of people among various positions in a social system, and not in a lack of definition as to how people filling particular positions should behave (1975: 62).

Guemple's views perhaps might have gained a wider following if more Inuit groups constructed their models of social reality in the way he proposed; i.e., that kinship, adoption, spouse exchange, name-sharing, etc. were equally symbolic conferrals that emerge out of the need to convert a person who is in the local group into a relative of some sort (1979: 94). However, few Inuit groups outside of Guemple's Belcher Island experience accord kinship such a minor role in organizing social relationships—the Copper Inuit being another—at least in my experience and that of other anthropologists. In fact, it appears that Guemple's conclusions were drawn from an acculturated situation whereby many Belcher Islanders were exploring alternative means of forming productive relationships in the context of decreased mobility and increased centralization of the population (M. Freeman, personal communication, 1993). A few years prior to Guemple's fieldwork, when Islanders were distributed throughout the islands in several small camps, Freeman (1967: 154) found that kinship was indeed the basis of group composition.

Certainly, the kinship model more closely approximates our own 'typically Western' way of thinking about social life (Guemple 1979: 92). But I do not believe that it ought to be discarded on this account. In fact, many features of Inuit social life across the Arctic directly point to the importance most Inuit groups attach to kinship. Some of the more obvious ones include (a) the way strangers are treated, (b) the use of kin terms over personal names in forms of reference and address, and (c) the strength of parental and sibling bonds over spousal ties.[8] Moreover, if kinship was as unimportant as Guemple suggests, then we might expect to find groups who trace virtually no genealogical connections among families to be as common as groups composed of families who are all related through blood and marriage. We do not. To say that kinship was just another way of rationalizing relations of production is to underestimate its role and to ignore many facts of Inuit social life.

Guemple's perspectives on Inuit social organization appear to have been influenced as much by traditional anthropological definitions as by his own work among the Belcher Islanders. Guemple appreciated the fact that the opposition of kinship versus alliance, which was rooted in the distinction between ascription and achievement, is totally unacceptable when applied to the study of Inuit social structure (1972c: 57). Indeed, it can be argued, following Needham (1974: 39), that anthropological constructs such as marriage, kinship, descent, exogamy, bride-service, etc., may have little utility in the study of Inuit social organization because so many disparate relationships and categories are brought together under each of these terms that they lose all precision in meaning and distinctiveness (Trott 1982). However, what Guemple failed to recognize was that kinship is not a rigid, invariant structure of behavioural directives. Instead, kinship is a model, if you will, that exists in the conscious and unconscious minds of individuals for organizing relations of production and reproduction. Locality does not attain its importance at the expense of kinship. Rather, biological and geographical propinquity were simply the 'hooks' upon which Inuit hung their hopes for emotional security and economic well-being (Hickey and Stevenson 1990). Which was ascendant, or came first, are likely to be unproductive exercises in metaphysical gymnastics. Trott (1982: 101) and Turner (1978: 245) addressed this issue when they observed that, if one sees that '(a) models of relations of various orders are constructed out of relations of reproduction, within a cultural definition of reproduction, (b) these are in part defined relations between men and women, one aspect of which can be called marriage, and (c) both these factors are further mediated in terms of territory and proximity', then neither 'kinship' nor 'locality' are objective conditions, but are a part of a constructed 'given' human condition. Productive activity, thus, structures and is itself structured by relations between people commonly subsumed under 'kinship' and 'locality' (Trott 1982: 101).

The distinction between kinship and locality, as espoused by Guemple, may have stemmed initially from the emic difference between 'looking up' and 'looking across and down' generations (Trott 1982: 102). Graburn (1964) was cognizant of this distinction when he observed that, while Takamiut camp arrangements of previous generations were already established, those of Ego's generation and the next were in the process of 'becoming'. Perhaps this is why terms in the first ascending generation of most groups seem to be more invariant than those in Ego's and the first descending generation. Structural unity in many groups appears to have been accomplished in Ego's generation by emphasizing co-operative behaviours in same-sex relationships, i.e., 'brotherhoods' (Trott 1982). Opposite-sex relations, on the other hand, often become part of a group of 'outsiders' from which potential marriage partners emerge. The production and reproduction of 'brotherhoods' and 'sisterhoods' from one generation to the next through symbolism and ritual exchange, I believe, are what Fienup-Riordan's structural analysis of Nelson Island social organization sought to address. It is also a fundamental problem in Central Inuit social organization.

An emphasis on either kinship or locality at the expense of the other is likely to be unrewarding: the former, because only partial answers will be provided; the latter, because no answer will be forthcoming at all—an exclusive focus on locality assumes *a priori* that there is no structure, that all social relationships are negotiable. Having rejected the latter view, where does the search for structure in Central Inuit society begin? First and foremost, it begins with a comprehensive analysis of the arrangement of social features within specific groups, including the behavioural content of kinship systems. Are these features complementary or contradictory? What is the relationship of territoriality to leadership, leadership to marriage, marriage to adoption? Do these features form an integrated structure? And, how do their systemic associations inform us about relations of material and social production and reproduction in preceding and succeeding generations?

Questions of this nature, of course, cannot be answered by synchronic data alone. Towards this end, I trace in Chapter 5 the social composition of numerous local groups in Cumberland Sound, Baffin Island, over a 50-year period from 1920 to 1970. However, prior to this study, the effects of various acculturative forces on traditional or aboriginal Cumberland Sound Inuit social organization must be assessed. For example, how did prolonged participation in commercial whaling, the impact of foreign diseases, or the early adoption of Christianity affect traditional social relationships during the late 19th and early 20th centuries? Did Euroamerican influences fundamentally alter or transform the structural basis of Inuit society in Cumberland Sound? And, are we justified in utilizing the Cumberland Sound Inuit to inform us about the structural principles of Central Inuit social organization?

# Part Two

# The Cumberland Sound Inuit in History and Prehistory

# Chapter 2

## Cumberland Sound Before Qallunaat

The Cumberland Sound Inuit played an integral, though largely unappreciated, role in the development of anthropology and thinking about Inuit social organization. Their terminology was the cornerstone of the 'Eskimo' type kinship system, and they were the first Native culture to be studied by Franz Boas, the 'father of American anthropology'. The Cumberland Sound Inuit had a profound effect on Boas' views on racial equality, perspectives which helped shape the discipline of anthropology. However, Boas' upper-middle class German background failed to prepare him for life among the Inuit. Initially appalled by the living conditions in which he found himself on Kekerten Island in the fall of 1883 (Cole 1983: 21), Boas was soon to write:

> I often ask myself what advantages our 'good society' possesses over that of the 'savages' and find, the more I see of their customs, that we have no right to look down upon them. Where amongst our people could you find such hospitality as here? Where are people so willing without the least complaint, to perform every task asked of them? We have no right to blame them for their forms and superstitions which may seem ridiculous to us. We 'highly educated' people are much worse, relatively speaking. (Cole 1983: 33)

Considering the unique position that the Cumberland Sound Inuit occupy in anthropology, it is ironic that we do not know more about their social organization, especially when compared with that of other Inuit groups. In this chapter I endeavour to draw a clearer picture of precontact social organization in Cumberland Sound. My ultimate objective is to lay the groundwork that would enable us to determine whether commercial whaling and other Euroamerican influences during the historic period significantly altered the structural basis of Cumberland Sound Inuit social organization. Are the kinship and local group composition data presented in Chapter 5 representative of the aboriginal period? Or did contact with Qallunaat fundamentally transform the structure of Inuit social organization in Cumberland Sound? Before these questions can be addressed, however, some understanding of the environment in which these Inuit lived and of their precontact social organization must be sought in order to appreciate the extent and magnitude of any subsequent changes.

## The Environment of the Cumberland Sound Inuit

### Tides, Rock, Ice, and Climate

The Cumberland Sound Inuit inhabit the shores of the largest and deepest indentation on the east coast of Baffin Island (Figure 5). This body of water is approximately 230 km long and ranges in width from 40 km at its mouth to nearly 65 km mid-way up its length. Variously named Cumberland Straits (Markham 1880), Hogarth's Sound (W. Penny 1840), and Northumberland Inlet (Wareham 1843), Cumberland Sound has become its official designation. At contact in 1840 local Inuit knew this inlet as 'Tenudiakbeek', which according to M'Donald (1841: 7) and Sutherland (1852: 229) referred to 'the number of whales frequenting it'—*akbeek*, or more correctly, *arvik* is the Inuit word for bowhead whale (*Balaena mysticetus*). Although Cumberland Sound became one of three major commercial whaling grounds in the eastern Arctic, Boas' (1964: 27) interpretation of the name of the Sound, Tiniqdjuarbing, i.e., 'the place of great tides', is more accurate: *tinu* is the root word for tides. Indeed, the Sound is well known for its extreme tides; tidal ranges of *c.* 7.5 to 8 m are common at the head of the Sound and the entrance to Nettilling Fiord (Haller et al. 1966: 19). Compared to most of Baffin Island, where mean tides average no more than 1 or 2 m (Haller et al. 1966: 18), tides in Cumberland Sound are extreme.

In island-congested waters at the head of the Sound the great range in tides produces strong currents. Although these tidal rips present hazards for small craft in summer—Oleetivik at the mouth of the McKeand River is known as 'the place where one waits for the tides'—in winter they frequently prevent the formation of ice. These open water areas or *sarbut*, in turn, attract ringed seals (*Phoca hispida*) and other marine mammals, and thus their exploitation by humans. *Sarbut* are found at the mouths of most fiords near the head of the Sound, particularly Kangiloo Fiord and Clearwater Fiord, and near the island of Nunaata (Figure 6). Tidal and ice conditions are among the most important factors influencing animal distributions in the Sound and therefore human settlement. However, ice conditions near the shore are also affected by physiography and climate.

The terrain surrounding the Sound can be divided into three physiographic zones (Haller 1967: 4), (1) Chidlak Hills, (2) Nettilling Uplands, and (3) Cumberland Fiords. The Chidlak Hills are located on the southeast shore of the Sound between Popham Bay and Brown Inlet, and consist largely of rolling hills and plateaus dissected by deep valleys. Southeast of Chidlak Bay cliffs rise 500 m or more above sea level. Here long, narrow, steep-sided inlets

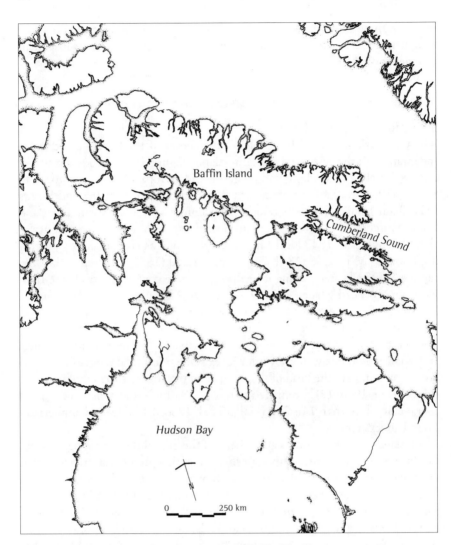

FIGURE 5. Map of eastern Canadian Arctic and Baffin Island.

are common. Northwest of Chidlak Bay the land is less eroded and broad valleys and shallow lakes predominate away from the coastline. While the shore is still rugged and indented, the increased frequency of offshore islands make these characteristics less apparent (Haller 1967: 5).

Adjoining the Chidlak Hills is the Nettilling Uplands, which extends northward from Brown Inlet and crosses the head of the Sound to Shimilik Bay. This is a rocky terrain intersected by linear scarps that form the sides of many lakes, islands, and inlets (Haller 1967: 5–6). Innumerable islands are found in most large bays of this zone. These islands, in turn, have reduced the

## Cumberland Sound Before Qallunaat  35

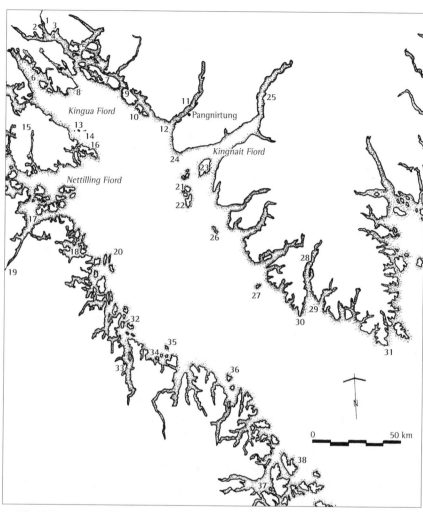

1. Milurialik
2. Millut Bay
3. Shimilik Bay
4. Issortuqjuaq (Clearwater) Fiord
5. Shark Fiord
6. Qaggilortung (Kangiloo) Fiord
7. Kekertelung
8. Anarnitung (Bon Accord) Harbour
9. Ussualung
10. Sunigut Islands
11. Pangnirtung Fiord
12. Peroeetok
13. Kilautang (Drum) Islands
14. Pujetung
15. Tarrionitung Peninsula
16. Imigen and Saunirtung Islands
17. Irvine Inlet
18. Brown Inlet (Kanagatlook)
19. McKeand River (and Oleetivik)
20. Nuvujen
21. Kikistan Islands
22. Kekerten
23. Kekertaqjuaq
24. Umana
25. Niutang
26. Miliakdjuin
27. Midlikjuaq
28. Ugjuktung Fiord
29. Abraham Bay
30. Nuvuk Point
31. Cape Mercy
32. Kingmiksoo
33. Kangertlung (Chidlak) Fiord
34. Naujateling
35. Umanaqjuaq
36. Kaxoudlin
37. Popham Bay
38. Leybourne Islands

FIGURE 6. Cumberland Sound with selected geographical place names.

effects of wave action in some inlets, while allowing the accumulation of extensive sand deposits near the mouths of rivers in others. These inlets often serve as access routes to the interior. In this regard, Irvine Inlet and Nettilling Fiord are the two most important—the former because it is the main link between Frobisher Bay and Cumberland Sound, the latter because it is the main route through the uplands to the caribou hunting grounds around Nettilling Lake.

Northeast of Nettilling Fiord, submersion of the uplands has created many deep inlets and isolated islands. East of Kingua Fiord the Nettilling Uplands give way to the Cumberland Fiords, a series of broad inlets intersecting the Penny Highlands, a rolling, upland plateau rising 800 m or so from the shoreline. The majority of these fiords are glacially-scoured, U-shaped valleys which have been flooded by sea water. Northwest of Pangnirtung Fiord, silt and sand from the Penny Highlands has filled and shortened these valleys, while towards the southeast most fiords are considerably longer and broader, and bordered by rugged cliffs and scree slopes. The largest of these is Kingnait Fiord, the head of which is the principal overland route between the Sound and Davis Strait. Except for a group of islands at the mouth of Kingnait Fiord, few islands occur along the northeast shore of the Sound, the major ones being Miliakdjuin and Midlikjuaq—the latter apparently being the place where death, and thus spiritual rebirth, originated in the mythology of the Iglulingmiut and south Baffinlanders (Boas 1907: 173, Rasmussen 1929: 92).

A strong low pressure area brings cold northern air to Baffin Island throughout much of the winter. Combined with radiative cooling, a cold prolonged winter is assured (Haller 1967: 19), moderated only by occasional invasions of Atlantic maritime air. For example, the mean January temperature at Pangnirtung in 1935 was -33° C, while the following year it was 10° warmer (Haller 1967: 21). From March to May anticyclones accompany clear skies throughout the district. By June, radiative heating and an increase in cyclonic activity heralds the onset of summer in the form of thicker cloud cover and greater precipitation. July and August represent the period of greatest annual rainfall. Though mean daily temperatures during these months may rise to 15° C, temperatures average about 7.2° C. Cloud cover and the presence of cold surface waters, which never exceed 4.5° C, however, normally prevent warming beyond these temperatures. By mid-August, the disappearance of ice from the Sound is met with an increase in cyclonic activity. Following a short autumn, there is a rather rapid reversion to winter conditions. The heaviest snowfalls are recorded in October and November, with totals averaging 575 mm for these months (Haller 1967: 22). By early November winter has taken hold.

Variable topography and expanses of cold water combine to modify this pattern in certain embayments of the Sound. For example, the steep walls of many fiords and valleys are effective in reducing exposure to sunlight, and heating takes place at a slower rate in these areas. In addition, the Penny Ice Cap and the heavily indented coastline on the northeast shore of the Sound often produce violent winds in many fiords. However, these winds, which frequently exceed 120 km per hour, usually disperse and subside when they reach the Sound. This pattern occurs so frequently in Kingnait Fiord that two distinct climates may be found within this inlet, one at its head, the other at its mouth (Haller 1967: 19–20). Thus, unequal exposure to wind and solar radiation creates a number of local micro-climates in the Sound.

Sea ice conditions in Cumberland Sound are rather complex and variable. Areas surrounding *sarbut* are the first to become ice-free, sometimes as early as April because of tidal currents. Along the shoreline of most fiords solar heating and tidal action create large areas of open water as early as mid-May. Ice in most of the larger and deeper fiords begins to break up a month or so before that in the Sound, and by mid-June most fiords are ice-free. Although boat travel across the Sound is usually possible by mid-July, heavy polar ice entering this body of water on currents from Davis Strait prevents the outward movement of ice until early August.

This pack-ice is subject to currents and winds, and will often jam ice at the mouths of fiords for considerable periods of time. In 1987, for example, while Pangnirtung Fiord became navigable in late June, a strong southwest wind jammed pack-ice against the mouth of the fiord until mid-August. Conversely, northeast winds will often block off the southwest shore of the Sound. However, when north and northwest winds prevail the entire region will be cleared of pack-ice. The timing of this latter event is thus quite variable. In 1859, for instance, several American whaling vessels were able to enter the Sound as early as mid-June, while in 1903 ice jammed the Sound throughout the summer (Low 1906). By early October sea ice begins to form in shallow and protected inlets. However, tidal currents and winds usually delay the formation of solid ice in these areas until early November. In 1923 Pangnirtung Inuit were hunting on the sea ice as early as November 12th (Haller 1967: 17), while in 1877 a continuous ice cover did not form over the Sound until late December (Howgate 1879).

During the early winter, ice consists of two types, land-fast shore ice and an agglomeration of rafted pans and icebergs made fast by the formation of new ice among them (Haller 1967: 17). The former surrounds the latter, and the junction of the two is marked by broken blocks of ice caused by tidal movements and wind conditions. Until mid-January the *sina*, or edge of the land-fast ice, rarely extends more than 10 km from the shore, and significant cracks appear in the central ice owing to tides and winds (Figure 7). The Inuit respect this phenomenon, and travel is restricted to the land-fast ice. After January, however, the central ice stabilizes and becomes fast to the shore ice, creating a less extensive floe edge closer to the entrance to the Sound. While sled travel across the Sound is usually possible after late January, the writer has witnessed on several occasions large expanses of open water off the southwest shore in March. Boas (1964) observed that the *sina* normally ran from Kekerten at the

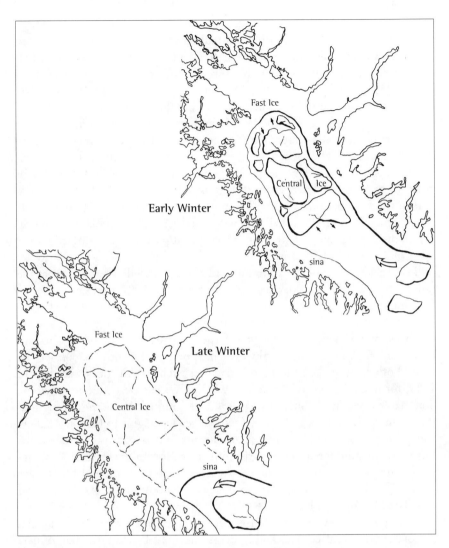

FIGURE 7. Early and late winter ice conditions in Cumberland Sound.

southern entrance to Kingnait Fiord to Nuvujen, although it had been known to extend as far north as Pujetung and Umana Islands. Alternatively, in severe winters the *sina* often forms much farther south. In the winter of 1853–54, for example, the floe edge was located over 32 km southeast of W. Penny's whaling quarters at Kingmiksoo (Penny 1854b), and in 1966 it extended from Umanaqjuaq across to Nuvuk Point (Haller 1967: 19). A strong current between Nuvuk and Kaxoudluin Island on the southwest shore, however, normally prevents the floe edge from forming beyond these points.

## Animals Hunted

Seasonal abundances and distributions of animals hunted by the Cumberland Sound Inuit are directly affected by all the above variables, especially tidal currents and ice conditions. Throughout their history the Cumberland Sound Inuit depended principally on several species of marine mammals to fulfil their nutritional, social, and other needs. These included ringed seal (*nettik; Phoca hispida*), bowhead whale (*arvik; Balaena mysticetus*), bearded seal (*ugjuk; Erignatus barbatus*), harp seal (*qairulik; Pagophilus groenlandicus*), beluga whale (*qilalugaq; Delphinapterus leucas*), narwhal (*qirniqtuq* or *qilalugaq tuugaalik; Monodon monoceros*), walrus (*aivik; Odobenus rosmarus*), and polar bear (*nanuq; Thalarctos maritimus maritimus*), although not necessarily in that order. While caribou (*tuktu; Rangifer tarandus arcticus*) may have been more intensively hunted in the past by some groups (Boas 1964: 22–3), this species and char (*iqaluk; Salvelinus alpinus*) appear to have always been secondary food sources; the former were hunted primarily for winter clothing, the latter for a change in an otherwise steady diet of sea mammal meat (*nerqri*) and blubber (*uqsuq*). The inclusion of Arctic fox (*tiriganirk; Alopex lagopus*) in the economy of the Cumberland Sound Inuit is largely a product of the fur trade.

The ringed seal inhabits the Sound year round. Weighing as much as 80 kg, *nettik* remains the staple food in the diet of the Cumberland Sound Inuit. In winter this species maintains breathing holes through the land-fast ice, and its distribution and abundance is determined principally by the thickness and extent of this ice (McLaren 1961). As adult females require stable ice with suitable snow cover in order to pup, adult seals tend to move well away from the *sina* into the fast ice, where their dens remain undisturbed by wind and tidal action. Consequently, the majority of seals maintaining breathing holes in the fast ice are mature, while those at the *sina* are immature (one to five years), a fact born out by Kumlien's (1879) observation that no immature seals were killed in the tide rips at the head of the Sound during the winter of 1877–78. By early March the females select their dens and pupping takes place from early March to mid-April. In April ringed seals can be found on the ice basking in the sun. During this period feeding activity is reduced and a corresponding thinning of the blubber layer occurs (McLaren 1961). If not quickly retrieved, seals killed at this time will often sink. As the ice breaks up, many immature seals move to leads and cracks. During this time, seals born in the spring, known as 'silver jars' or *netsiavinik*, tend to congregate around Nuvujen and Brown Inlet (Boas 1964: 26, Haller 1967: 53), where they were frequently hunted by the inhabitants of Keemee and Kipisa.

After the ice leaves the Sound, many ringed seals are found within 5 km of the shore, where they remain throughout the summer. However, a significant number of seals apparently migrate out of the Sound with the ice, leaving a scarcity of seals in most areas and forcing hunters to travel further afield

(Haller 1967: 56–7). Local Inuit assert that as October approaches there is, in turn, an influx of immature seals into the Sound from Davis Strait. Haller (1967: 57–60) provides support for both the out- and in-migration of ringed seals in the summer and fall. The theoretical population of this species in Cumberland Sound has been estimated to be c. 74,000, of which an average of 9700 were taken each year between 1962 and 1965 (Haller 1967: 61, 65)—a figure three times the number of sealskins traded annually at the Hudson's Bay Company's Pangnirtung Post between 1924 and 1936 (Goldring 1986).

Attaining weights of 40,000 kg or more, *arvik* was the largest animal hunted by the Inuit. In former times, the bowhead whale was a frequent visitor to Cumberland Sound. While at anchor among the islands at the head of the Sound in 1585, Davis observed three or four whale skulls on a nearby island. In the same area two years later, a whale swam by his anchored ship (Markham 1880: 13, 46). Europeans did not enter Cumberland Sound again for another 253 years. However, when W. Penny sailed into Tenudiakbeek in 1840 with his Inuit guide, Eenoolooapik, abundant evidence of whales was once more encountered. Freshly killed whales were observed near the traditional village of Noodlook in Bon Accord Harbour (M'Donald 1841: 89), and further up Kingua Fiord, which was considered by Inuit to be 'the principal resort of the fish' (M'Donald 1841: 82). Here, Penny found:

> a very large fish which had been killed about ten days before. It was supposed that there were not less than twenty tuns of blubber piled upon the beach at this point. Near the same spot there were also the remains of former victims in great abundance. (M'Donald 1841: 93)

And near Ussualung Penny met several Inuit whose kayaks were loaded with 'whalebone' (baleen) from a very recently killed whale (M'Donald 1841: 86). Just prior to Penny's arrival, numerous whales had been sighted by Inuit near Kingmiksoo and in Kingua Fiord, where Penny was told he 'could still find them in abundance' (M'Donald 1841: 80–2). While Eenoolooapik stated that the tribes occupying the shores of Tenudiakbeek 'were in the practice of killing considerable numbers of whales for the sake of their flesh, which forms a staple article of food' (M'Donald 1841: 7), Penny (1840) observed that the Sound's inhabitants killed 'annually from 8 to 12 whales', a custom he felt 'worthy of notice as it seem[ed] to be peculiar to these Esquimaux'. While anchored at Nuvuk Point in early September Penny saw a large number of whales entering the Sound from the south, as local Inuit assured he would. In 1839 Penny's guide, Eenoolooapik, drew a map of Cumberland Sound recording the locations of various settlements and principal whaling grounds (Figure 8). In regard to the latter, females and calves were to be found at the mouth of Kingua Fiord, whereas other whales were numerous on the east side of the Sound between Miliakdjuin and Midlikjuaq.

# Cumberland Sound Before Qallunaat

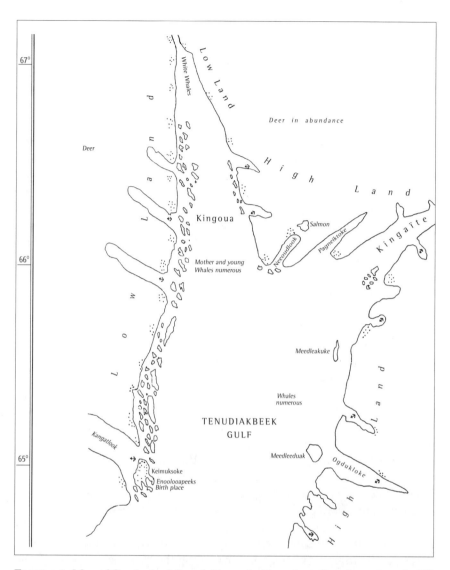

FIGURE 8. Map of Cumberland Sound. Drawn by Eenoolooapik and associates for W. Penny in 1839, showing locations of settlements and principal whaling and hunting grounds (adapted from NMC-59335).

Local knowledge supports not only a spatial separation of the stock, but a temporal one as well. According to my informants, as soon as the ice permits, females and calves enter the Sound from the southwest, while during the fall males enter the Sound along the northeast shore, where they remain until freeze-up. In late August of 1841, an elder hunter of some importance from Anarnitung informed Captain M. Wareham (1843: 24) that:

no whales were to be found at . . . (their) present location, except when land-ice occurred, and when the ice in the more southerly parts of the inlet was breaking up; but that the route of the whales was round Queen's Cape (Cape Mercy), and along the east land as far as Gambier Island (Midlikjuaq), where they crossed the inlet and remained among the islets near Kimmock-sowick (*sic*) till the ice set in for the winter.

Prior to their decimation by commerical whalers, bowhead whales were also encountered at the floe edge throughout the spring (M'Donald 1841: 128, Wareham 1843: 24). It is difficult to estimate the aboriginal population of bowheads in the Sound. However, if the average whale was valued at $5000, and if C.F. Hall's estimate is correct that commercial whalers took $15,000,000 worth of oil and 'whalebone' alone from Cumberland Sound between 1840 and 1871 (Davis 1876: 40, see Holland 1970: 41), the precontact bowhead population may have numbered more than 1500 whales. Although intensive whaling from 1840 to 1920 all but decimated the bowhead in Cumberland Sound waters, as early as 1927 its numbers began to increase.[1]

Bearded seals sometimes grow to 300 kg and are regarded, along with the walrus and various species of whales, as large marine mammals in Inuit ideology. Bearded seals were formerly taken when and wherever encountered; this animal's durable hide was highly valued for boot soles, boat covers, lashings, and whaling and sealing lines. Occasionally *ugjuk* will maintain breathing holes through the land-fast ice. However, normally they live in open water during all months of the year. Throughout the winter, *ugjuk* may be found at the floe edge, especially below Kaxoudluin Island where a strong current prevents the water from freezing over (Haller 1967: 73–4). At break-up bearded seals move further up the Sound to Bon Accord and Kangiloo waters, where they remain until the formation of land-fast ice in the fall. Haller (1967: 76) estimates the bearded seal population in the Sound to be *c.* 6000, of which no more than 150 were taken annually during the early 1960s.

The harp seal is larger than the ringed seal, but considerably smaller than the bearded seal. Unlike the latter two seals, *qairulik* is only a seasonal visitor to the region. During the winter and early spring the harp seal is found off the Labrador coast, where its pups are born. In May, huge herds of adults migrate north through Davis Strait, with an undetermined number of these animals taking up residence in Cumberland Sound. A few weeks later the immature seals arrive. Harp seals are especially numerous at the mouth of Pangnirtung Fiord and in Kangiloo Fiord from Bon Accord to Imiyoomee, where an abundance of their principal food, Arctic cod (*uugaq, Boreogadus saida*), may be found. *Qairulik* also occurs in lesser concentrations around Kingmiksoo and Kekerten. The harp seal is a sociable animal and is found in herds of several hundred or more throughout the Sound during the summer. When the ice begins to form in the fall, *qairulit* leave for southern waters, although Haller (1967: 68, 70) speculates that a small percentage of non-breeding adults and immature seals remain throughout the year. Interestingly, the occurrence of this animal in the past, both spatially and temporally, appears to have been less extensive. Kumlien (1879: 61) noted that the harp seal was only occasionally found as far north as Anarnitung and that it returned in the spring for only a short time. Until 1965, the harp seal was killed only for domestic purposes, usually for sleeping and ground sheets. However, with the increase in the value of its skin and the advent of faster boats, which reduced the risk of losing a sinking seal—the harp arrives in the region with very little fat content—this seal became a popular item of trade during the late 1960s and 1970s.

As the ice begins to break up in late May and early June, the *qilalugaq* enters the Sound in groups of varying size, eventually making its way to its head amongst the ice pans. Milurialik, 'the place where stones are thrown', a small tidal flat near the mouth of the Ranger River in Clearwater Fiord is the favourite gathering place for the beluga from late June to early August. While federal government scientists believe that less than 500 beluga remain in this herd (Department of Fisheries and Oceans 1994), many more beluga apparently gathered in the vicinity of Millut Bay in the past.[2] As its name implies, Milurialik was the principal hunting grounds of the beluga in precontact and historic times. It was also the location where most commercial beluga whaling took place during the late nineteenth and early twentieth centuries. After July, beluga are dispersed throughout the Sound. With the onset of cooler temperatures, beluga begin to congregate near the mouths of Nettiling and Kangiloo fiords. With the formation of ice in the fall, they leave for more open waters. Although the meat of this 500 to 1000 kg whale is no longer used for dog food—snowmobiles replaced dogs as the major form of winter transportation in the mid-1960s—they are still taken for their *maqtaak*, which is one of the most sought after foods of the Cumberland Sound Inuit, as is the *maqtaak* of the narwhal.

The latter whale is also a seasonal visitor to Cumberland Sound, although it is sometimes taken at the *sina* in the spring. In the past few years the narwhal has been seen and hunted in increasing numbers between break-up and freeze-up, perhaps exploiting the ecological 'elbow-room' created by increased dispersal of the beluga population.

During the winter the polar bear can be found among the ice-congested waters off the *sina* in pursuit of its favourite prey, the ringed seal. After break-up *nanuq* is usually confined to shores on either side of the entrance to the Sound. Although the polar bear is rarely seen north of the Kikistan Islands, in August of 1988, 30 or more bears were found within a 2 km radius of a 13 m (40,000 kg) bowhead whale that had washed ashore at Pujetung. Formerly, the polar bear was hunted at the *sina* with spears and dogs, and much prestige was associated with its capture.

Walrus are no longer plentiful in the Sound north of Nuvuk Point. However, based on the occurrence of place names at the head of the Sound referring to this animal (e.g., Anarnitung translates as walrus 'shit') *aivik* apparently once frequented this area in considerable numbers. Today, as in the mid-nineteenth century, walrus occur primarily at the mouth of the Sound between Nuvuk Point and Cape Mercy on the northeast shore and between the Leybourne Islands and Abraham Bay on the southwest shore. In 1840 M'Donald (1841: 115) visited the village of Togaqjuaq ('big tusk') near Cape Mercy, reporting that 'it was a place favourable for the capture of walrus'. Inuit hunters stationed at Kekerten and Umanaqjuaq (Blacklead) Islands at the turn of the century hunted walrus for commercial purposes. However, by 1902, after several annual kills of around 150 animals, the walrus harvest fell off dramatically (Lubbock 1955). Like most whales, the walrus only visits the Sound during the open water season.

Before the introduction of the snowmobile and permanent housing in Pangnirtung in the mid-1960s, caribou were hunted primarily for their skins, even though Euroamerican clothing had been worn for decades. Traditionally, Inuit travelled inland every summer for several weeks or more to procure enough skins for winter clothing. The main hunting grounds were located between the head of the Sound and Nettilling Lake. In addition, caribou were pursued from the heads of most fiords. Prior to the early nineteenth century a division of the Talirpingmiut, the 'tribal' or regional division on the southwest shore of the Sound, apparently spent the greater part of the year inland living off caribou and other resources in the vicinity of Nettilling Lake (Boas 1964: 22). From the shores of Nettilling Fiord this group made its way inland about the beginning of May, returning to the coast in December—a pattern of transhumance which Boas considered unique among all Baffinlanders and indeed most Central 'Eskimo' groups. Formerly, most Inuit occupying the shores of Cumberland Sound hunted caribou only in the late summer and early fall. However, with the introduction of the repeating rifle in the late nineteenth century, caribou soon became an important supplementary food source throughout the winter. While many Inuit still wear caribou skin clothing when out on the land in winter, caribou is now hunted primarily for its meat. As Stenton (1989) has clearly demonstrated, the South Baffin caribou herd is subject to extreme local fluctuations in population. For example, during the early 1920s herds of several hundred caribou were observed each year in Pangnirtung Fiord. By the end of the decade few caribou were seen in this fiord and none were killed for the next 40 years or so (see Haller 1967: 82). Nowadays caribou appear to be making a comeback in this area; Pangnirtung hunters rarely travel more than a few hours during the winter before caribou are encountered.

Both anadromous (ocean-going, river-spawning) and landlocked char are found in the vicinity of Cumberland Sound. Even though char was never an important staple among the Cumberland Sound Inuit, *iqaluit* were usually taken with spears, and later with nets, at the mouths of rivers and streams in early August where they school before starting their upstream migration. Camps were often established at the heads of inlets in summer in order to exploit local anadromous fish runs and caribou herds. Some of the more important fishing localities in the Sound include Iqaluit near Opinivik, Iqalugaqdjuin Fiord, and Avatuktoo, where char can be taken in a chain of nearby lakes throughout the winter.

Fox were not actively pursued by Inuit in Cumberland Sound until after the turn of the century, when the depletion of whales forced station managers at Kekerten and Umanaqjuaq to begin harvesting alternative resources. Throughout the twentieth century, the local hunter was known as a 'pretty fair sealer and whaler, but a very poor trapper'.[3] While Cumberland Sound was never regarded as a productive trapping ground, and Arctic fox is subject to extreme cyclical fluctuations, the general failure of trapping rested on cultural factors: 'They are more or less content to hunt seals, and the fur (fox) hunt is becoming of secondary importance. They appear to have little ambition to secure anything but ammunition and tobacco.'[4] In good fox years, however, seal returns were lower, not because fewer seals were killed, but because of 'the disinclination of the natives to clean and bring in the skins while they can obtain their requirements much more easily with fox skins.'[5]

The above species do not represent all the animals traditionally hunted by the Cumberland Sound Inuit; a variety of migratory birds, for example, were taken in the summer. Nevertheless, they do include the major species upon which important economic and social decisions were based in historic times, and presumably late prehistoric times as well.

## Reconstructions of Cumberland Sound Inuit Prehistory

The prehistory of the Cumberland Sound Inuit was largely unknown until Schledermann (1975, 1979) undertook three seasons of archeological fieldwork there in the early 1970s. Based on excavations at over two dozen winter village sites in the upper half of the Sound, Schledermann constructed a model of Cumberland Sound Inuit prehistory which remains to this day. Although artifacts belonging to various phases of Dorset Inuit culture were recovered, Schledermann was concerned primarily with documenting the development of Thule Inuit culture in the Sound. Toward this end, he recorded 264 sod houses, of which about 10 per cent and a number of associated midden deposits were excavated, producing over 2000 artifacts and at least ten times as many identifiable bones. Changes in climate, house form, and various species in refuse deposits over time, led Schledermann to suggest that the Thule occupation of the Sound was divided into three periods: (1) an initial and major period of occupation beginning in the thirteenth or fourteenth centuries A.D., and lasting until about A.D. 1650, (2) a second period of very sporadic settlement, which lasted about 100 years, and (3) a third period of major occupation which began around A.D. 1750 and terminated with the establishment of Euroamerican shore-based whaling stations in the mid-nineteenth century.

The arrival of the first Thule Inuit in Cumberland Sound was considered to be a continuation of the original migration of Thule culture-bearing Inuit out of Alaska. Schledermann felt that the first Thule Inuit to arrive in the Sound followed the north coast of Hudson Strait from the Foxe Basin area to Frobisher Bay, from whence they entered Cumberland Sound. A comparison of precontact material culture from several regions of the central Canadian Arctic suggested a close tie between Cumberland Sound and the Iglulik area (1975: 275). During the first occupational phase, which is ascribed to the middle Thule period elsewhere in the central Arctic (1975: 247), there was an economic emphasis on the bowhead whale, as indicated by an abundance of baleen in midden levels located between 20 and 25 cm below the ground surface. However, at the height of the 'Little Ice Age' in the sixteenth and seventeenth centuries, land-fast ice increased markedly in extent and duration. While this cold period produced more favourable habitat for ringed seals, it resulted in a significant reduction in the amount of open water and thus bowhead whale habitat. The effects of climatic cooling became particularly evident around A.D. 1650, and the amount of baleen in refuse deposits after this time period fell off sharply (1975: 256). Schledermann recorded a 10-cm hiatus in cultural material immediately above his 'baleen period' levels which he interpreted to be evidence of sporadic occupation between A.D. 1650 and 1750. In the upper 10 cm of Schledermann's middens, contact materials and faunal refuse, particularly ringed seal bones, increased dramatically in volume. This 'great increase in seals and other animals must certainly have been indicative of a shift in hunting pattern, and was probably directly related to the decreasing availability of baleen whales' (1975: 259).

Attendant with this climatically induced shift in economy, other changes were also taking place, most notably a change in harpoon head and winter house styles. New varieties of harpoon heads, which may have been related to an increase in seal hunting, appeared in levels stratigraphically associated with the last major period of occupation (1975: 248–9). While there was a greater number of single-room houses in use during the first occupational period, a proportional increase in multiple family platform structures in later occupation periods was noted. Of the five three-room communal dwellings tested or excavated by Schledermann, all were placed within the early contact period (1975: 262). The tendency towards communal living during the later part of the Little Ice Age was considered to be a direct function of an increasing dependence upon seal hunting and a corresponding decline in whaling, whereby:

> the less fortunate family would likely experience more frequent food shortages, as well as problems in obtaining an adequate supply of fuel for cooking

and for heat. (In turn) the lack of an adequate fuel supply for heating the winter house could certainly be offset to a large degree by use of a few large dwellings, rather than attempting to heat many smaller ones. The use of large communal dwellings can be seen, then, as a positive adaptive mechanism which served to stabilize the economic situation, and the sharing practices helped to conserve fuel for heating and cooking. (1975: 266–7)

Cold conditions and 'hard times' during late prehistoric times were also held responsible for the gradual abandonment of land-based winter villages (and thus a hiatus of cultural materials in refuse deposits) and greater mobility, which resulted in an increased use of snowhouses and *qammat* (less substantial fall/early winter dwellings) . . . 'when the more permanent settlements were abandoned' (1975: 269).

This interpretation of Cumberland Sound prehistory has been accepted by most archaeologists, while providing support for the view that climatic cooling throughout the central Arctic during the Little Ice Age resulted in an economic shift away from bowhead whaling to sealing. However, the historical evidence provided above clearly suggests otherwise. In fact, perhaps more than any other Central Inuit group, the Cumberland Sound Inuit possessed a specialized whale hunting economy at contact.[6] Boas recorded numerous stories among the Cumberland Sound Inuit which referred to various aspects of aboriginal whaling (e.g., 1907: 249–51, 255–6, 276–9). Indeed, according to Boas (1964: 32) Inuit whaling in Cumberland Sound was:

formerly carried on in . . . bulky skin boats. They pursued the monstrous animal in all waters with their imperfect weapons, for a single capture supplied them with food and fuel for a long time. I do not know with certainty whether the natives used to bring their boats to the floe edge in the spring in order to await the arrival of the whales, as the Scotch and American whalers do nowadays, or whether the animals were caught only in the summer. On Davis Strait the Padlimiut and the Akudnirmiut used to erect their tents in June near the floe edge, whence they went whaling, sending the meat, blubber, and whalebone to the main settlement. In Cumberland Sound whales were caught in all the fjords, particularly in Kingnait, Issortuqdjuaq (Clearwater Fiord); and the narrow channels of the west shore. Therefore the Eskimo could live in the fjords during the winter, as the provisions laid up in the fall lasted until spring.

### An Alternative View

How, then, does one reconcile Schledermann's interpretations with historical and ethnographic fact? Assuming that Schledermann's midden profiles accurately reflect the depositional histories of such deposits, the absence of baleen in levels where contact materials and ringed seal bone predominate may be most parsimoniously explained by the possibility that Inuit were no longer discarding baleen because they were trading it to the whalers. In this light, the apparent hiatus in cultural materials in levels assigned to the period A.D. 1650 to 1750 begs re-examination. We know that as early as 1835 Inuit from Cumberland Sound were travelling to Durban Island on Davis Strait to trade baleen and to interact with whalers (M'Donald 1841: 94–5). In 1837, for example, several Inuit travelled overland from Durban Island in five days to fetch 'whalebone' from the Sound to trade with two Scottish whaling captains (Goldring 1986: 159). By 1840 a 'lively trade' existed between Cumberland Sound and Davis Strait (Boas 1964). It is also noteworthy that the Inuit who migrated to Durban Island from Anarnitung and Kingmiksoo appear to have been the more influential and 'well-to-do type of native' (M' Donald 1841). The temporary absence from the Sound of more productive families could possibly account for a marked decrease in artifacts and faunal refuse in levels occurring between deposits dominated by baleen on the one hand, and contact materials and ringed seal bone on the other.[7] But this explanation does not fit with Schledermann's chronology of events as he speculated that baleen virtually disappeared from refuse deposits around A.D. 1650, not nearly two hundred years later.

Schledermann's stratigraphic sequence is not suspect. Rather, his chronological and associated cultural interpretations should be re-examined. Given the historical evidence presented above, Schledermann's model of Cumberland Sound Inuit prehistory becomes tenuous when he implicitly assumes a constant and uniform rate of deposition, not only within individual deposits, but between middens and sites as well. The possibility that the rate of refuse deposition in household middens in Cumberland Sound was not so much a function of time but of intensity of occupation, was recently demonstrated to the writer during excavations of an early twentieth century *qammaq* on Kekerten Island. An elderly informant in Pangnirtung revealed that she had lived in this dwelling with her husband for a little less than two years around 1917–18. Yet excavation of the midden deposit near the entrance to this feature produced over 20 cm of historic materials and faunal refuse (mostly seal), which, if we were to employ Schledermann's methodology, would place the initial occupation of this feature squarely at the end of the 'baleen period' around A.D. 1650. The fact that this individual, Qatsu, was the daughter of the leader of the Qikirtarmiut, Angmarlik, and that she was married to the son of another prominent whaler (Keenainak) probably contributed to the overall productivity of the household and thus thickness of the deposit. Nonetheless, the lesson is instructive.

The likelihood that the upper portions of Schledermann's midden deposits, including his 'baleen period' levels, were produced much later than he estimated, while difficult to substantiate without additional research, is indirectly supported by faunal data provided in his paper (1975: 97–103). Schledermann suggests that the marked increase in ringed seal bone during the late precontact and early contact periods was the result of an economic shift towards sealing. However, while he documents an increase in the number of ringed seal bones, there was also a proportionate increase in the number of bones of other seals. What changed, then, was not the relative importance of individual species or the economic focus, but rather the intensity of exploitation, or, more accurately, the intensity of occupation at specific locations as reflected in increased rates of refuse production.

Based on evidence and discussion presented to date, an alternative model of Cumberland Sound Inuit prehistory may be considered. From the time of their arrival in the Sound to a century or two prior to contact, Thule Inuit living on the shores of the Sound were generalized in their economic orientation, exploiting a variety of available species on the basis of their encounter frequency. During this period, settlement was characterized by small, single family dwellings arranged in linear fashion along the same contour lines of sites (Schledermann 1975). However, during later prehistoric times, bowhead whaling increased in importance. This specialization may have been socially motivated (see Chapter 6), or it may have been driven by environmental factors whereby a spatial and temporal expansion of land-fast ice in island-congested waters at higher latitudes (e.g., the Iglulik area) during the Little Ice Age may have resulted in an increase in the number of bowheads visiting the Sound. Alternatively, intensive commercial whaling in Baffin Bay throughout the 1820s may have forced more whales into Cumberland Sound than previously. Whatever the case, this development resulted in an increase in the use of specific locations where whales and seals could be taken (e.g., islands at the head of the Sound), and communal living arrangements. The latter developed, not because of the need to conserve resources in 'difficult times', but because greater numbers of men had to be assembled into co-operative work units in order to hunt bowhead whales effectively. Double- and triple-room houses simply helped to co-ordinate, maintain, and reinforce the solidarity of these production units. The possibility that some of these production units were based on either same or adjacent generation partnerships among extended family members seems likely; even the most productive households probably could not have formed a large enough unit to hunt bowhead whales.[8]

The intensification of whaling and occupation at the head of the Sound reached its peak in the century prior to contact, resulting in an abundance of baleen in refuse deposits. The absence of baleen after this period in middens marks the onset of trade with British whalers, while the hiatus in cultural materials observed below those levels in which contact items were abundant could represent a temporary migration out of the Sound by more productive families in order to establish trading relations with the whalers in Davis Strait.

This model accounts for both the archaeological and historical evidence, while adopting an informed view of the formation of household middens in Cumberland Sound. Even so, it too should not be accorded too much credibility until more research is carried out; Schledermann's stratigraphic sequence was based largely on the excavation of one refuse deposit at Anarnitung (MbDj-1) and two house middens at a large village site 8 km west of Imigen Island (LlDj-1). The possibility that we still know very little about the complexities of Cumberland Sound Inuit prehistory is evident in Salter's (1984) study of two late precontact burial populations from Tasaiju (McDi-1) on the northeast side of the Sound near Nunaata and Niutang (MbDc-1) in Kingnait Fiord (Figure 9).

Salter (1984) undertook an intensive analysis of 46 and 48 human skeletons, respectively, from Niutang and Tasaiju. Her analyses indicated that both burial populations dated to late precontact times, or from about A.D. 1750 to 1840. No European pathologies nor genetic traits were found in either population, and associated artifacts appeared to be late prehistoric in age. While marine mammals were placed as grave offerings in burials at Tasaiju, burials at Niutang contained a much higher proportion of caribou (1984: 302)—an economic focus likewise found by Schledermann's (1975) excavations at Niutang. However, Salter's most significant finding was that there was relatively little biological similarity, and presumably genetic exchange, between Inuit living at Niutang and Tasaiju:

> The biological distances generated using cranial discrete and cranial metric data were examined. The results from these two differing techniques both demonstrated a dramatic phenetic distance between people from Niutang and Tasioya (*sic*). They were not closely related phenetically, as one would expect based on their geographic propinquity. (1984: 304)

In fact, the biological distance between burial populations at these two sites was equivalent to that between the Silumiut (Sadlirmiut) of Southampton Island and Inuit of the Labrador coast—a distance of more than 1440 km (1984: 291). Yet the Niutang and Tasaiju burial sites are only 90 km apart.

## Aboriginal Social Groups and Interaction

Boas was fairly explicit in differentiating between aboriginal patterns of land-use and those altered by Euroamerican contact, even though his data may have been derived from postcontact sources. The following descriptions are abstracted from Boas (1964: 16–32) and may be considered accurate for that period immediately prior to contact.

Inuit inhabiting the shores of Cumberland Sound and the southern shore of the Cumberland Peninsula, or Saumia, were known to other Inuit groups as Oqomiut, 'those living on the lee side' (Boas 1964: 16). Formerly, the Oqomiut

## Cumberland Sound Before Qallunaat    51

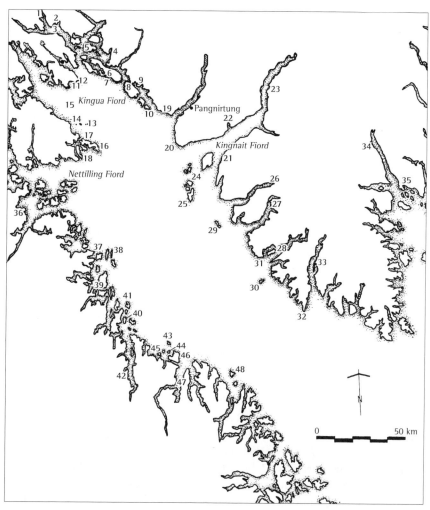

1. Milurialik
2. Shimilik
3. Issortuqjuaq Fiord
4. Tasayu (McDi-1)
5. Nunaata
6. Kekertelung
7. Tulukan ? (MbDi-1)
8. Ussualung
9. Avatuktoo (MbDg-1)
10. Aupalluktung
11. Anamitung (MbDj-1)
12. Idlungajung (Noodlook?)
13. Pujetung
14. Kilauting
15. Sakiaqdjung
16. Imigen Islands
17. Sauniqtuajuq
18. LiDj-1
19. Inuguaarulu (MaDg-2)
20. Tesseralik
21. Kitingujang
22. Kingnait (Torgnait)
23. Niutang (MbDc-1)
24. Tuapait
25. Kekerten
26. Iqalugaqdjuin (Exaluqdjuin)
27. Iqaluqdjuaq (Exaluqdjuaq)
28. Kangertukjuaq
29. Miliakdjuin
30. Midlikjuaq
31. Shaumia (Saumeer)
32. Nuvuk (Nuvukjaluin)
33. Ugjuktung
34. Togaqjuaq
35. Aukadliving
36. Auqardneling
37. Nuvujadlu
38. Nuvujen
39. Opinivik
40. Kingmiksoo
41. Idjorituaqtuin
42. Kangertlung (Chidlak) Fiord
43. Umanaqjuaq (Blacklead Island)
44. Sagdluaqdjung
45. Naujateling
46. Qeqertaujang (Kekertaluk)
47. Qasigidjen (Ptarmigan Fiord)
48. Kaxoudlin

FIGURE 9. Map of Cumberland Sound with selected historical and archaeological sites.

were divided into four regional subdivisions:

—Talirpingmiut ('the people of the right side') on the southwest shore of the Sound,
—Kinguamiut at the head of the Sound,
—Kingnaimiut ('the people of the mountains') on the northeast shore,
—Saumingmiut ('the people of the left side') on the southeastern slope of the highland of Saumia.

Interestingly, two of these divisions, the Talirpingmiut and the Saumingmiut, utilize the head of the Sound as their reference point.

The most southerly settlement of the Talirpingmiut was Naujateling. In the fall, seals were hunted in narrow channels and fiords from this location. When winter arrived and the sea ice froze, usually about December, this group moved to Umanaqjuaq. In March, Inuit left Umanaqjuaq in search of young seal. This hunt was apparently pursued with much energy throughout the entire Sound as the white coat of the young animal was of prime importance for inner garments (Boas 1964: 20). The principal summer settlements of the Naujateling natives were located at the head of Qasigidjen and Kangertlung Fiords, which are situated near Idjorituaqtuin and Kingmiksoo, from which they ascended the level highlands in search of caribou.

Further up the Sound was the settlement of Idjorituaqtuin. The same relationship that existed between Naujateling and Umanaqjuaq pertained between Idjorituaqtuin and Kingmiksoo; in the fall, after caribou hunting, Talirpingmiut gathered at Kingmiksoo until freeze-up, whence they moved to Idjorituaqtuin. Kingmiksoo was regarded as the principal settlement in the Sound at contact (M'Donald 1841, Sutherland 1856, Wareham 1843), as over 110 Inuit in 16 houses were stationed here (see below). Summer settlements were located at the heads of numerous fiords to the west (Boas 1964: 21). North of Idjorituaqtuin the winter site of Nuvujen was found along with the fall settlement of Nuvujadlung, which is located near a high cliff at the entrance to Nettilling Fiord. Boas regarded the inhabitants of Nettilling Fiord as by far the most interesting group of Talirpingmiut as 'among all the tribes of Baffin Island . . . it is the only one whose residence is not limited to the seashore' (1964: 22). The Talirpingmiut apparently once occupied three or four settlements near the south end of Nettilling Lake and the outlet of the Koukjuaq River. As seals are permanent residents of this large freshwater lake, Boas (1964: 22) speculated that some members of this 'tribe' stayed here year-round and rarely descended to the coast. In November of each year this branch of the Talirpingmiut gathered at Isoa on the easternmost bay of the lake and made their way to the entrance to Nettilling Fiord, where, in the same manner as other Oqomiut, they pursued seals at their breathing holes, and later at their dens. However, in spring this group travelled west towards Amitoq and Isoa, and finally to the Koukdjuaq. Here, they split, with the older men and women staying behind and the others descending the river in search of game. In August the group reunited at Qarmang, where they stayed until sled travel across the lake to Isoa was possible.[9]

Nettilling Fiord and its numerous islands formed the northern boundary of the Talirpingmiut, whereas the Kinguamiut inhabited the area from Imigen around the head of the Sound to Ussualung on the northeast shore. Formerly, the Kinguamiut were associated with three principal settlements: Imigen, Anarnitung, and Tulukan. The former site is apparently 'situated in the midst of one of the best winter grounds' in the Sound as the 'southern portion of the island, on which the huts are erected, projects far out to sea' (Boas 1964: 27). However, strong tidal currents frequently create rough ice conditions which hamper breathing-hole and open water sealing. As spring approached the natives of Imigen often moved to the largest island in the Pujetung group, which is even closer to open water. At other times, late winter settlements were established on the land-fast ice further north, where the natives remained until late April. In summer, the natives of Imigen usually went caribou hunting either to Issortuqjuaq Fiord, where they lived at the popular fishing locations of Exaluaqdjuin, Shimilik, or Midlurielung (Milurialik), or to Exaluqdjuaq Fiord near Ussualung. In the fall, the natives resorted to Saunirtung or Saunirtuqdjuaq, two islands northwest of Imigen, staying until January, whence they returned to the sea (1964: 27–8).

The second settlement of the Kinguamiut, Anarnitung, together with its neighbouring point of land in Bon Accord Harbour, Idlungajung, were, outside of Kekerten, the principal seat of occupation in the Sound during the early 1880s (Boas 1964: 28). It is also apparent that it was an important settlement in 1840 when Penny visited it. Here, in the village of Noodlook, M'Donald (1841: 89, 91) found 40 people living in 'seven huts . . . of very portable description' (i.e., skin tents). However, he was assured that 'during the winter their number would be much increased—the majority of their tribe having gone to the lakes (of which there were many not too far distant) for the purpose of catching salmon', while still others were inland hunting caribou (M'Donald 1841: 91). Caribou were hunted at the head of Issortuqdjuaq as well as around Nettilling Lake, which they reached by crossing Tarrionitung Peninsula. If the ice in the upper parts of the Sound was smooth, families from Anarnitung moved to Kilauting in the Drum Islands just north of Imigen, where they hunted seal at breathing holes (Boas 1964: 28). If the ice was rough they remained near Anarnitung to hunt seals at *sarbut*. During the young sealing season most families left for Sakiaqdjung and other small islands near the entrance to Qaggilortung (Kangiloo) Fiord. However, heavy snowfalls often compelled them to forego this region for the open sea. While tidal rips often concentrated seals at *sarbut* in winter, the size of these tide holes increased markedly in spring and mild winters, forcing travel over undesirable routes and passages. As Boas (1964: 29) does not mention the annual movements of the Tulukan community, it may be inferred that he could not find anyone who had lived at this settlement.

In the early 1880s, the Kingnaimiut tribe resided entirely on Kekerten Island. However, prior to contact their principal settlements appear to have been situated at the mouth of Pangnirtung Fiord, Miliakdjuin, and at Niutang and Kitingujang in Kingnait Fiord. In summer, the Kingnaimiut hunted caribou from the heads of Nirdlirn, Pangnirtung, Kingnait, Exaluaqdjuin, and Kangertlukdjuaq fiords, where fishing was also carried on. The favourite settlement of the Kingnaimiut was Kitingujang in Kingnait, as large numbers of char were taken in the river at this spot, and the gentler lay of the land afforded ample opportunity for long caribou-hunting excursions. Although Boas did not detail the subsistence and settlement patterns of the Kingnaimiut during the rest of the year, these may be inferred from archaeological and informant data. In the absence of numerous *sarbut*, hunting was divided between breathing-hole sealing and floe edge hunting during the winter. While the former pursuit was carried out between the Kikistan Islands and the mainland, and in the islands northwest of Pangnirtung Fiord, the location of the latter activity depended upon the position of the *sina*, which may have been located as far south as Midlikjuaq or as far north as Umana. Young seals were taken all about Kekerten in the early spring as well as off the mouth of Pangnirtung Fiord south of Aupalluktung and the Sunigut Islands.

The Saumingmiut inhabited the inlets of Cumberland Sound southeast of Midlikjuaq, including Ugjuktung Fiord, where the winter settlement of Qeqertaujang was located. Walrus were taken just before freeze-up in the fall by Inuit from Ugjuktung, and during the winter seals were hunted at the entrance to this fiord. In March these Inuit either went polar bear hunting or moved up the Sound to join the Kingnaimiut during the young sealing season. Later during the spring, the Ugjuktung natives joined 'others of their kind' on Davis Strait, where another principal winter village of this regional division, Aukadliving, was located. 'Here walrus are hunted in the summer and in the fall and a great stock of provisions is laid up . . . (while) in winter the floe (edge) offers a good hunting ground for sealing and in the spring the bears visit the land and the islands to pursue the pupping seals' (Boas 1964: 31). In summer, Togaqjuaq on Davis Strait was a favourite settlement as caribou were easily hunted from this location. Another important summer station was Qarmaqdjuin, which was also used by the more northerly Padlimiut.

Boas (1964: 17) was of the opinion that no great difference ever existed between the above 'tribes'. However, in light of differences in the Tasaiju and Niutang burials, as well as numerous stories collected by Boas himself, it is clear that substantial barriers to social interaction, if not outright hostility, existed between some local groups, and perhaps larger regional groupings. In

## Cumberland Sound Before Qallunaat    55

fact, according to Boas, the last feud to occur in the Sound took place between the Kingnaimiut and Kinguamiut during the early 1820s:

> At that time a great number of Eskimo lived at Niutang, in Kingnait Fiord, and many men of this settlement had been murdered by a Qinguamio of Anarnitung. For this reason the men of Niutang united in a sledge journey to Anarnitung to revenge the death of their companions. They hid themselves behind the ground ice and killed the returning hunter with their arrows. (1964: 57)

The Kingnaimiut appear to have been particularly prone to feuding. Boas (1907: 290–1; 294–5; 299–301), for example, recorded at least two instances of feuding between villages in Kingnait Fiord and another with an Aggomiut village north of Padli on Davis Strait, the latter occurring in the early 1800s. Feuds also apparently took place within the same settlement; Boas (1964: 231) relates a story of two enemies in the village of Niutang who vowed to kill each other, with only one succeeding. A burial cairn at Niutang contains the remains of a middle-aged male whose skull and mandible had been cut in half sagittally, and subsequently burned in driftwood (Salter 1984: 112). Considering the value of driftwood in the Sound (Boas 1964: 61), the treatment of this individual undoubtedly reflects some very strong sentiments in real life. The migration of a group of Oqomiut from Cumberland Sound under the leadership of Qitdlarssuaq around 1835 also appears to have been motivated by a blood feud (Boas 1907: 535, Mary-Rousselière 1991).

Feuds, however, were generally restricted to individual extended families and warfare between 'tribal' groupings was rare (Boas 1964: 57). The only real instance of regional group warfare that Boas was able to record occurred sometime prior to the 19th century at Sagdluaqdjung, near Naujateling. Here, the remains of several huts are found on top of the island, which:

> are said to have been built by Eskimo who lived by the seashore and were attacked by a hostile tribe of inlanders. The tradition says that they defended themselves with bows and arrows, and with bowlders (*sic*) which they rolled down upon the enemy. (1964: 57)

Kumlien (1879: 12) also reported that 'numerous traditions exist among (the Cumberland Sound Inuit) of the time when they warred with the other tribes, and old men, now living have pointed out to us islands that were once the scene of battles.' Interestingly, Barron (1895: 89) in 1857 was told by some Kingmiksormiut that the Nugumiut of Frobisher Bay formerly 'made raids upon them, killing the men and taking the women away'. Barron, not surprisingly, observed that when the natives of these groups met they were very distant: 'if both tribes happened to encamp for a short period of time in the same place, they always left a large space between their huts' (Barron 1895: 89). Although tales are still told today of murderous inland warriors living in the interior west of the Nettilling Uplands, who occasionally descended to the coast to attack people (also see Warmow 1859: 90), it is not known if there is any connection between these and the last stories. Boas (1964: 210–12) relates another incidence of warfare between local groups in Cumberland Sound. But this story, 'The Emigration of the Sagdlirmiut (*sic*)', while perhaps having some grain of truth, possesses mythic elements. Interestingly, this story involves two camps near Tasaiju (a Kinguamuit site) and Ussualung (a Kingnaimiut site), and the extermination of the men of one group by the other.

The feuding that Boas recorded between the villages of Anarnitung and Niutang was undoubtedly symptomatic of ongoing hostilities between prominent individuals in these villages; historically, only men with substantial means and influence dared to eliminate their enemies. Whether this hostility characterized social interaction between all Kinguamiut and Kingnaimiut, however, is not known. In this regard, Burch and Correll's distinction between warfare and feuding in Inuit society may offer some insight:

> Feuding involved only close kinsmen on a side, and was usually pursued actively by only one individual, or at most a man and his adult sons (or perhaps a set of adult brothers) at any given time. Direct confrontation was avoided, the objective being to shoot the enemy in the back in some isolated part of the district. This was regarded as murder, and had to be followed by an extensive series of ritual observances. Warfare, on the other hand, involved anywhere from ten to several dozen men from several different families.... Although ambush was often sought, the direct confrontation of the opposing sides was frequent. Killing in this context was not regarded as murder, and no ritual observances were necessary. (1972: 34)

While we can see that the Niutang-Anarnitung example contained elements of both definitions—i.e., a large number of men, probably from different families, seeks revenge on one man for killing their own kinsmen—the archaeological record is perhaps more informative. If interregional hostilities did exist between the Kingnaimiut and Kinguamiut then we might expect settlements near their boundary to assume more defensive, organized arrangements. In this connection, it is illuminating that the three most spatially concentrated village sites recorded by Schledermann (1975) are located in the presumed boundary area between these two groups; MaDg-2 (Inuguaarulu), MbDi-1 (Kekertelung, Tulukan?), and MbDg-1 (Avatuktoo). The fact that the spatial concentration of houses at these sites differs markedly from either that at Niutang or Anarnitung is evident from Figure 10. In light of his evidence, the possibility that the biological distance between Tasaiju and

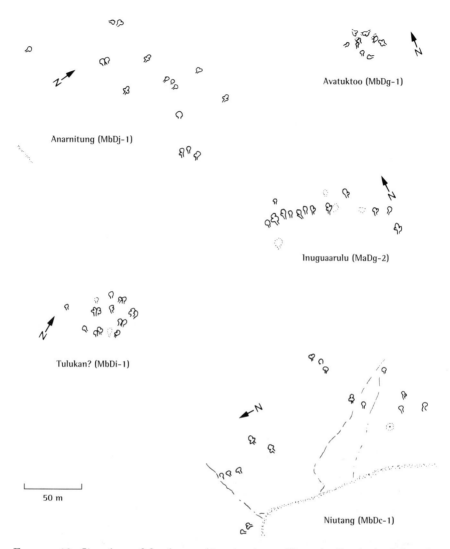

FIGURE 10. Site plans of five late prehistoric winter villages in Cumberland Sound. Redrawn from Schledermann (1975: 39, 41, 52, 54, and 67).

Niutang skeletal specimens was, in part, a function of hostilities between the Kinguamiut and Kingnaimiut is enhanced.

Salter (1984) remarked on the geographical and social isolation of the Kinguamiut relative to the Kingnaimiut and Talirpingmiut. Not only was the former group hemmed in by mountainous terrain, but it appeared to have had no regular intercourse with either neighbouring group. Conversely, the Kingnaimiut had regular contact with the more southerly Saumingmiut and easterly Padlimiut (Boas 1964), while the Talirpingmiut maintained relations, though perhaps not always friendly, with the Nugumiut, and to a lesser extent, with the Iglulingmiut.[10] More favourable economic conditions at the head of the Sound, especially the occurrence of numerous *sarbut*, may have reduced the necessity to form economic or social alliances outside the Kingua region, thus promoting the development of regional group endogamy.

The formation of a marriage isolate under such circumstances would seem to be a reasonable explanation for the lack of shared genetic traits between Niutang and Tasaiju's burial populations. However, Burch and Correll's (1972) study of north Alaskan warfare clearly indicates that groups that fight one another also stick together. In other words, groups who feud also participate in other sorts of socioeconomic alliances, including trade and marriage. Certainly, the Niutang-Anarnitung case could be interpreted in this light. After all, there must have been some grounds for hostility between these villages. Whether the Niutang and Tasaiju skeletal data are indicative of broader patterns of Kingnamiut-Kinguamiut interaction, or whether the Tasaiju skeletal specimens simply represent a localized development within the Kingua area, will be addressed in a later chapter. Nonetheless, until further burial population studies are undertaken, it seems obvious that this issue will continue to challenge our understanding of precontact social group organization and interaction in Cumberland Sound.

## CUMBERLAND SOUND INUIT SOCIAL ORGANIZATION: TRADITIONAL FEATURES

On the basis of the foregoing discussions, it can be argued that some local groups in Cumberland Sound prior to contact possessed a fairly specialized whale-hunting economy, and that they displayed a certain amount of insularity as well as hostility towards other groups. While these traits are more commonly associated with historic Inupiat in northwest Alaska than Inuit in the eastern Canadian Arctic, we still know virtually nothing about the basics of Cumberland Sound Inuit social organization. For example, what was the average size of local groups? What were their marriage practices, residence rules, authority and leadership patterns? And did these features differ between local groups, between regional divisions? We cannot answer these questions with the degree of resolution we would like, but we can examine the historical record in order to arrive at a more complete picture of traditional Cumberland Sound Inuit social organization.

### Local Group Size

Boas calculated that, when the whalers first over-wintered in the early 1850s, the population of the Sound may have amounted to about 1500. Between Nuvujen and Naujateling there were apparently three settlements totalling about 600 people, and at Kekerten enough Kingnaimiut assembled to man 18 whaleboats: 'Assuming five oarsman and one harpooner for each boat, the steersman being furnished by the whalers, and for each man one wife and two children, we have in all about 400 individuals'. Boas estimated that the inhabitants of Nettilling Fiord numbered almost as many, and that the number of Inuit living at Anarnitung and Imigen were 200 and 100, respectively. In total, Boas thought that 'probably eight settlements, with a population of 200 inhabitants each . . . would be about the true number in 1840', though he later considered this figure to be 'too large rather than too small' (1964: 17). Indeed, 200 individuals per settlement would seem remarkable given our understanding of local group size among other Central Inuit societies. Boas' estimates are suspect not only because they were derived from 'conjecture and hearsay' from the whalers, but because many Inuit from outlying districts likely immigrated into the Sound after the whalers began to over-winter. Surely, a more accurate estimate for the aboriginal population of the Sound is Penny's (1840) figure of 1000. However, this estimate did not provide a breakdown by local groups. For this, we must look to M'Donald (1841) and other sources.

As pointed out above, M'Donald found 40 Inuit at Anarnitung in seven tents in the autumn of 1840. However, this figure was somewhat less than the total number of people that lived there during the winter. At Togaqjuaq M'Donald (1841: 115) recorded another 30 Saumingmiut, while at Kingmiksoo he was visited by 'about sixty of the natives,—great numbers of who were related to Eenoo(looapik)' (1841: 101). Again, given the time of the year—September—many inhabitants of this site may have been inland, caribou hunting. Kingmiksoo was considered by the whalers to be the principal settlement in the Sound, and the possibility that its fall population may have been double M'Donald's figure is provided by Sutherland (1856). In the fall of 1846 Sutherland spent two months at Kingmiksoo, during which time he conducted a detailed census of this settlement, enumerating 111 individuals in 16 huts (1856: 213). It might be argued that Kingmiksoo's population in 1846 had grown beyond aboriginal levels owing to the presence of Europeans. However, Eenoolooapik's map of the Sound clearly shows the presence of 16 huts here in 1839 (Figure 11).

Eenoolooapik and other Inuit boarded the *Neptune* at Durban Island in 1839 to produce a tolerably accurate map of Cumberland Sound for Penny, detailing principal whaling grounds, caribou hunting areas, as well as occupied settlements, denoted by accumulations of dots at specific locations (Figure 8). The latter appear not to be haphazard accumulations of points, however. Rather, at least some of them appear to represent the actual number and layout of dwellings at various locations in 1839. This interpretation finds support not only in the different numbers and arrangements of dots at various locations, but in the correspondence between the dots and actual sod houses at specific sites.

A cluster of eight dots at the head of an inlet named 'Neeoudlook' on Eenoolooapik's map is clearly Schledermann's MbDg-1, or Avatuktoo, a site in which nine Thule Inuit winter houses were recorded (Figures 8 and 10). Similarly, the cluster of nine dots that Eenoolooapik located at the north entrance to the first inlet north of Kangatlook is probably Schledermann's LlDj-4, a site of 11 winter houses. The spatial arrangement of dots at Kingmiksoo provides additional support for this interpretation as it very closely approximates the actual spatial distribution of winter houses at this location. Even though Gardner (1979: 382) recorded some 25 dwellings at Kingmiksoo, including the house foundation inhabited by the crew of the American whaler *McLellan* in the winter of 1851–52, the bipolar to circular arrangement of dwellings on his map is similar to the distribution of 16 dots on Eenoolooapik's map (Figure 11).

Assuming, then, that approximately the same number of people lived in Kingmiksoo's 16 houses in 1839 as 1846, we may estimate the aboriginal population of Kingmiksoo to be *c.* 115, not 200 as Boas suggested, but still large by Central Inuit standards. From M'Donald's information it is clear that local group size varied considerably, and that Boas' figure of 200 in eight settlements, while perhaps having some factual basis when the whalers began to over-winter in the early 1850s, is unfounded for the period before 1840.

A more concrete, though still speculative, estimate of local group size may be obtained from Eenoolooapik's map. If we interpret the clusters of points at various locations on this map to represent settlements, and individual dots to be occupied dwellings, 182 huts in 29 sites are obtained. Averaging M'Donald's (1841) and Sutherland's (1856) figures for the number of people occupying dwellings at Anarnitung (40/7) in August of 1840 and Kingmiksoo (111/16) in the fall of 1846 we arrive at a figure of 1151.3 people (6.326 x 182) for the aboriginal population of the Sound. If we exclude the cluster of dots in the bottom right corner of the map (Figure 12), which is not in the Sound—it appears to be the Saumingmiut settlement of Togaqjuaq—the total population is reduced by 38 to 1113. Allowing for a 10 per cent margin of error, a total of 1001.7, a figure closer to Penny's, is obtained. Employing the same methodology, the theoretical range of the size of local groups varies from 19 (3 x 6.326) to 101 (16 x 6.326).

A potentially illuminating picture of subregional variation in local group size emerges if we overlay Boas' three Cumberland Sound Inuit 'tribal' divisions onto Eenoolooapik's map (Figure 12). Not only do the Kinguamiut demonstrate the most sites (n=13) of any regional group, but they would appear to have had the largest population (n=434), assuming that an average of 5.71 (40/7) people occupied each of the 76 dwellings in Kingua Fiord. However, contrary to expectations, they also demonstrated the lowest average group size (n=33.4). Given that the head of the Sound was more productive than either adjacent shore, one might have anticipated not only more sites, but a corresponding increase in the size of local groups. Yet, assuming that an average of 6.94 (111/16) people lived in each of the 54 Talirpingmiut dwellings plotted on Eenoolooapik's map, the population of the average Talirpingmiut site is

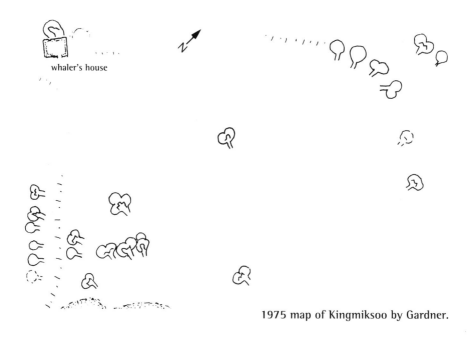

FIGURE 11. Modern and 1839 maps of Kingmiksoo. Redrawn after Eenoolooapik (NMC-59335) and Gardner (1979: 382).

considerably larger (n=46.8). Although the reason for this finding is obscure at this point, a following chapter will explore the possibility that it may be related to the ways in which local groups within each area of the Sound were organized.

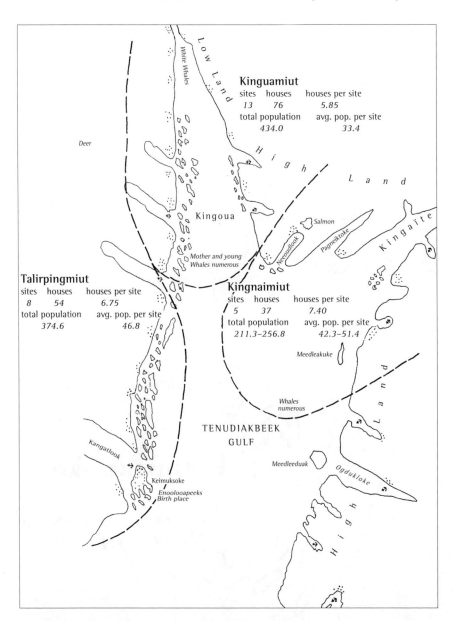

FIGURE 12. Boas' (1964: 16–32) 'tribal' divisions or subregional groups superimposed over Eenoolooapik's 1839 map of Cumberland Sound, with associated statistics.

No comparable estimate is available for the average number of people that occupied dwellings among Kingnaimiut. However, based on the above estimates, we find that the average size of a Kingnaimiut site ranged from 42.3 to 51.4 people. In light of the foregoing discussion, it is interesting that the

Kinguamiut and Kingnaimiut potentially display the greatest differences in local group size, 33.4 vs 51.4.

The potential information contained on Eenoolooapik's 1839 map has gone unnoticed by researchers. Yet, if the above assumptions are valid, this map may contain a wealth of data about Cumberland Sound Inuit social organization. If this source has been used too uncritically here, it was only because there seemed to be too good a fit between it and the archaeological data to ignore. Nonetheless, until further research establishes a firmer correlation between Eenoolooapik's map and the late prehistoric archaeological record of Cumberland Sound, our interpretations, though potentially informative, must remain speculative.

Assuming that both Boas' reconstructions and Eenoolooapik's map have some factual basis, how do we explain their differences? Eenoolooapik plotted 29 settlements of various sizes, while Boas estimated that the Cumberland Sound Inuit were divided into eight principal settlements. However, Boas recorded the use of many more settlements; for the coastal Talirpingmiut alone, 11 sites were reported. It is obvious from Boas' descriptions that his eight settlements represent not so much actual geographical locations, but social groups within larger subregional configurations. For example, although the coastal Talirpingmiut between Nuvujen and Naujateling may have occasionally occupied three principal locations during the winter, they also inhabited several other sites at different seasons.

The fact that Boas recorded three functionally identical pairs of settlements habitually used by the same groups each fall and winter on the southwest shore points to the existence of three subdivisions of coastal Talirpingmiut. In contrast, Kinguamiut settlements appear to have been associated principally with two major groups centred about Imigen and Anarnitung, and there appears to have been no systemic pairings of fall and winter sites, such as those on the southwest shore. Thus, it appears that three rather distinct social configurations were present in Cumberland Sound: (1) the local residential group or *nunatakatigiit*, (2) a larger group which likely consisted of all resident and non-resident kinsmen or *ilagit*, and (3) the regional division or 'tribal' group, with recognized 'miut' designations.

## Leadership

The exploration of leadership is hampered by a lack of information on the subject. Boas' (1907, 1964) descriptions refer specifically to leaders and/or high status individuals, usually within the contexts of whaling, feuding, and migration. In this connection, he noted that there was a kind of chief in every settlement, called the *pimain* or *issumautang* (more correctly, *isumataq*) or he who knows everything best, whose authority is limited:

> to the right of deciding on the proper time to shift the huts from one place to the other, but the families are not obliged to follow him. At some places it

seems to be considered proper to ask the pimain before moving to another settlement and leaving the rest of the tribe. He may ask some men to go deer hunting, others to go sealing, but there is not the slightest obligation to obey his orders. (1964: 173)

For reasons which may be related to Boas' own particular biases or lack of experience, he appears not to have been too concerned with authority and decision-making in traditional Cumberland Sound Inuit society. Nor does he describe the criteria upon which leaders were chosen. In the latter regard, we may assume that age, wisdom, experience, hunting skills, personality, and support from kin were all important factors in the emergence of local group leadership.

A potentially more useful treatment of traditional leadership in Cumberland Sound is provided by M'Donald (1841). At Togaqjuaq, M'Donald (1841: 115) could not discover 'any chief or superior' among the 30 people he met at this settlement. However, at Anarnitung (Noodlook) and Kingmiksoo, he encountered recognized leaders. At Bon Accord Harbour, M'Donald encountered his old acquaintance, Anniapik, at the helm of an *umiaq*—M'Donald first met the elderly *angaquk* at Durban Island in 1835. While aboard Penny's ship Anniapik was given officer's dress, which he wore proudly, while openly helping himself to anything he desired—a behaviour that did not exactly endear him to Penny's crew. Anniapik also had many long conferences with Eenoolooapik as the latter wanted the old man's daughter's hand in marriage. The result of these negotiations 'was that . . . Eenoo(looapik) was to give his green-painted canoe (whaleboat) for the beautiful Coonook, and this canoe was to become the property of Anniapik's youngest son, he himself being unable from the infirmities of age to manage it' (M'Donald 1841: 88). While this anecdote raises questions about inheritance and residence—Boas (1964: 172–3) noted that the first inheritor of a man's possessions was his eldest resident son—it clearly establishes the substantial age and status of Anniapik. Alternatively, at Kingmiksoo M'Donald (1841: 101) was informed that Eenoolooapik's cousin was 'chief', or as Eenoolooapik expressed it, 'captain of the tribe'. Although M'Donald thought that there was 'but little difference between the chief and the others', what is interesting about this individual is his apparent age. As the latter was a member of the same generation as Eenoolooapik, and as Eenoolooapik was only about 20 at the time, it is obvious that Kingmiksoo's leader was much younger than Anniapik. Whatever qualities Eenoolooapik's cousin possessed, we may assume that there were older men in the settlement with considerably more experience, wisdom, and acumen.

**Marriage, Residence, and Descent**
Boas (1964: 170–1) briefly listed a number of social organizational features of Cumberland Sound and Central Inuit society. Individual families were held

together by ties of consanguinity and affinity, implying that descent was more bilateral than unilateral. And while child betrothal was common, this arrangement was not strictly binding between the families involved. After marriage the young man normally went to live with his wife's parents, although as Boas (1964: 171) noted 'it happens frequently that the young man's parents are unwilling to allow him to provide for his parents-in-law...'. While there was no definite period of bride-service,[11] Boas observed that 'not until after his parents-in-law are dead (was the groom) entirely master of his own actions.' By this remark Boas probably meant that the son-in-law did not escape the 'grasp' of his parents-in-law until after their death, rather than the strict observance of matrilocal residence. Regardless of whether marriages were prearranged or not, brides were normally acquired with gifts. Consent of the bride's parents, or if the latter were dead, brothers, was always necessary. As noted previously, marriages with close relatives was strictly forbidden, though a man was permitted to have two wives or to marry two sisters.

Despite Boas' cursory examination of social life, a number of tendencies are apparent. The exclusion of close relatives from the universe of marriage partners, for example, obviously indicates a preference towards kin and local group exogamy, as does the grouping of parallel and cross-cousins under one term in Morgan's (1870) Cumberland Sound Inuit kinship terminology.

More revealing are the customs of bride-price, bride-service, and polygyny, which suggest an emphasis on the patri-line. The apparent ascendancy of one line over the other may also be reflected in Morgan's terminology. Of particular interest are the terms for FZ and FZH (Figures 2 and 13). Whereas the latter was simply not provided for either gender, the former, *atchunga*, is enigmatic. This term is not commonly used today on Baffin Island, though, interestingly, *achun* was used to refer to both MZ and FZ among some Mackenzie Delta groups. Rather, for both male and female Ego *attak* is used for FZ, while the term for FZH is most often either *ningauk* (i.e., in-marrying male) or *akka* (FB, or male related through the father). In Pangnirtung today the term for FZH and MZH is *airaapik*. In this light, it is curious that neither of Morgan's informants appear to have known the term(s) for FZH. This seems to imply that neither individual maintained close relationships with nor lived in the same camp as his/her FZH, and presumably FZ. While this perhaps indicates an emphasis on the patri-line, an emphasis on the matri-line, especialy since MBW was equated with MZ, cannot be discounted. Whatever the case, the existence of dual exogamy is obviated since one would expect that either one or the other would have known the term for FZH, as he/she would have lived in the same village. Yet, the fact that the terminology appears to be affinal-including on at least one side, would seem to suggest the existence of cousin marriage in the past; systems of cousin marriage often use consanguineal terms for affines in the first ascending generation (Murdock 1949: 122). The ascendancy of the male line in Morgan's terminology may also be reflected in the greater development of terms for males. For example, for male Ego, older sister and younger sister are the same (Figure 2), while for female Ego, older brother and younger brother are differentiated (Figure 13).

Two additional sources that may be of some use in our attempt to derive a clearer picture of traditional Cumberland Sound Inuit social organization are Boas' presentation of numerous stories and legends from Cumberland Sound, and Sutherland's 1846 census of Kingmiksoo. Central Inuit mythology contains many lessons and guidelines for living in a social context. For example, the story of the 'Sun and the Moon' establishes the incest taboo between brother and sister, while the story of 'Sedna' warns of the dangers that befall the woman who disobeys her father and rejects all suitors known to her group, while accepting a foreigner as a husband. Some of the more specific legends that Boas was able to record deal with the themes of:

- local group exogamy, 'Emigration of the Sagdlirmiut' (1964: 210–12),
- sister or female solidarity, 'Origin of Agdlaq' (1907: 171–2, 261–5),
- brother or male solidarity, 'Qaudjaqdjuq' (1907: 288–9, 1964: 220–2),
- mistreatment of daughters by fathers, 'Origin of Agdlaq' (1907: 171–2),
- mistreatment of boys by (grand)mothers, 'Origin of the Narwhal' (1964: 217–19),
- mistreatment of younger brothers by older brothers (1907: 283–5),
- dangers of forming economic relationships with brothers-in-law (1907: 282),
- the consequences that result when men want to marry their wives' sisters (1907: 261–5),
- rejection of marriage with women inside one's own group, 'Ititaujanq' (1964: 207–10),
- dangers of marrying strangers, 'The Girls Who Married Animals' (1907: 217–18),
- the pitfalls of taking wives from foreign groups, 'Ititaujanq',
- patrilocality, 'The Girls Who Married Animals', 'Ititaujanq'.

Of particular interest in these stories is the continual reference to the logistical problems of finding eligible marriage partners outside one's own group (local group exogamy) who are not strangers (subregional group endogamy).

A more concrete, though still far from ideal, portrait of traditional Cumberland Sound Inuit social organization can be drawn from Sutherland's 1846 census of Kingmiksoo. For reasons discussed above, the size and organization of this settlement was considered to be relatively unchanged from 1839, i.e., largely unaffected by European contact. Sutherland spent two months among the Kingmiksormiut in the fall of 1846, during which time he conducted a detailed census of 16 households. Here he recorded (a) the number of independent nuclear families within each household, (b) the names, sexes, and ages

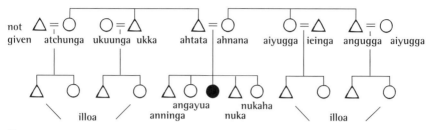

FIGURE 13. Cumberland Sound cousin terminology for female Ego, *c.* 1860 (from Morgan 1870).

of family members, as well as (c) the type and number of watercraft owned by individual family heads. While age determinations are not likely to be exact in all cases, they are nonetheless educated guesses by a medical expert with some familiarity with Inuit culture and society. Unfortunately, Sutherland did not systematically record kinship connections between various male or female adults either within or between households—only in few cases were kinship relationships supplied. Sutherland's census data are summarized in Figure 14.

This figure reveals several interesting patterns, perhaps the most obvious being the high incidence of polygamy. Of the 24 marriage arrangements documented, 25 per cent represent polygynous unions. Conversely, no instances of polyandry were recorded. As Kjellstrom (1973) observed, these data document one of the highest occurrences of polygyny ever recorded in the Canadian Arctic. An excessive death rate among adult males, owing to hunting accidents and exposure (Sutherland 1856: 213), undoubtedly favoured the formation of polygynous unions by creating a surplus of socioeconomically productive females. Indeed, there were 47 adult women as opposed to 30 men at Kingmiksoo. This tendency is even more apparent when one considers that, while seven widows were enumerated, no widowers were documented. Unmarried adult males may, in fact, have been a contradiction in Kingmiksormiut society.

Also contributing to the frequent occurrence of polygyny may be the tendency for first wives of productive hunters to bring in a second wife to assist in domestic chores. Of the five polygamous unions where the ages of both wives are given, three have an average age difference of 15 years, while the remaining two have an average age differential of 6.5 years. The possibility that adult male productivity may account for at least some of Kingmiksoo's polygynous marriages is suggested by the fact that, of the six *umiat* recorded, three are owned by men with two wives.

Sutherland's census data also reveal a patrilocal slant. Five married children living with either one or both parents were recorded, all of whom were males that had taken wives from other households or settlements. In a society that supposedly practised bride-service, it is remarkable that no matrilocal living arrangements were documented. Given the large size of this settlement, however, in some cases both sets of parents may have been co-resident,

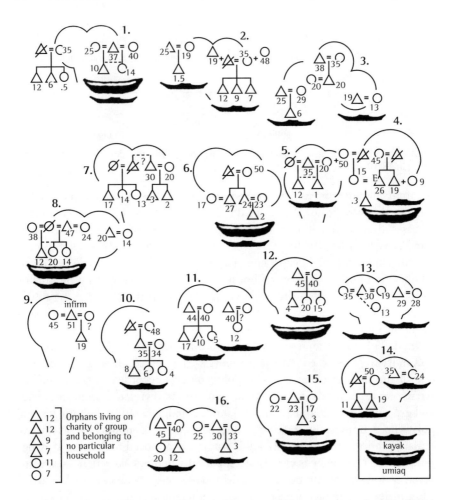

FIGURE 14. Graphic depiction of household census of Kingmiksoo, in the fall of 1846. Adapted from Sutherland (1856: 210–12). Census by individual nuclear families within households. Numbers indicate ages of individuals. 'E' refers to Eenoolooapik. Shapes of dwellings estimated on the basis of Sutherland's separation of nuclear families within households. Arrangement of households follows Sutherland's numerical presentation, and are assumed to be accurate sequentially, but not spatially.

suggesting perhaps the acceptance of local group endogamy. Whatever the scenario, a preference for local group and/or kin group exogamy may be indicated by the number of eligible unmarried males (n=5) and females (n=7) present in this population.[12]

Another interesting pattern that emerges from Sutherland's census data is the very low incidence of three-generation families within households. Only three out of the 26 individual families documented contained members of three generations. Albeit uncertain, the high rate of adult male death and, to a

TABLE 2. Occurrence of same generation and intergenerational relationships between nuclear family heads in 'multi-family' houses at Kingmiksoo, fall 1846. Single-family dwellings with adult children, which are, by definition, vertically structured, are not considered. Table based on age data provided by Sutherland (1856: 210–12).

| HOUSEHOLD NO./PAIRING | PROBABLE (KNOWN) RELATIONSHIP | LESS LIKELY ARRANGEMENT |
|---|---|---|
| 1. ab | Same generation | — |
| 2. ab | Same generation | Intergenerational |
| 3. ab | Intergenerational | — |
| 3. ac | Same generation | — |
| 3. bc | Same generation | Intergenerational |
| 5. ab | Intergenerational | — |
| 6. ab | Intergenerational (M-S) | — |
| 6. ac | Intergenerational (M-S) | — |
| 6. bc | Same generation (B-B) | — |
| 7. ab | Same generation | Intergenerational |
| 8. ab | Intergenerational | Same generation |
| 11. ab | Same generation | — |
| 13. ab | Same generation | — |
| 14. ab | Intergenerational | Same generation |
| 16. ab | Same generation | Intergenerational |

lesser extent, the incorporation of productive widows into polygamous unions operating under a patrilocal tendency, may account for this finding. Individual households probably represented extended families of some sort. However, whether these households were held together by sibling, parent-child, or other types of kin relationships cannot be determined, though we might surmise that most were founded on either *nukariik* or *irniriik* ties. Even though Sutherland did not document kinship relationships between nuclear family heads, he did estimate the ages of individuals, which, in turn, should allow us to determine whether nuclear families within households were bound primarily by same generation (e.g., B-B, B-Z, male Ego-FBS, etc.) or intergenerational (e.g., F-S, M-D, female Ego-MZ, etc.) relationships. Table 2 is constructed on the basis of two assumptions: (1) that an age difference greater than 14 years between nuclear family heads indicates that a household was held together by intergenerational ties, while age differences less than 14 years reflect same-generation relationships,[13] and (2) that male relevant ties played a larger role in structuring relationships than either opposite-sex or female relationships.

Based on this table we can see that multi-family households at Kingmiksoo were held together largely by same-generation relationships (n=9) as opposed to intergenerational relationships (n=6). Although speculative, the high death rate among adult males may likewise have contributed to the relatively low frequency of vertical relationships within multi-family households, while encouraging the formation of horizontal relationships among nuclear family heads. When single-family dwellings with adult children (married and unmarried) are considered, however, a more even balance between vertically (n=10) and horizontally (n=9) structured households is obtained. Still, in a society which supposedly emphasized parent-child relationships above all others, the number of same-generation ties seems excessive.

The question that this analysis raises is whether the unexpectedly high frequency of horizontal relationships within multi-family dwellings is characteristic of Cumberland Sound Inuit society in general, or unique to Kingmiksormiut or Talirpingmiut? As suggested above, perhaps an increased dependence on bowhead whaling during late prehistoric times encouraged the formation of larger production units among kinsmen. Whatever the case, without comparable data from other settlements from equivalent time periods, it is very difficult to resolve this issue, though a following chapter will present data to inform this pattern.

## A Cumberland Sound Type of Social Organization?

Based on the above interpretation and discussion, is it possible to distinguish a Cumberland Sound type of social organization? In spite of the inherent limitations of the evidence, it seems apparent that both similarities and differences existed among aboriginal social groupings in Cumberland Sound. For example, with an abundance of whales and *sarbut* at the head of the Sound, the Kinguamiut appear to have lived in a more productive environment than either the Kingnaimiut or Talirpingmiut. Conversely, although the southwest side of the Sound was blessed with more tidal rips than the northeast shore, Inuit in both these districts normally spent the winter hunting at seal breathing holes or at the *sina*. From Eenoolooapik's map we saw that, while the Kinguamiut lived in more settlements than either the Talirpingmiut or Kingnaimiut, their villages were generally smaller in size. The Kingnaimiut not only appeared to have had the largest and the fewest settlements, but on the basis of Boas' descriptions they also seem to have exhibited the greatest insularity and hostile tendencies. Nonetheless, it was the Kinguamiut who appeared to be the most geographically and socially isolated of the three regional divisions. Salter's Tasaiju and Niutang burial data as well as Schledermann's site plans suggest the maintenance of social boundaries between Inuit living in Kingnait Fiord and Kingua Fiord. However, whether these data were characteristic of social interaction at the level of the local group or regional group could not be determined. While M'Donald observed that both Kingmiksoo and Bon Accord Harbour had recognized leaders in 1840, leadership seems to have been far more developed in the latter settlement.

Most of our knowledge of local group organization in Cumberland Sound was derived from either Boas' general descriptions or Sutherland's census data from Kingmiksoo. Both sources suggested that the Cumberland Sound Inuit possessed a social structure that could be classified as exhibiting a patrilineal tendency on a bilateral structure. For example, the kinship system, while basically bilateral in character, tended slightly to favour the male line. Moreover, bride-price and bride-service were common, as was patrilocal residence. Local and kin group exogamy prevailed. Yet, regional group endogamy seems to have been preferred. Polygyny appears to have been both a goal and an accepted fact of life, though sororal polygyny may have been less socially accepted.

I began this chapter with the objective of drawing a clearer picture of aboriginal social organization in Cumberland Sound. Whether or not I have succeeded in this endeavour is probably less important than acknowledging the complex nature of the results obtained. We may not have obtained a more concrete understanding of Cumberland Sound Inuit social organization, but we have certainly gained a better appreciation of the problems involved in reconstructing Inuit socioeconomy in this region, and of the potential dynamism of local group organization owing to a multiplicity of factors. Perhaps all that can be said at this point is that the ideal social arrangement within local groups was one which sought to impose some male bias on a bilateral structure. However, differences between subregional groups as well as consistent arrangements of social features within these configurations lead us to anticipate that not all Cumberland Sound Inuit achieved this ideal in the same way.

# Chapter 3

## Cumberland Sound
## Inuit-Qallunaat Interaction to 1970

No Inuit on Baffin Island participated longer in a foreign economic system than the Cumberland Sound Inuit. Nor did any other regional group adopt Christianity as early, or arguably suffer so many losses to foreign diseases during the historic period. These factors alone might suggest that Cumberland Sound is a poor place indeed to undertake a search for structure in Inuit social organization. But this presumption assumes *a priori* that the Cumberland Sound Inuit underwent irreversible structural changes as a result of contact.

Some Arctic researchers have dismissed the study of Inuit groups having lengthy contact with Qallunaat as if contact somehow destroyed the underlying fabric of Inuit society or Inuit were in some way immune to change before Euroamericans arrived. We must reject this view not only because of its questionable validity, but because it fails to acknowledge the complex, richly textured nature of Inuit society prior to and after the coming of the white man. In turn, we must accept the fact that the Inuit possessed a vibrant, integral culture irrespective of our own biases and of the harsh environment in which they lived. At the very least, we must regard the assumption of significant social and cultural change in the context of Inuit-Qallunaat interaction as a working hypothesis to be tested rigorously in each region. Ultimately, we may discover that structural changes did occur as a result of contact. However, did these transformations render these groups any less Inuit? Alternatively, fundamental social change may not have followed in the wake of Qallunaat. Whatever the case, in the process of examination we will surely learn more about Inuit responses to contact, and thus aboriginal social organization, than we knew before.

This chapter outlines the historical background required to assess whether the Cumberland Sound Inuit underwent significant structural changes in social organization during the late nineteenth and early twentieth centuries. How did population decimation by foreign diseases, participation in commercial whaling, the adoption of Christianity, and other Euroamerican influences affect traditional group composition, intergroup relationships, leadership roles, aboriginal subsistence patterns, socioeconomic partnerships and kinship relations? Examining these questions should allow us to determine whether local group composition in Cumberland Sound between 1920 and 1970—the period remembered by my informants—was representative of precontact social organization, and whether this knowledge might inform us about the structure of Central Inuit social organization. A brief history of Inuit-white relations will help assess the extent and magnitude of various Euroamerican influences on Cumberland Sound Inuit social organization.

## PRECONTACT, 1820–1840

Between 1820 and 1840 the extensive ice-free areas off Lancaster Sound and Pond Inlet were the principal whaling grounds of the British whale fishery. During these two decades over 1,300 voyages took more than 13,000 whales from these waters (Ross 1981: Table 1). However, the industry paid a heavy price for its success. Particularly disastrous years were experienced once in every five, when dozens of whaling ships and as many men were lost.[1] If the Baffin Bay whale fishery had a devastating impact on its own resources and the bowhead whale, it certainly did not have a detrimental effect on the Inuit. Throughout the 1820s and 1830s, contact with Inuit on the east coast of Baffin Island remained sporadic and ephemeral:

> Along much of the Baffin Island coast the movement of the whalers was unpredictable and irregular . . . depending on the southward migration of whales and the state of sea ice and weather. Accordingly, contact with vessels was largely fortuitous and of short duration, and systematized trade and employment could not develop. (Ross 1979a: 251)

Although the odd whale carcass or ship's wreckage may have occasionally washed ashore, providing a windfall for some lucky group, the only real opportunity for social interaction occurred when ships, caught in the northern ice pack, were forced to over-winter.

As these whaling grounds were 'fished out',[2] whalers expanded southward into uncharted waters. By 1830, it was common practice to 'fish' off Pond Inlet and Lancaster Sound in summer, and then proceed south along the east coast of Baffin Island to Cape Dyer before freeze-up (Ross 1979b, 1981). By the mid-1830s this pattern had become so routine that Inuit began to gather at Durban Island to trade and interact with the whalers before the latter returned home in late September.

It was during this time that William Penny first came into contact with Inuit from Cumberland Sound. For several years British whalers had realized that if their industry was to survive new whaling grounds would need to be found and a colony established to serve as a permanent whaling base and refuge for ship-wrecked whalers (Holland 1970: 26–7). Penny, almost single-handedly, took up the task of carrying out these plans. And in 1839 he met an Inuk, Eenoolooapik, who originated from the shores of the now legendary 'Tenudiakbeek', a large bay to the southwest that purportedly abounded in whales.[3] Penny returned to Scotland with Eenoolooapik in an attempt to obtain funding for an expedition to these waters. A year later he and Eenoolooapik entered Cumberland Sound in the company of three other vessels.

## EARLY CONTACT, 1840–1857

Penny's discovery did not receive the response he expected,[4] and in the ensuing years only Wareham (1843) visited the Sound. Penny returned to Cumberland Sound in 1844 and again in 1846, where Eenoolooapik assisted in the capture of several whales, marking the first time Inuit were employed in the industry. The year 1846 also heralded the beginning of a routine which was to characterize the fishery for the next half dozen years. The season was a particularly good one for Penny,[5] and the Sound finally lived up to its reputation as a productive whaling ground:

> I mind in 1846, which was the first year I went to Cumberland Gulf, in the old *Alexander*, we was terrified to go out in the boats, the whales was that large and numerous they raised quite a heavy sea with their fins and tails. (cited in Lubbock 1955: 346)

Four vessels were successful at whaling in the Sound that fall, and in subsequent years varying numbers of ships delayed their return home from Baffin Bay to 'fish' in Cumberland Sound.

Increased interaction between Inuit and whalers after 1846 appears to have resulted in a concomitant rise in (a) the population of the Sound, (b) settlement size, and (c) deaths owing to starvation. The whalers brought with them metal knives, needles, pots, and a host of other useful and not so useful articles that the Inuit found irresistible. The sheer excitement created by interaction with the whalers also enticed Inuit to settle at whaling harbours (Ross 1985b: 172). As Eenoolooapik remarked in 1846 'a tribe of Esquimaux does not soon get over the visits of the whalers in the autumn' (Sutherland 1852: 327). During the late 1840s an unknown number of Inuit from outlying areas migrated into the Sound, reversing the flow of population to Davis Strait begun more than a decade earlier. In 1848 Captain Parker of the *Truelove* reported that 'many more natives than usual were in Northumberland Inlet (*sic*) this autumn, in the expectation of meeting with the whalers and obtaining useful articles from them.'[6] As much as any factor, the immigration of Inuit into the Sound during the late 1840s may account for Boas' (1964: 17) inflated estimate of 1600 Inuit inhabiting its shores at contact.

Autumn settlements also increased in size during the late 1840s and early 1850s. Captain J. Parker, for example, found a camp of 160 Inuit at Naujateling in 1848,[7] while Penny spent the fall 'fishing' season of 1853 at Kingmiksoo among 270 Inuit (Penny 1854a). As early as 1847, large aggregations of Inuit appear to have over-taxed available resources in the vicinity of settlements, resulting in starvation. When Parker reached Naujateling in 1848 he found that 20 of the 160 people at this settlement had died of hunger the preceding winter, 'of whom several, horrible to relate, had gnawed the flesh from their own arms'.[8] Failure to build up winter stores and obtain enough skins for clothing and shelter in favour of lingering around the ships may have also contributed to numerous deaths (Barron 1895: 43, Harper 1981: 46–7, Warmow 1859: 89).

Winter aggregations on the southwest shore expanded in size and duration after the winter of 1851–52, when the crew of the American whaling ship, *McLellan*, was left at Kingmiksoo in order to pursue whales at the floe edge in the spring.[9] For years, the Inuit had reported an abundance of whales at the *sina*, and had tried to convince the whalers of the benefits of over-wintering (Goldring 1984, 1986; Ross 1985a: 189). The experiment marked the first time whalers had purposely wintered in Arctic Canada, and despite the inexperience of the crew, it proved a success in every way; 16 small whales, yielding 16,000 pounds of 'bone' (Clark and Brown 1887: 95), were taken under native guidance and very friendly relations were established with the Inuit (Barron 1895: 37–43). After 1852, American and British whalers routinely over-wintered in Cumberland Sound and contact with the Inuit increased markedly in frequency and duration. Spring whaling was, in fact, impossible without Inuit assistance, as Penny (1854b) discovered:

> My first trip to the water with my dog-sledge was on the 25th of March, twenty-five miles distant. Pitched three tents at the water's edge, where eighteen men managed to kill eighteen whales, and to drag up to the ship seventeen (lost one). I had sometimes as many as twenty-one sledges on the ice! The distance, in a straight line was twenty-one miles, or about twenty-two miles and a half with traverse course. The dogs went to the water's edge and back every day, making a daily journey of forty-five miles; the distance put upon end, would have amounted to 14,000 miles.

Inuit were also hired to crew on whaleboats, flense whales, and 'try-out' whale blubber.[10] In addition, they hunted seals and caribou to feed and clothe wintering whalers. At Naujateling in 1853 Penny hired as many as 50 Inuit in the fall fishery (Goldring 1986: 160), and when the *Truelove* called at Kingmiksoo in the summer of 1852, all the Inuit of this settlement were engaged in whaling with the crew of the *McLellan* at Nuvujen (Barron 1895: 39). Soon most ships carried less than half their normal complement of 40 to 50 men (Barron 1895: 44). In exchange for their services, Inuit were given food, clothing, tobacco, alcohol, and a host of more useful items. Most prized of all, however, were whaleboats and muskets, and by 1852 a number of Inuit on the southwest shore had begun to acquire these goods (Barron 1895: 43). Access to vast amounts of meat and *maqtaak* from whales flensed on the shore during the fall whaling season was another benefit that the Inuit derived from their participation in the whale fishery, as Captain G. Tyson (Ross 1985a: 190) observed in 1852:

> the natives would seize upon the latter (flensed whale) and strip off all the meat. What they could not eat, they put in sealskin 'drugs' or bags, and they stowed these away for future use, hiding the bags by covering them up on the various islands in the gulf. . . .

Although the Inuit were better off materially than ever before, increased interaction with the whalers had serious consequences. Overpopulation and neglect of daily duties resulted in some deaths. But foreign diseases claimed even more lives. Eenoolooapik succumbed to 'consumption' (pulmonary tuberculosis) in 1848 (Holland 1970: 35), and in December of 1853 'cholera broke out, and carried off a third of the Esquimaux who formed the little colony at Newacktoolick (Naujateling) Harbour' (Penny 1854b), including the 'chief' and seven of his relatives (Goldring 1986: 160). Two years later Penny (1856: 143) wrote to the Moravian Church that many had died since his last visit. The Inuit were always susceptible to colds, fevers, and respiratory ailments in the late fall when food reserves were at their lowest, the ice was not yet firm, stormy weather prevented hunting, and snowhouses could not yet be built. Exposure to wintering whalers and a host of new infectious diseases only served to exacerbate the situation. In November of 1857 Mathias Warmow, a Moravian missionary, found several natives 'dangerously ill of pleurisy, and (afflictions) of the chest' near Kekerten (1859: 91). More remarkably, however, Warmow counted no more than 350 Inuit inhabiting the shores of Cumberland Sound (Holland 1970: 40).

If Penny's estimate of 1000 Inuit in Cumberland Sound at contact is correct, Warmow's figure would seem to indicate that at least two-thirds of the population lost their lives to influenza, tuberculosis, syphilis, and various other foreign diseases in a period of less than ten years. Warmow's population estimate, however, is suspect because it was derived at a time of year (i.e., August and September) when many Inuit were inland hunting caribou. For example, although Warmow (1859: 88) observed a number of recently occupied settlements on a voyage from Tornait in Kingnait Fiord to the head of the Sound in late September of 1857, not a living soul was found. In addition, it also seems likely that the formation of stable, enduring relationships between whaling masters and prominent resident Inuit may have forced less fortunate, non-local Inuit headmen and their kinsmen to look for 'greener pastures' outside the

Sound. Finally, we must consider the possibility that Warmow's count may have been less a reflection of reality than of his own unconscious biases and possible motives. Both Parker and Penny were well aware of the negative impact that the whalers had on the Inuit, and had tried unsuccessfully to land a Moravian missionary in Cumberland Sound for several years. However, it was not until 1857 that they finally succeeded. Appalled by the living conditions of the Inuit and their association with the whalers, Warmow (1859: 89) wrote to his superiors:

> I am always sorry to see the Esquimaux wearing European clothes, and, in short, imitating the Europeans in all respects. They were undoubtedly better off in their original state, and more likely to be gained for the kingdom of God. But when they begin to copy our mode of life, they are neither properly Europeans nor Esquimaux, and will speedily die out in consequence of the change.

But Warmow's plans back-fired. Owing to the small size of the population and the difficulties created by the presence of the whalers, the Moravian Mission Board concluded that a mission was just not justified (Holland 1970: 40). Cumberland Sound would not be visited by another missionary for three and a half decades.

The whalers brought diseases against which the Inuit had no immunity. However, while scores of Inuit undoubtedly lost their lives to epidemics in the mid-1850s, the number is likely not nearly as high as Warmow's population estimate would seem to suggest.

## CUMBERLAND SOUND WHALING, 1857–1870

By 1850 Inuit and whalers began to congregate each fall at places such as Naujateling and Kingmiksoo where they pursued whales among the many islands and inlets of this shore and intercepted whales on their way out of the Sound before freeze-up. Contact, however, remained opportunistic up to 1851–52, when over-wintering created a more solid foundation for trade and intercourse. Inuit now found regular employment in the whale fishery and stable relationships developed between individuals of both cultures. The itinerary of the whalers also changed; it was now common to engage in two seasons of fall whaling before returning home. With the advent of spring whaling, activities expanded north along the coast to Nuvujen. Although this location appears never to have been a popular winter harbour (Goldring 1984: 478), Nuvujen provided immediate access to the floe edge in most years.

In 1857 the nature of whaling in Cumberland Sound changed once again when Penny finally realized his long-standing ambition of establishing a shore-based whaling station in the Arctic. At Nuvujen and at Kekerten near

TABLE 3. Number of ships wintering in Cumberland Sound, 1851 to 1880. From Goldring (1986: 152; personal communication, 19 June 1986); based on: Dennis Wood, 'Abstracts of Whaling Voyages', manuscript in New Bedford Free Public Library; Dundee University Library, 'Kinnes manuscripts', Printed Annual Returns of Whaling Voyages, *History of the American Whale Fishery* ... (Starbuck 1964); and *Returns of Whaling Vessels Sailing from American Ports, 1876–1928* ... (Hegarty 1959).*

| | | | |
|---|---|---|---|
| 1851–52: 1 | 1858–59: 2 | 1866–67: 7 | 1874: 1 |
| 1852–53: 0 | 1859–60: 4 | 1867–68: 9 | 1875: 2 |
| 1853–54: 4 | 1860–61: 12 | 1868–69: 5 | 1876: 2 |
| 1854–55: 0 | 1861–62: 2 | 1869–70: 10 | 1877: 5 |
| 1855–56: 3 | 1862–63: 3 | 1870–71: 6 | 1878: 3 |
| 1856–57: 3 | 1863–64: 6 | 1871: 4 | 1879: 1 |
| 1857–58: 5 | 1864–65: 7 | 1872: 2 | 1880: 0 |
| | 1865–66: 9 | 1873: 1 | |

*Note: Because of access to native labour, the size of over-wintering crews during the 1850s and 60s likely rarely exceeded half the normal complement of men aboard a whaling ship, i.e., 40 to 50 men (see Barron 1895: 44). In this regard, the 18 men employed by Penny at the floe edge in the spring of 1854, and the 19 and 14 men, respectively, of the *Sophia* and *Union* who participated in a cricket match at Union Harbour on 4 January 1860 (Ross 1985b: 161) probably approximated the average number of men carried aboard most over-wintering ships relying on Inuit labour. Wintering ships intending to make up deficiencies in the crew by hiring Inuit were also obliged to provision the family of each native hired so that to secure a crew of seven men, a ship had to provide for 30 or more people (Barron 1895: 162–3, Kumlien 1879: 21).

the mouth of Kingnait Fiord, Penny erected buildings to serve as permanent quarters for wintering whalers and storage for equipment and supplies. The floe edge normally ran between these islands in the late spring, and in the fall whales were more abundant around Kekerten than along the southwest shore (Barron 1895: 194, Goldring 1984: 499). In 1857–58, Penny over-wintered at Kekerten in the company of 150 Inuit, mostly Kingnaimiut, all of whom were retained within the service of the whaling ships (Harper 1981: 46). Kekerten, Nuvujen, and Naujateling were the principal Inuit settlements and whaling bases from which the joint Inuit-white assault on the bowhead was undertaken—Kingmiksoo's role as a rendezvous for whalers had declined steadily since the advent of over-wintering (Goldring 1984: 470–2).

Five ships over-wintered in Cumberland Sound in 1857–58, more than in any previous year. Over the next four years, whaling in Cumberland Sound reached its peak. In the fall of 1859 more than 30 vessels may have entered the Sound,[11] while a year later one dozen ships over-wintered in the vicinity of Kekerten and Naujateling (Table 3).

By 1860 American whalers had erected buildings at Kekerten, Umanaqjuaq (Blacklead Island), and Nuvujen (Hantzsch 1977: 98). Although Nuvujen's population may have exceeded 150 during this time, its buildings were dismantled within a few years owing to the site's poor anchorage (Hantzsch 1977), and the possibility that young whales may have been rapidly exterminated from its waters. For the next several years, winter harbours at Naujateling and in the Kekerten Islands, together with their associated spring floe whaling bases at Umanaqjuaq and Miliakdjuin,[12] were the principal whaling centres in the Sound.

During the 1860s, whalers wintering at Union and Penny's harbours in the Kekerten Islands engaged in a routine that was described in detail by A.C. Whitehouse, mate of the Hull whaler *Emma* (Ross 1985b: 155–73). This 1859–60 journal warrants detailed summary here, if only because it illustrates the nature of Inuit-Qallunaat contact during this important time period.

In late September of 1859 the *Emma* pulled up in winter quarters at Union Harbour in the Kekerten Islands.[13] While preparing for winter, the crew ventured forth in stormy seas and constructed whaling lookouts on nearby islands for the spring 'fishing' season. November 5th was the last day of whaling and by the end of the month the ships were frozen in for the winter, though gale force winds buffeted the ships for several more weeks. Throughout December, crew members hunted, hauled ice from icebergs, played cricket, and celebrated Christmas. A small community of 50 Inuit gathered about the *Emma* and *Isabel*, and almost every night was spent singing and dancing:

> Several fresh Yacks came today from Kingaway (Kingua). We have got about fifty men, women and children now. All the Yacks on board, several from the other ships, men and women. Weather fine but very cold. Night, all dancing and singing. (16 and 17 December 1859, Ross 1985b: 159).[14]

By early January drinking had become a problem among the crew and scurvy began to appear.[15] Over the next two months Inuit spent the days shooting seals at the floe edge, while the whalers continued to hunt ptarmigan, trap foxes, shoot targets, and play rounders and cricket. The nightly ritual of gathering aboard ship for various forms of entertainment continued well into mid-May, when all the Inuit engaged by the ships moved to Miliakdjuin for the spring whaling season. Whales were spotted off the floe edge in late February, suggesting that they wintered near the mouth of the Sound, and preparations for the spring whaling began in early March. A month later the transfer of whaleboats, tents, casks, and whaling equipment to Miliakdjuin (and possibly Midlikjuaq [Yankee Island?]) had begun. Throughout the spring, Inuit continued to hunt seals for the *Emma*'s crew and haul supplies to the spring whaling base.

Over the winter, whalers and Inuit attached to the four British vessels wintering in the 'Kekertens' got along well and freely associated with each other. However, on April 23rd a fight broke out between the masters of the *Union* and *Emma* over the use of the latter ship's natives and their dogs for the spring whaling season. The competitive nature of spring whaling had surfaced, and several natives belonging to the *Emma* were posted at Miliakdjuin to guard the boats and equipment. There was also apparently a shortage of Inuit labour in the 'Kekertens', and several *qamutiit* (sleds) were sent to Nuvujen, Kingnait, and Kingua to fetch more Inuit for the spring whaling season. Inuit with whaleboats and *umiat* appear to have been in particularly high demand as the *Sophia* and *Emma* had lost a number of boats during an early winter storm. Disagreements and fights among the whalers, and with natives, continued into late April as each whaling master tried to secure as favourable a position as he could for the spring whaling season.

By the beginning of May the spring whaling season had started. At the floe edge near Miliakdjuin Inuit and whalers attached to the *Emma* took turns whaling, and on May 15 the native shift made fast to a 'fish'. Although the Inuit were quick to adapt to the use of the whaleboat and new whaling methods, they retained certain elements of their aboriginal technology, and on May 1st all hands aboard the *Emma* spent the day 'making drogues (*niutang*) for the natives' boats to drogue fish with' (Ross 1985b: 165). The whalers flensed the whale at the floe edge, while the Inuit hauled the blubber back to the island. By mid-May the ice began to 'rot' and snow-blindness had become a common ailment among both whalers and Inuit. Whales were spotted almost every day throughout May, though only a handful were actually taken. In addition, beluga and narwhal were seen in great numbers. Spring floe-edge whaling was just as dangerous a pursuit as fall whaling, and on May 23rd a whale capsized a boat, killing the master of the *Sophia* and another sailor. Towards the end of the month a prominent Inuit whaler, Tesuwin, came across from Naujateling with three or four native boats and crews to participate in the spring whaling. Tesuwin appears to have run his own independent whaling operation, bartering 'whalebone' to the highest bidder and negotiating the blubber on the side. When not engaged in whaling or hauling, the Inuit hunted seals, beluga, and narwhal, with most of the blubber going to the whalers.

Whaling, sealing, caribou hunting, and hauling casks, ice, and blubber to Miliakdjuin and various other duties associated with the fishery continued well into June, when sawing and blasting the *Isabel* out of its ice-bound harbour was added to the daily routine. Although the floe edge began to break up in mid-June, young whales were still plentiful in the waters off Miliakdjuin and Kekerten,[16] and three 'fish' were taken by crews belonging to the *Emma* and Tesuwin. Most of the equipment and whaleboats were brought back to the ship in late June, while four boats were sent up to Kingua. By the end of the month whaleboat crews began to cruise more extensively, sailing up Kingnait Fiord and to the head of the gulf. The *Isabel* 'broke harbour' in early July to work in the company of the whaleboats, providing food and accommodation for the crews and standing ready for flensing and the storage of 'whalebone' and blubber (Ross 1985b: 171–2). But the whaling was not very successful. With the breaking-up of the ice and the arrival of six American ships at Penny's Harbour (Ross 1985b: 178), whales had become scarce. On September 4th the *Emma* set sail for home after spending time in Frobisher Bay.

By 1860 most Inuit were acquainted with the use of firearms, and hunters hired to procure seals and caribou for the whalers either owned a rifle or were in the process of obtaining one. Several Inuit may have also owned whaleboats by this time.[17] Some Inuit boat owners and crews 'fished' independently of the whalers, while Inuit and sailors manning boats belonging to the ships took alternate shifts whaling.

Another benefit for Inuit involved in the whale fishery was access to vast quantities of wood and metal in the form of shipwrecks. Between 1859 and 1870 at least eight ships were wrecked in Cumberland Sound (Stevenson 1984, Ross 1985b: 156), mostly at Naujateling and in the Kekertens, where they were run aground and stripped, first by the whalers, then by the Inuit.[18]

Over the winter of 1860–61 whaling crews may have numbered more than half the Inuit population of the Sound (Table 3). In one harbour alone aggregations of 200 people—natives and sailors—attended theatrical events on board the *Antelope*.[19] The heyday of Cumberland Sound whaling, however, was short-lived; only two ships appear to have over-wintered the following year (Table 3). The decreasing abundance and availability of bowhead whales in Cumberland Sound waters,[20] and the discovery of productive whaling grounds in Roes Welcome Sound by American whalers in 1860 resulted in a flurry of activity in Hudson Bay over the next decade (Ross 1975, 1979a, 1979b). While Hudson Bay attracted the attention of most American ships, many soon returned to the Sound to over-winter, either on their way into or out of Hudson Bay. Double winter voyages in both whaling grounds also became routine for American whalers.

American whalers soon returned to Cumberland Sound only to find that it had been 'fished out'. A decade of intensive whaling had reduced the Sound's bowhead population to such an extent that the large capital investments of the late 1850s and early 1860s were no longer warranted. By the mid-1860s, Penny's participation in the Cumberland Sound whale fishery had come to an end and his former employer, the Aberdeen Arctic Company, had sold out to the Crawford Noble Company of Dundee. Although there was an increase in over-wintering towards the end of the decade (Table 3), perhaps in response to the depletion of whales in Roes Welcome Sound, only a few ships wintered each year in Cumberland Sound after 1870. Moreover, most of these ships were not the large, appropriately equipped vessels of the 1860s. Rather, they were smaller vessels, permanently attached to the one or two American whaling operations left in the Sound.[21] The banner years of the Cumberland Sound whale fishery were over, and a new era of Inuit-Qallunaat relations had begun.

## ECONOMIC DIVERSIFICATION: WHALING AND SEALING, 1870–1894

As whales were depleted from Cumberland Sound the number of wintering ships declined proportionately.[22] Nonetheless, those few vessels that sailed these waters during the 1870s offered a more stable foundation for interaction than ever before. By 1868, the American company of Williams and Haven began to send a heavily freighted tender to deposit stores at its holdings on Kekerten and Blacklead Islands, and to take back the blubber and 'whalebone' collected during the previous season (Colby 1935). The supplies would be given to the Inuit for their participation in the whale fishery and used by skeleton staffs manning the company's schooner. The Noble Company appears to have adopted the same routine, though its managers preferred to operate from shore-based facilities at Kekerten and later Umanaqjuaq.[23]

By the early 1870s the whale population appears to have been decimated to the point where those few whaling companies remaining in the Sound were forced to diversify in order to sustain their operations at a profitable level. For example, the Williams and Haven Company made two systematic attempts to capture beluga with nets at the head of Kingua Fiord. In spite of the fact that over 800 whales were taken, these efforts barely paid expenses and large beluga whaling operations were quickly suspended (Clark and Brown 1887: 247–8). By far the most common quarry, after diversification of the economy, was the ringed seal. Even though the blubber of small whales and seals was collected by American and British whalers during the 1850s and 1860s to 'top off' their tanks before returning home (Clark and Brown 1887: 147), these species were not routinely exploited until the 1870s.

However, as bowheads grew scarce, seals began to attract the attention of the owners of the American and Scottish interests at Kekerten and Blacklead. In addition to the blubber, the skin of the seal, particularly the 'silver jar', possessed a market value, at least to the Scots. Although the Americans were taking seals in Cumberland Sound as early as 1852 (Clark and Brown 1887: 95), they appear never to have become involved in the blubber-skin trade to the same extent as the Noble Company. Rather, the main response of the Americans to the depletion of bowheads in Cumberland Sound after 1870 was to gradually scale down operations and establish shore-based whaling stations elsewhere (e.g., Hudson Strait). In 1878 the American building(s) at Kekerten were partially dismantled by the owner, C.A. Williams (Howgate 1879: 159), and by 1883 Blacklead Island was no longer a whaling base (Boas 1964: 59).

Throughout the 1870s whale oil prices gradually decreased on world markets owing to the large supply of cheaper substitutes and to the fact that seal oil, which is of inferior consistency, was often mixed with whale oil (Clark and Brown 1887: 147, 156). Despite the reduction in oil prices and the decreasing availability of bowhead whales, Arctic whaling remained profitable because of an increase in 'whalebone' (baleen) prices during the same period (Table 4).[24]

TABLE 4. Prices for Arctic whale oil and 'whalebone' (baleen) on American markets, 1868–1880 (year open, year high, year close). Extracted from Clark and Brown (1887).

| | ARCTIC WHALE OIL ($ PER US GALLON) | | | ARCTIC 'WHALEBONE' ($ PER POUND) | | |
|---|---|---|---|---|---|---|
| YEAR | OPEN | HIGH | CLOSE | OPEN | HIGH | CLOSE |
| 1868 | .65 | 1.25 | 1.00 | .70 | 1.40 | .80 |
| 1869 | 1.00 | 1.20 | .90 | .70 | 1.30 | .85 |
| 1870 | .725 | .80 | .65 | .85 | .85 | .65 |
| 1871 | .65 | .80 | .80 | .65 | 2.00 | 2.00 |
| 1872 | .73 | .73 | .68 | 1.90 | 1.90 | 1.18 |
| 1873 | .68 | .68 | .63 | 1.15 | 1.20 | 1.10 |
| 1874 | .61 | .675 | .675 | 1.10 | 1.25 | 1.25 |
| 1875 | .675 | .70 | .70 | 1.20 | 1.30 | 1.30 |
| 1876 | .70 | .70 | .70 | 1.30 | 3.50 | 3.50 |
| 1877 | .70 | .70 | .60 | 3.50 | 3.50 | 2.00 |
| 1878 | .60 | .60 | .39 | 2.00 | 3.25 | 3.25 |
| 1879 | .38 | .60 | .60 | 3.25 | 3.25 | 2.25 |
| 1880 | .60 | .60 | .50 | 2.25 | 2.50 | 1.30 |

Thus, even the capture of one good-sized whale a year was usually enough to cover expenses. In this regard, the three whales taken by the American and Scottish operations at Kekerten in the fall of 1877 must have been considered a good catch indeed (Howgate 1979: 61). Yet, whales were not only less numerous, they were considerably smaller.[25] Consequently, even though the bowhead remained the most commercially important species during the late nineteenth century, its occurrence was simply too unpredictable and its yield too small to justify an exclusive emphasis on this species. By the early 1870s the seal had become a major commodity.

As seals assumed greater economic value, Inuit began to return to their former camps. Seals were more easily obtained in the vicinity of traditional hunting grounds than around Naujateling, Umanaqjuaq, and Kekerten, though the latter area was always known as a productive sealing ground (Etuangat Aksayuk, personal communication, 1983). After nearly two decades of living much of the year in the company of commercial whalers, most Cumberland Sound Inuit returned permanently to their original camps, travelling to the stations only to trade and participate in spring and fall whaling activities.

In 1877 Kumlien (1879: 12, 15) estimated that about 400 Inuit lived in 10 villages from Nugumiut at the north entrance to Frobisher Bay around to Saumia. While Naujateling and Kekerten were the principal settlements, only a few elderly couples lived at Kingmiksoo. Six years later, Boas (1964: 18) reported that the Cumberland Sound Inuit inhabited eight settlements (Table 5). The Talirpingmiut (n=86) lived at Umanaqjuaq/Naujateling, Idjorituaqtuin, Nuvujen, and Qarussuit; the Kinguamiut (n=60) at Imigen and Anarnitung; the Kingnaimiut (n=82) at Kekerten; and the Saumingmiut (n=17) at Ukiadliving.

An examination of Table 5 reveals several interesting features, the most obvious being an apparent reduction in the population of the Sound since

TABLE 5. Cumberland Sound and Davis Strait census by Boas (1964: 18), December 1883.

| SETTLEMENT | MARRIED | | UNMARRIED | | | | TOTALS |
|---|---|---|---|---|---|---|---|
| | MEN | WOMEN | WIDOWS | WIDOWERS | ADULTS | CHILDREN | |
| Naujateling | 6 | 6 | | 1 | 1 | 6 | 20 |
| Idjorituaqtuin | 3 | 3 | 1 | | 1 | 3 | 11 |
| Nuvujen | 8 | 8 | 2 | 1 | 1 | 6 | 26 |
| Qarussuit | 10 | 10 | 2 | | | 7 | 29 |
| Imigen | 6 | 6 | | | | 5 | 17 |
| Anarnitung | 12 | 12 | 1 | 1 | 1 | 16 | 43 |
| Qeqerten | 26 | 26 | 6 | | 4 | 20 | 82 |
| Ukiadliving | 6 | 6 | 1 | | 1 | 3 | 17 |
| Padli | 11 | 13 | 2 | 2 | 1 | 14 | 43 |
| Akudnirn | 8 | 12 | | | 2 | 18 | 40 |
| Totals | 96 | 102 | 15 | 5 | 12 | 98 | 328 |

1857 (n=245), and the relatively low number of children compared to adults. As Boas (1964: 60) notes, emigration in response to an increase in commercial whaling and sealing in Davis Strait likely accounts for the population decline:

> As the whale catch in Cumberland Sound has fallen off during the last fifteen years, a reimmigration of the population of Davis Strait has occurred, ships visiting these shores every fall and a regular traffic being kept up. Therefore many Oqomiut now travel as far as Qivitung (Kivitoo) in order to trade there. As Nugumiut is still frequently visited by the whalers, there is no inducement for the inhabitants to leave their country.

The ratio of adults to children, 3:1 (179/66), however, clearly suggests a high incidence of infant mortality compared to the early contact period—the ratio of adults to children at Kingmiksoo was about 2:1 (see Figure 14). Only two settlements, Anarnitung and Naujateling, appear to have approached Kingmiksoo's adult/child ratio. It is possible that the low survival rate of children may have been due to nutritional stress brought about by the decline in employment opportunities in the whaling industry and/or a readjustment to traditional patterns of living. As early as the 1850s, various observers (e.g., Barron 1895: 44, Warmow 1859) expressed concern that the Cumberland Sound Inuit were fast losing their ability to hunt in the manner of their forefathers. More likely, however, disease may have been the primary cause of infant mortality as well as infertility, particularly among the Kingnaimiut at Kekerten. While Boas (1964: 18) reported that diphtheria, a disease that he himself may have introduced, killed five children at Kekerten in 1883, he felt that of all the foreign diseases syphilis had made the 'greatest ravages' among the Cumberland Sound Inuit.

A combination of social, economic, nutritional, and pathological factors is most likely responsible for the decline in population from 1857, though contact with the whalers seems to have had little effect on the high rate of adult male death; widows outnumbered widowers 13 to 3 (Table 5). Contact may have also affected marriage patterns as 6 of the 19 marriages recorded by Boas on Davis Strait (i.e., Padli and Akudnirn) were polygynous unions, whereas he found no such cases in Cumberland Sound (Table 5). It is possible that some of the Sound's Inuit purposely adopted European ideals of marriage and fidelity for social and economic gain within the context of the whale fishery. Alternatively, access to imported clothing as well as a high rate of infant mortality may have alleviated the socioeconomic need for two productive women in one household.

Certain customs may have suffered as a result of contact. However, many features likely remained unchanged from the early contact period. Boas may not have been aware of it, but Cumberland Sound Inuit social organization in 1883 was probably more similar to that of the precontact situation than of any time within the last 30 years.

In the 1880s most Cumberland Sound Inuit who returned to their former settlements pursued seals and other marine mammals for domestic purposes and the blubber-skin trade. Yet, every spring after the young sealing season and every fall after the caribou hunting season, Inuit gathered at Kekerten, and later Umanaqjuaq, for whaling. Boas (1964: 59–60) has described the economic round of the Cumberland Sound Inuit during the early 1880s. As this cycle characterized the annual routine of most Inuit in the Sound from the mid-1870s to the demise of commercial whaling around 1920 (see Low 1906: 9–10, Wakeham 1898: 74–5), it warrants presentation here:

> When the Eskimo who have spent the summer inland return at the beginning of October they eagerly offer their services at the stations, for they receive in payment for half a year's work a gun, a harmonium or something of that nature, and a ration of provisions for their families, with tobacco every week. Every Saturday the women come into the house of the station, at the blowing of the horn, to receive their bread,[26] coffee, sirup, and the precious tobacco. In return the Eskimo is expected to deliver . . . a piece of every seal he catches.
>
> The time for the fall fishing commences as soon as the ice begins to form. If the weather, which is generally stormy, permits it, the boats leave the harbour to look for the whales which pass along the east shore of the sound toward the north. During the last few years the catch has been very unprofitable, only a few whales having been seen. As the ice forms quickly the boats must be brought back about the end of October or the beginning of November. Since the whale fishery has become unprofitable the stations have followed the business of collecting seal blubber and skins, which they buy from the Eskimo.

A lively traffic springs up as soon as the ice becomes strong enough to allow the sledges to pass from shore to shore. The sledges of the stations are sent from one settlement to another to exchange tobacco, matches, coffee, bread, &c. for skins and the spare blubber which the Eskimo have carefully saved up. On the other hand, those natives who require useful articles, such as cooking pots, lamps, &c., collect quantities of hides and blubber and go to Qeqerten to supply their wants. The winter passes quickly amid the stir of business, till everything comes to a stop at the end of March, when the young sealing season fairly opens.

When the sun has reached such a height that the snow begins to melt in favored spots, a new life begins at the stations. The skins which have been collected in the winter and become frozen are brought out of the store room and exposed to the sun's rays. Some of the women busy themselves, with their crescent shaped knives, in cutting the blubber from the skins and putting it away in casks. Other clean and salt the skins, which are likewise packed away. The men also find enough work to do after the young sealing is over, for the whale boats must be got ready for the spring fishing. Strangers whose services have been engaged by the station for the next few months arrive daily with their families and all their goods to take up their abode on Qeqerten. The boats are dug out of the deep snow, the oars and sails are looked after, the harpoons are cleaned up and sharpened, and everything is in busy preparation. The boats are made as comfortable as possible with awnings and level floors, for the crews are not to come to the shore for about six weeks.

By the beginning of May, the arrangements having been completed, the boats are put upon the sledges, which, under the direction of native drivers, are drawn by dog teams, with their crews, to the floe edge. The sledges being heavily laden and food for the dogs having to be provided by hunting, each day's stage is rather short. Arriving at the floe edge the sledges are unloaded and the boats are launched. Seal and birds of all kinds are now found in profusion and the chase is opened without delay upon everything that is useful and can be shot. Sledges are regularly sent back to Qeqerten with skins and meat for the families of the Eskimo, while the blubber is packed in casks and kept ready on the spot.

The most important object of the expedition is the whale. Harpoons and lines are always in readiness for the contest with the mighty monster. The boats return to the north with the breaking up of the ice and the fishing ends in July. The Eskimo are paid off and dismissed and resume their reindeer hunting, while the whites are glad to enjoy some rest after weeks of exhausting labor.

Throughout most of the 1880s few ships intentionally wintered in the Sound, and the American operation at Kekerten appears to have been without a permanent manager much of the time (Mutch 1886). Shrinking personal

TABLE 6. Returns of whales and seals from Noble's stations at Kekerten and Umanaqjuaq, 1883–1903. From Goldring (1986: 163) and Lubbock (1955). Table does not include returns of American or other British vessels.

| Year | Whales | Seals | Year | Whales | Seals |
|---|---|---|---|---|---|
| 1883 | 1 | 4300 | 1895 | 3 | 4500 |
| 1885 | 2 | 5000 | 1896 | 3 | 3890 |
| 1886 | 2 | 1200 | 1897 | 1 | 5750* |
| 1888 | — | 3300 | 1898 | 2 | many |
| 1889 | 3 | 2480 | 1899 | 2 | 2900 |
| 1890 | — | 2227 | 1900 | 1 | 3400 |
| 1891 | 1 | ? | 1901 | 2 | 3400 |
| 1892 | 1 | 8618 | 1902 | — | 1750 |
| 1894 | 1 | 7000 | 1903 | 3 | 3044 |

* 70 tons of seal oil, which, based on statistics provided in the 'Kinnes Lists', amounts to 4000 to 7500 (or c. 5750) seals, were returned to Scotland in this year.

contact with Qallunaat, however, was temporarily postponed in the late 1880s when the Noble Company and C.A. Williams Co. established shore-based operations on Blacklead Island (Goldring 1984: 489–90). A decline in whaling in more northern waters[27] and an increase in the price of 'whalebone' apparently stimulated a rise in both whaling activity and the Sound's population during these years:

> In the present winter—1887–88—one American and two Scottish stations are in operation in Cumberland Sound ... and the Scottish steamers which used to fish in Baffin Bay ... are beginning to visit Cumberland Sound and Hudson Strait. The whaling in Baffin Bay shows a sudden falling off. ... This cannot be without influence upon the Eskimo, who will probably begin again to flock to Cumberland Sound and Nugumiut. (Boas 1964: 259)

By the mid-1890s most Cumberland Sound Inuit had returned to Kekerten and Umanaqjuaq, their populations rising, respectively, to 140 (Wakeham 1898: 24) and 170 (Harper n.d.: 34). Even so, the seal skin trade remained fairly constant throughout the late 1880s and 1890s, with an average of 3900 seals being traded each year to the Noble Company's stations at Kekerten and Blacklead (Table 6). The fact that the seal harvest remained unchanged during these years, despite an increase in the concentration of the population, would suggest a concomitant rise in individual mobility. The latter, in turn, may have been facilitated by access to wood for sleds created by the wrecking of three ships in the Sound in 1886–87 (Stevenson 1984: 23).

While Inuit continued to find seasonal work throughout the 1890s, employment opportunities were reduced from previous years. After operating more or less permanently in the Sound for over three decades, the Williams Co. sold out to the Noble Company (Wakeham 1898: 75). The effects of the American withdrawal from Cumberland Sound, however, were eased somewhat by an increase in the procurement and processing of beluga whales by other British interests. This species seems to have been more actively pursued in the early 1890s than previously,[28] for it was customary that if a 'whaling voyage to Baffin's Bay, or Lancaster Sound, had not been a profitable one, for the whaler to call in at Cumberland Gulf on his way home and if possible fill up his tanks with the oil of the white whale' (Wakeham 1898: 73).

Against this background of economic diversification and population concentration, a new harbinger of Euroamerican culture arrived in Cumberland Sound.

## A Clash of Ideologies: A New Way of Believing, 1895–1906

In 1894 the Anglican Church of England sent E.J. Peck and J.C. Parker to Blacklead Island to set up a mission. At first, these missionaries were well received by the 170 Inuit living at this settlement; the latter were quick to realize that Parker's medical training could help them in times of sickness (Peck 1922: 28). However, Peck soon ran into opposition from Umanaqjuaq's three leading *angaqut*, and principally Kanaaka, their chief shaman (Greenshield 1914: 13):[29]

> The magicians, who learnt by this time, that their business was in danger, should the truths of God prevail, tried in every possible way to undermine the good work. They arranged, chiefly in the dwellings of the minor conjurors gatherings, to oppose our teachings and where even amidst the howling of the wind, could be heard their unearthly yells. (Peck 1922: 30)

Realizing that the religious opinions and beliefs of most men were too strongly held to sway easily, Peck turned his attention to women and children, setting up daily bible study classes. To attract the adults, coffee and biscuits were served after Sunday morning and evening services.[30] Peck's efforts to convert the Umanaqjuarmiut, however, did not begin to bear fruit until 1897, when he received his first shipment of newly translated Inuktitut bibles. Peck had begun to teach syllabic literacy soon after he arrived. And these bibles 'proved . . . a mighty faith creating and life giving force' that Peck assumed the 'magicians could not in wise withstand'. Although the *angaqut* continued to fight the missionaries at every step, after the introduction of the bible their influence apparently 'became less potent among their former supporters' (Peck 1922: 31). By 1900 many women could read the gospels.[31] Nevertheless, most men continued to reject Peck's teachings and rarely attended church meetings.[32] The introduction of a slide projector or 'Magic Lantern' in the fall of 1900, however, had great drawing power,[33] and a few men began to attend the services.[34]

By 1901, 24 women and two men 'had publicly confessed their faith in Jesus',[35] while Peck (1922: 35) believed that the influence of the 'magicians' was nearing its end. However, 'heathen incantations' were still being said over the dying, and women, including some of Peck's candidates for baptism, were still in the habit of visiting the ships for tobacco and other favours.[36] The latter habit caused Peck much concern for he believed that 'the extermination of the whole of the Eskimo of Cumberland Sound and some other regions is only a matter of time if some check (was) not put to these dreadful practices'. Resistance to Peck's teachings appears to have been even more widespread at Kekerten. Here, some older women would not attend Peck's annual spring services unless they were paid.[37] Opposition to Peck's new order came to a head at Kekerten and in Cumberland Sound in 1902 when Angmarlik, an aspiring *angaqok* and the best seal hunter in the Sound (Hantzsch 1977: 32), propounded a syncretic religion which fused traditional religious concepts with elements of the Christian doctrine. In March, Peck and his new assistant, E.W.T. Greenshield, travelled to Kekerten to counter Angmarlik's revelation, which he had received from the goddess 'Sedna', and which had been known far and wide (Lewis 1904: 318–19). Every day for the next three weeks Angmarlik and Peck took turns proclaiming the benefits of their respective ideologies in front of gatherings in the old American whalers' bunkhouse (Stevenson 1984). Finally, a conclusion to the matter was reached, at least as far as Peck was concerned:

> A wonderful day. The church was packed morning and evening. Hardly any of the men had gone away hunting, and the attention and reverent behaviour of the people was quite remarkable. I naturally inquired what these things meant. This is the answer which I received—an answer which gave me great joy. . . . They told me that having considered the new doctrine propounded by Angmalik (*sic*), and having also considered the words they heard and read, viz., the words of Jesus, they had come to the conclusion that His words were in every way preferable, and therefore they had determined to cast away their heathen customs and come to the place of prayer. . . . (Lewis 1904: 319)

Peck's focus of disgust was the 'Feast of Sedna', and especially the ritual exchange of spouses presided over by the chief *angaquk*, the *qailertetang* (Boas 1964: 196–8). However, like the Umanaqjuarmiut, women at Kekerten appear to have accepted the Christian doctrine much sooner than the men. In a particularly symbolic act, Angmarlik's wife, Ashivak, and several other prominent women of the settlement made the ceremonial costume of the *qailertetang* and cast it, along with the old religious order, into the water:

> In the summer my mother got some women together to make caribou clothes. The clothes were at least two times as big as the ones Eskimos wore, or maybe bigger than anybody could use on earth. They made everything,

the parkas, the kamiks (boots), the mittens, the pants, in fact the whole works. After they finished, everything was thrown into the water because that is what my mother wanted them to do. They made these clothes so that they could throw them into the water and no longer be followers of this god. (Eevic 1976: 79)

This symbolic act and Ashivak's protestations aside, it was not until Angmarlik gradually began to profess to accept Christian beliefs a year or two later that the men of Kekerten began to adopt the new religious doctrine (Qatsu Eevic, personal communication, 1983; Eevic 1976: 79).

Leadership also played an important role in the acceptance of Christianity by men at Umanaqjuaq. Even though nine women and two men had been baptized by February of 1902,[38] it was not until after the prominent hunter and whaler Tooloogakjuaq began to lead prayer meetings and instruct congregations in the late fall of 1903 that men began to attend evening services.[39] Within a few months (Peter) Tooloogakjuaq was baptized and several hunters began to keep the 'Lord's day'.[40] However, without a permanent missionary, the situation at Kekerten worsened: 'the influence of the conjurors during our absence seems almost to drag the people back to their former state of heathen degradation'.[41] Nonetheless, Tooloogakjuaq began to instruct Inuit whalers from both Umanaqjuaq and Kekerten during the spring whaling season. A letter to Peck from Tooloogakjuaq at the floe edge records the progress he was making: 'I am greatly pleased with these men from the Kikkerton (*sic*) Station. ... Okittok, their chief, in particular desires to believe in God'.[42]

By 1904, a total of 30 Inuit had been baptized in Cumberland Sound, mostly women and all but one or two from Umanaqjuaq.[43] Although few Inuit appeared to have grasped 'a real sense of sin or true repentance' by the time the mission station was moved to Lake Harbour in 1906,[44] towards the end of the decade 12 men and six women—the former acting as lay preachers, the latter as teachers of children—were spreading the 'word' in Cumberland Sound. Among these, Tooloogakjuaq and (Luke) Kidlapik, who were known and respected for hundreds of miles around, played the principal role (Greenshield 1914: 13–14). By the late 1910s most Inuit appear to have embraced the new faith in some degree or another, or were 'coming up Jesusy' as Angmarlik's followers described it (Munn 1932: 274).

## Decreasing Expectations, Continuing Adjustments, and General Trading: 1906–1921

Commercial beluga whaling continued to be carried out in a desultory manner well into the 1900s. Although Low (1906: 11) reported that the stations had as yet made no systemic attempts to exploit this resource, the *Nova Zembla* took 418 white whales in the Sound in 1901 (Lubbock 1955: 440). Walrus, however, appears to have been intensively hunted around the turn of the century (Lubbock 1955). By 1903, Inuit were trading fox, bear, and other fur-bearing animals (Low 1906: 10–11; Qatsu Eevic, personal communication, 1984). Although 'whalebone' fetched over $5.00 (US) per pound (Hegarty 1959: 51) during the early 1900s, it was the broad nature of the resource base that kept the Noble Company's operations 'afloat' for the remainder of the decade.

The number of Inuit at Umanaqjuaq and Kekerten remained fairly constant from the early 1890s to 1910. However, in 1897–98 the population of Umanaqjuaq swelled to over 260, when the entire native settlement of Singnija (Cape Haven) arrived to look for work (Wakeham 1898: 75). The latter station was temporarily abandoned after its manager, Captain Clisby, drowned in a boating accident along with four Inuit and two other whites (including J.C. Parker). Although most Nugumiut soon returned to Singnija, a few appeared to have stayed, and by 1900, 40 dwellings at Umanaqjuaq housed 194 people.[45] The fact that the missionaries provided ammunition and gun powder as a source of relief in times of stress likely accounts for this situation.[46] But overpopulation created unsanitary conditions, which contributed to the death of 18 people at this settlement in the fall of 1899.[47] Though no such problems existed at Kekerten, sickness was still prevalent during the fall, and 20 per cent of the adult population (n=12) died over the winter of 1900–01.[48]

Hantzsch (1977: 31, 39) estimated that only 250 Inuit lived at Kekerten and Blacklead in 1909. While disease may have contributed to this decline (e.g., Fleming 1932: 102), emigration probably accounts for most of this reduction. After the Noble Company's supply vessel was wrecked in Cumberland Sound in 1902, poorly outfitted vessels with equally inadequate crews were sent out to supply the stations. As often as not, these vessels failed to reach their intended destinations,[49] and station managers, lacking adequate supplies, were unable to pay the Inuit their normal wages (Hantzsch 1977: 40, 61). By the end of the decade, many Inuit had returned to their former settlements. For example, over the winter of 1909–10 a 'great many natives' from Kekerten in a caravan of nine sleds and over 100 dogs migrated to Durban Island (Fleming 1932: 143), while Inuit attached to Blacklead 'scattered their winter dwellings more widely than was customary' in groups of two to four families, where they ate well and 'suffered no want' (Hantzsch 1977). In 1910 the Kinnes Company of Dundee established a post at Durban Island (Munn 1932, Usher 1971: 129) under the direction of William Duval, a German-American whaler/trader who had lived more or less continuously in the region for the last 30 years. The following year the newly formed Sabellum Company of Peterhead headed by James Mutch, a manager of Noble's Kekerten and Blacklead stations for some 35 years, established a small trading depot at Cape Mercy (Usher 1971: 129). The same company also purchased the old American station at Singnija and set up a small post at Kivitoo on Davis Strait. The era of general trading had begun.

Greenshield continued to provision Inuit at Blacklead in times of want so long as he was present.⁵⁰ However, after 1910, a crash in the 'whalebone' market—this commodity droped to $.10 a pound in 1912 (Hegarty 1959: 51)—resulted in a decline in whaling activity and seasonal employment opportunities. By 1913, no more than three whaleboats were maintained at each station (Munn 1932: 183), about half as many as a decade earlier (Low 1906: 10). While Umanaqjuaq's population remained fairly stable during these years (Fleming 1932: 13–14), the population of Kekerten fell markedly as most Inuit returned to their former camps or settled at other small trading depots. As Noble's operations seem to have been without resident managers much of the time, whaling and trading at Kekerten and Umanaqjuaq, respectively, were left in the hands of Angmarlik and Pawla (Etuangat Aksayuk, personal communication, 1983).⁵¹

In 1914, after nearly half a century in Cumberland Sound, the Noble Company sold out to Kinnes' Cumberland Gulf Trading Company. About the same time, the strangely named Arctic Gold Exploration Syndicate arrived on the scene, buying out Kinnes' Durban Island post in 1914 (Munn 1932, Usher 1971). Even though fox and polar bear furs had been collected along with whale and seal products for more than a decade, the emphasis shifted from systematic whaling and sealing to general trading in furs, skins, blubber, and ivory after 1915. Whaling was still kept up in the spring and fall at each station. In fact, Inuit at Kekerten appear to have been remarkably successful, taking three whales in one year (1917?) alone (Etuangat Aksayuk, personal communication, 1983). Nonetheless, an increasing emphasis on fur-bearing animals, together with the attraction of other independent trading operations elsewhere, resulted in a general decline in both whaling activity and the size of Umanaqjuaq and Kekerten. Since vessels frequently were unable to sail from Scotland during these years (Goldring 1986: 165), there was little reason to remain at the stations. Ironically, the lone white man in the Sound depended entirely on the charity of the Inuit for whom he had nothing left to trade (Akulujuk 1976: 75, Kilabuk 1976: 35–6).

As in the whale fishery, Inuit leaders figured predominantly in the general trade. While Angmarlik and Pawla continued to direct whaling and sealing operations, they also bartered for furs on behalf of white station managers, who were more often absent than not. The Sabellum Company also appears to have relied extensively on natives to manage its trading operations. Near the mouth of the Sound, Kanaaka managed two small depots (Parmi n.d.: 22), while Kingudlik and Niaqutsiaq operated posts at Durban Island and Kivitoo, respectively. The old post at Singnija and a new one in Frobisher Bay were, in turn, run by Michiman and Godiliak (Goldring 1986: 166). When feasible, small vessels belonging to each company would be sent to Cumberland Sound to deposit stores at their respective depots and collect the previous season's harvest of furs, oil, hides, and ivory. Towards the end of the decade, beluga whaling once again appears to have attracted attention as Inuit boat crews attached to both the Cumberland Gulf Trading Company and the Arctic Gold Exploration Syndicate pursued this species at the head of the Sound each summer. The latter company even went so far as to establish a small trading post at Ussualung in 1918 under the direction of William Duval (Munn 1932). However, it was the arrival of the Hudson's Bay Company in 1921 that heralded yet another era of Inuit-white relations in Cumberland Sound.

## PANGNIRTUNG AND THE HUDSON'S BAY COMPANY, 1921–1962

### Economic Concessions and the Fur Trade

As part of its expansion into the Arctic fur trade, the Hudson's Bay Company erected a post in Pangnirtung Fiord in 1921. Over the next few years an intense rivalry for Inuit labour and produce characterized relations between the HBC and its competitors:

> Kanaka is the man in charge of Jimmy Mutch's station at Shaumia and just came here on a visit—we have three . . . men from him and they are now trapping for the Co. Evidently the opposition Companys (*sic*) here intends to put up a good fight, Kanaka is selling a Columbia Gramaphone for five foxes and a 303 British rifle for two foxes and ten seal skins. Whether he is working from the instructions received from Mutch last fall is more than I can state but in any case it knocks the price of our rifles all to hell and it seems that all his other goods he has for trade is priced in proportion to his rifles and Gramaphones. As is the usual custom with Munn and Kinniss (*sic*) Kanaka treats his trappers as servants and they are issued rations either once or twice a week and besides they are allowed to trade their hunt of furs and it seems to me, the only way to fight the opposition here is to give them a dose of their own medicine as I have been doing up to now. . . .[52]

This competition benefited the Inuit, as they played one company off against the other:

> The Opposition at Ooshoolook (Ussualung) are doing every thing in their power to secure some of the men they lost last year, but so far we are holding our own. How long this will last I cannot say as these people are evidently used to the custom of changing masters every year and cannot be depended upon to stick to any one company, especially when there is another company at hand to issue free grub, free rifles and ammunition as well as pay $25.00 for one fox. . . .[53]

The high price of Inuit labour and produce was not the only problem faced by the HBC. Strong allegiances and kinship obligations existed between native traders and their followers, a fact which the post manager continually bemoaned:

> Both Angmalee (Angmarlik) and Kanaka seem to have a wonderful hold on the natives working with them and it means a lot of hard labor and long pow-wows and a great deal in the expense line to break their hold. The devil or some of his fools seems to have fallen possession of the opposition men here and compared with any of my former trading with the natives, Hell seems to be let loose.
>
> Evidently it is the men that the opposition has the greatest hold on that is visiting us and when we ask them to work for us they say that they are brother or brother-in-law to the man in charge of opposition and that they love him too much to leave him yet. . . . If you gave them the whole store and all thats (*sic*) in it you would not get them to come with us. . . .'[54]

Inuit hunters were not so much loyal to the company, but to the native trader to whom they were usually related in some way. By enticing a few prominent individuals such as Tooloogakjuaq with new rifles and other possessions, the HBC managed to secure the services of 33 hunters within a short period of time.[55] However, the HBC trader, Nichols, soon discovered that these hunters were poor trappers, a fact which he blamed, mistakenly, on the whalers:

> (They) had always taken every (pound) of fat or oil these natives would get and never allow them to keep any for themselves. Consequently, instead of hunting foxes in winter they have to hunt seals at the floe edge to get enough fat to keep their houses warm.[56]

In fact, quite the reverse was true as 'the older traders always used to give out seal-meat during times of hunger and distress in the winter from the stock of whole seals which had been brought in during better hunting.'[57] After a succession of post managers, the HBC was still trying to eradicate this perceived defect from the character of the Cumberland Sound Inuit:

> The Cumberland Sound native seems to be a pretty fair sealer and whaler but he is a very poor trapper. We know there is nothing new in this observation, but we also add that the best of the hunters are poor trappers. Fifty per cent of the men do not try to make a hunt. After he has secured his one or two foxes to pay his debt he sits back and calls it a day. He will spend day in and day out sealing and live entirely off the country: He will then decide he is hungry for tea and biscuit and go and have a look at his traps and find a fox that has been dead probably for weeks, come any distance up to 100

miles to the Post to trade it and go back home and repeat the process all over again. He is also essentially a whaler type of native brought up entirely on the whaling tradition. This symptom was noted in the early days at the Hudson Straits Posts but fortunately has since almost disappeared. We have no doubt however that in time the Cumberland Sound natives will become better 'Hudson's Bay men'.[58]

The Cumberland Sound Inuit did not, of course, become 'better Bay men'. Well into the 1950s, various authorities remarked on the poor trapping abilities of the Cumberland Sound hunter.[59] This predisposition was not the result of whalers' influences, but to the maritime hunting traditions of the Cumberland Sound Inuit, which were continually being reinforced by Inuit leaders. Angmarlik, for example, told all those under his authority that, before they trapped for the Company, they had to put enough meat and blubber on the table to feed their families (Etuangat Aksayuk, personal communication, 1988). So strongly developed was the maritime orientation, that 'the trapping of fur (was) looked upon by (some) with disdain', and that, until about 1920, trapping was carried out predominantly by women.[60] However, with increases in the price of fox and trading competition during the early 1920s, men took up trapping.

For the first time in the history of Cumberland Sound, fox began to play a role in the selection of campsites, and indirectly the food and clothing supply, as the HBC tried to organize Inuit for fox fur production.[61] New settlements were created in Kingua Fiord (Shimilik) and Nettilling Fiord (Kaneetookdjuaq) over the winters of 1923–24 and 1924–25, respectively. The Company adopted the practice of advancing Inuit whaleboats and provisions in exchange for consenting to relocate to better trapping grounds, where they were to trap for the Company (Pauloosie Angmarlik, personal communication, 1988).[62] However, these experiments quickly proved a failure; neither site appears to have been occupied for more than a winter as they were located too far from the sealing grounds and fox was scarce.[63] Cumberland Sound was never regarded as a productive trapping ground. Moreover, Arctic fox is well known for its cyclical fluctuations in abundance, and between 1924 and 1938 fox returns were high only in one year out of four (i.e., 1927, 1931, and 1935).[64]

Eventually, some of the more pragmatic Company traders resigned themselves to the fact that the fox trapping was never going to be a lucrative enterprise in the Sound:

> He (Parsons) said the Company wished to encourage the Eskimos to hunt throughout the year and not concentrate on white foxes or any other form of wildlife. He thought that by providing whale and walrus drives, fishing expeditions, etc. the Eskimos would retain their hunting instincts and . . . add to their earnings during the low fur cycles.[65]

As early as 1923 the Company, realizing the hunter's reluctance to forsake sealing and whaling in order to trap fox, began to look for alternative ways to exploit Inuit labour and produce. One of the less imaginative schemes entailed the construction of a fox farm in Pangnirtung Fiord in 1927. Yet, this venture too proved a dismal failure—farming was not congruent with Inuit attitudes and beliefs towards animals—and attempts to raise fox were abandoned in 1932.

By far the most successful venture was the establishment of a beluga whale fishery in 1923. Although the beluga had been pursued by whalers and traders using Inuit labour in a desultory manner for decades, and whale drives had been conducted at Milurialik every season for the last five, beluga whaling remained largely unorganized and opportunistic until the HBC built a hide and blubber processing facility in Pangnirtung. The whale drive quickly became the 'high spot' of the year and 'annual picnic' for many Inuit,[66] and between 1923 and 1940 over 5100 belugas were taken by Inuit working for the HBC (Table 7). Even though some observers felt that the Inuit benefited very little materially from their participation in the whale fishery,[67] its temporary suspension in 1937 and 1938 raised real fears, particularly among Inuit boat owners, that the whale drive would be cancelled permanently.[68] Apart from being the social event of the year, the whale fishery was the primary means by which hunters replaced their aging whaleboats and fitted themselves out for the caribou hunt. After World War II, the beluga whale fishery was 'loosely organized by various camp bosses' whereby the HBC advanced gasoline, ammunition, and other supplies to whaleboat owners against their catch.[69]

The incorporation of this fishery into the annual routine of many Inuit resulted in a pattern that dominated their lives for the next three decades:

> Natives come into the post for the whale hunt as soon as the open water allows them, usually the end of July. They are employed for about three weeks. At the end of this time most of the natives are paid off and go away for the summer caribou hunt. Many of them are in a hurry to get away before this time and are paid off before all the whales have been handled at the Post. The number of natives employed for the whale drive amounts to about 40 men and 35 women. The Company also has the use of 9 native boats which belong to the men hired for the drive and which they are responsible for buying back the greater part of the whales obtained.
>
> About 20 families usually come to the post for the whaling, (while another) 20 families live at settlements around the sound who do not come into Pangnirtung for the summer whaling. These families live at Shalmea (Saumia), and around Bear Sound and Blacklead. (The latter) however, are employed for the annual summer work on the seal oil at the outpost. After the supply ship has sailed they return to their camps, and, until the commencement of the trapping season, are chiefly occupied in hunting seals.[70]

TABLE 7. Beluga whale and ringed seal returns, Pangnirtung Post, 1923–1940.

| Year | Whales | Seals | (Jar) | (Whitecoat)* | (Silver Jar) |
|---|---|---|---|---|---|
| 1923 | 600 | 961 | 920 | 41 | — |
| 1924 | 800 | 2169 | 1969 | 200 | — |
| 1925 | 422 | 3337 | 2195 | 1142 | — |
| 1926 | 248 | 4579 | 3672 | 907 | — |
| 1927 | 250 | 1796 | 1239 | 530 | 27 |
| 1928 | 350 | 1523 | 732 | 721 | 70 |
| 1929 | 240 | 3012 | 1238 | 1424 | 350 |
| 1930 | 272 | 2563 | 1050 | 930 | 583 |
| 1931 | ? | 1852 | 224 | 872 | 756 |
| 1932 | 160+ | 1595 | 240 | 663 | 692 |
| 1933 | 425 | 4353 | 675 | 1875 | 1803 |
| 1934 | 180 | 1757 | 55 | 750 | 952 |
| 1935 | 200 | 4205 | 81 | 2638 | 1486 |
| 1936 | 240 | 3187 | 454 | 1801 | 932 |
| 1937 | no drive | 4440 | 682 | 2313 | 1445 |
| 1938 | no drive | 2613 | — | 1500 | 1113 |
| 1939 | 300 | 3307 | — | 1855 | 1452 |
| 1940 | 424 | 1502 | — | 911 | 591 |

*The ringed seal retains a covering of white fur for 2 to 3 weeks after birth, while for the rest of the first year of its life, it possesses a silver coat of fur.

Sources: Goldring (1986: 171, Table 4); HBCA RG 3/26 B/36, 'Annual Report Pangnirtung Post, Outfit 271'; PAC RG85/1069, file 251–1, 4 Feb. 1925; 85/1044, file 540–3 (3A), 25 July 1925; 85/755, file 5648, 31 July 1928; 85/1045, file 540–3, pt. 3-c, 5 Aug. 1937. HBCA B455/a/3, 11 Aug. 1923; a/6, 14 Aug. 1925; a/7, 18 Aug. 1926; a/9, 20 July 1928, 28 July 1929, 11 Aug. 1929; a/10, 7 July 1930; a/11, 14 July 1932; a/13, 24 July 1933; a/14, 23 July 1935. HBCA RG 3/26 B/36, 'Annual Report Pangnirtung Post, Outfit 271', Thom; HBCA A97/6, Milling, 'Report on Visit to Pangnirtung, 1927–28'.

Except for a few times each winter to trade and for the mid-summer whale drive,[71] the majority of Inuit preferred to remain at their camps where they hunted seals most of the time and trapped fox in peak seasons. The relationship between sealing and fox trapping was such that when trapping was unsuccessful, hunters relied almost exclusively on the trading of 'blubber skins' for ammunition, tobacco, flour, and other requirements:

> In a poor season for the fox fur the natives will visit their traps less and consequently have more time to spend hunting seals at the floe edge, or open waters of the Gulf. The past season (1927) has been an abnormally good one for trapping and the natives have spent the whole of their time going to their trap lines with only a day or so between visits to hunt seals for food for themselves and their dogs. As a result no blubber skins were traded during the winter . . . .[72]

But good fox seasons seriously affected Inuit welfare, a fact recognized by the authorities:

> At Cumberland they had a very good year so far as food and clothing were concerned and were quite contented although they practically trapped no foxes. For the native welfare he must have plenty of seal for boots, food and fuel and deer for clothing. Both these abound at Cumberland hence a happy, healthy people without foxes, while at Pond Inlet although foxes were much more plentiful, food and clothing were hard to get, hence untold suffering and hardship.[73]

Thus, in good fox years the health of most camps worsened as fewer seals were procured and more whiteman's food was bartered from the trader. Fortunately, as much by tradition as design, good sealing years far outnumbered good fox seasons in Cumberland Sound.

It has been estimated that the number of common jar (adult) seals traded to the post in any one year amounted only to 10 per cent of the actual catch, and that the total number of seals taken annually in Cumberland Sound totalled about 10,000.[74] While it seems likely that only a fraction of the number of seals taken each year in the Sound were ever traded, the 1920s and 1930s appear to have been especially prosperous years for the Cumberland Sound Inuit. As early as 1923, authorities observed that families trading at the Pangnirtung Post were considerably larger than elsewhere—e.g., the latter averaged about five, whereas families at Pond Inlet averaged about two.[75] By 1925, 47 per cent of the population of the Sound was less than 15 years of age.[76] This rate of infant survival, which was much greater than that recorded for the 1880s (see Table 5), was likely the result of a combination of factors including (a) a continuing emphasis on sealing after the adoption of fox trapping, (b) the incorporation of systematic white whaling into the annual routine, which provided access to vast amounts of meat and *maqtaak*, (c) increasing immunity to foreign diseases, and (d) the dispersion of the population into small camps, which reduced interactions with foreigners and the spread of infectious diseases. An increase in the population of the Sound during the 1930s was subsequently observed (Table 8).

Nonetheless, by the end of the decade sustained interaction with Qallunaat led to deprivation and hardship. Throughout the late 1920s and 1930s the Pangnirtung Post purchased whitecoats and 'silver jars' while scaling down the trade in blubber skins (Table 7). The increased demand for seal fur products and concomitant decline in the value of seal oil on international markets appear to have placed undue pressure on Cumberland Sound's immature ringed seal population. By the late 1930s seals were not as plentiful as they once were and hunters were travelling further afield to keep up the supply.[77] Half a dozen years later, many of the older hunters reported that seals were much less abundant than formerly, and that excessive killing of whitecoats— 2000 to 2500 per year—was likely the cause.[78]

TABLE 8. Populations of Cumberland Sound settlements, 1923–1936.

| SETTLEMENT | 1923 | 1927 | 1930 | 1933 | 1936 |
|---|---|---|---|---|---|
| Saumia | 17 | 21 | ? | 19 | 16 |
| Aukadliving | — | 17 | — | — | — |
| Naulineaqvik | — | — | — | — | 12 |
| Kekerten | 35 | 14 | 10 | 9 | 14 |
| Tesseralik | 6 | 6 | 9 | — | — |
| Kingnait | — | — | 25 | 22 | 17 |
| Pangnirtung | 39 | 66 | ? | 54 | 54 |
| Ussualung | — | 12 | 9 | 16 | 12 |
| Nunaata | — | 27 | 34 | 29 | 30 |
| Kingua | 64 | — | — | — | — |
| Idlungajung | 21 | 19 | 21 | 30 | 42 |
| Sauniqtuajuq | 49 | 17 | 35 | 25 | 44 |
| Iqalulik | — | 9 | 13 | 12 | 23 |
| Nuvujen | 14 | 7 | — | — | — |
| Kingmiksoo | 12 | 52 | 40 | 39 | 53 |
| Opinivik | 20 | 14 | 12 | 18 | 13 |
| Umanaqjuaq | 48 | 24 | 7 | 12 | — |
| Koangoon | — | 7 | 7 | 12 | — |
| Neakunggoon | — | 11 | 12 | 12 | 9 |
| Illutalik | — | — | — | 23 | 25 |
| Aupalluktung | — | — | — | 8 | 11 |
| Etelageetok | — | — | 16 | — | — |
| Iglulik | — | — | 13 | — | — |
| Mamukto | — | — | 6 | — | — |
| Etalik | — | — | — | — | 16 |
| Totals | 325 | 323 | 269 | 340 | 391 |

SOURCES: PAC RG85/64, file 64–1, pt. 1, 3 March 1925, Burwash to Finnie; 15 Aug. 1927, Friel to 'HQ' Division; 85/1044, file 540–3 [3B], 30 June 1930, Petty to Off. Comm. 'HQ' Div.; 85/64, file 164–1 [1], 31 March 1933, McPhail; 85/815, file 6954 [3], 14 Sept. 1936, MacKinnon.

While the HBC heralded many changes in Inuit-Qallunaat relations in Cumberland Sound, none was greater than the shift from the 'collective rationing' system of the whalers to the barter system of the Company.[79] For decades, Inuit were advanced ammunition, tobacco, tea, biscuits, and the like by the whalers and the free-traders, regardless of how many whales were caught or blubber skins were traded.[80] At the end of each season, productivity in the fishery was rewarded in the form of bonuses.[81] Under this system, Inuit were not directly exposed to fluctuations in international markets and their relationship with station managers remained stable economically. However, under the barter system of the HBC 'articles . . . acquired a value they never had before'.[82] Many hunters who had been 'brought up under a system which guaranteed their simple wants in return for services . . . (failed) to understand why biscuits, etc. (had) a value'.[83] Although some Inuit may have at times exploited the willingness of the RCMP and other government agents to offer relief in the form of ammunition, 'those used to the old system of what seemed to them a free distribution of supplies (apparently) could not adapt to the new barter system':[84]

> All the natives of the Gulf find a difficulty in adapting themselves to the straight forward barter system of the 'H.B.Co.' from the old system of the whalers who assured them of their food supply in all seasons, irrespective of a successful hunt or not. It seems, and is, apparent that they are incapable of producing the overflowing products of the country, without someone giving them instructions....[85]

The whaling system mirrored a pattern of food distribution not uncommon to Inuit tradition, and provided a way of incorporating the white man into the local system of sharing. As such, barter was looked upon by the Inuit as a 'one-sided abrogation of local tradition' (Goldring 1989: 17). Nonetheless, as other contact agents including medical officers, Anglican missionaries, and police officers came to settle in Pangnirtung, an individual could visit all these institutions, securing the best price for his products,[86] and thus come away with all his needs and most of his wants satisfied. By this sort of manipulation, hunters could recover some of the customary benefits that they had lost when the whalers and free-traders departed (Goldring 1989).

### Law, Order, and 'No Loitering'

The increase in trading activity in Cumberland Sound during the early 1920s prompted the Canadian government to establish a RCMP detachment in Pangnirtung to look after the welfare of the natives and the animals upon which they depended. The RCMP were particularly concerned that the Inuit would suffer at the hands of the HBC and the free-traders. As long as there was trading competition and Inuit influenced the price of their produce and labour, this concern was misplaced. Nevertheless, the purchase of Kinnes' and Munn's trading stations by the HBC in 1923–24 and numerous complaints against the HBC trader rekindled interest in the welfare of the Inuit. Seeing that Inuit were treated fairly by the HBC was not the only duty of the RCMP; they also issued relief to the needy and elderly in times of hardship, a role that they inherited from the HBC.

The chief function of this detachment, however, appears to have been to prevent Inuit from settling in Pangnirtung and becoming too dependent upon the trader and other whites. By 1925 'twenty able-bodied men', most of them 'not good hunters', and their families had settled in Pangnirtung.[87] Within five years, Pangnirtung, which once contained 33 per cent of the population of the Sound, 'was greatly reduced in size'.[88] The RCMP continued to perform this function well into the 1950s, encouraging Inuit not to hang around Pangnirtung, while continually reminding them that they could not live in Pangnirtung unless they were employed by one of the 'White Concerns';[89] 'the natives do not loiter in the settlement, there are a few individuals who try to remain in the settlement and live off the employed natives, these natives are not tolerated and soon sent on their way.'⁹⁰ The police also took it upon themselves to establish new camps with individuals who appeared to serve no useful purpose in Pangnirtung, and to break up old ones if it was in the best interests of the Inuit as well as the government:

> In recent years . . . (Tesseralik) has become the gathering point for all the bums and scroungers in the district, all of which require relief assistance sometime during the year. They were informed during their visit to the settlement that unless they moved from this location they would receive no further family allowance.⁹¹

At the same time, the RCMP continually discouraged Inuit from purchasing 'store-bought' clothing and canvas tents, when seal skin tents and caribou skin clothing were much more practical and efficient for their purposes.

By 1955 the socioeconomic situation of most Cumberland Sound Inuit was perhaps closer to the aboriginal context than any time within the last century. Mechanization of the hunt had altered technological capabilities, to be sure. Every hunter owned a rifle and most camps possessed at least one whaleboat, sometimes two or three, often with engines, as well as four or five seal nets. However, skin-covered kayaks were still used at the floe edge in the winter and, among some camps, along the coasts in the summer. Moreover, virtually everyone outside of Pangnirtung still lived in seal skin dwellings and wore caribou and seal skin clothing. Sealing remained the primary occupation, while fox trapping continued its decline in importance. Finally, the population of the Oqomiut was beginning to approach precontact levels (Table 9). The Cumberland Sound Inuit, in fact, were considered to be so self-reliant that the RCMP divided the south Baffin region into two economic zones, one centred around the DEW (Distant Early Warning) line stations on Davis Strait, where Inuit lived off 'hand-outs', and the other in Cumberland Sound, 'where the Eskimo lives his normal social life and lives off the land as he has done in the past'.⁹²

Pangnirtung as a settlement existed largely to serve Inuit needs through white personnel (Goldring 1989). With perhaps the exception of the Anglican mission, and then only initially, all contact agents encouraged Inuit to live away from Pangnirtung and to remain independent of various sources of relief. As game was scarce around Pangnirtung, the HBC trader benefited by encouraging Inuit to live in the same small, dispersed camps that their forefathers had for decades. Health conditions were also better in the camps, as 'country food' was more abundant and sanitation was not a problem. Finally, and most importantly, a native family living off the land required no relief. Not only did this make life easier for various officials, it also conveyed a sense of fiscal responsibility to their southern superiors (Goldring 1989). Collectively, perhaps as much by parsimony as by design, various contact agents and the institutions

TABLE 9. Populations of Cumberland Sound settlements, 1944–1966. After Haller et al. (1966: 150, 156; Tables 22 and 29). With the exception of 1944, populations figures do not include Inuit temporarily absent from the Sound.

| Settlement | 1944 | 1951 | 1961 | 1966 |
|---|---|---|---|---|
| Saumia | — | 23 | — | — |
| Aukadliving | — | 13 | — | — |
| Sukpeeveesuktoo | — | — | 32 | — |
| Kekerten | — | 26 | — | — |
| Tesseralik | — | 16 | — | — |
| Kingnait | 31 | 4 | — | — |
| Tuapait | — | 16 | 31 | 30 |
| Pangnirtung | 45 | 75 | 98 | 342 |
| Ussualung | 14 | 32 | — | — |
| Avatuktoo | 20 | 23 | 47 | — |
| Nunaata | 39 | 31 | 38 | — |
| Idlungajung | 66 | 56 | 66 | 46 |
| Sauniqtuajuq | 39 | 46 | 40 | 37 |
| Iqalulik | 40 | 31 | 41 | 20 |
| Naujeakviq | 19 | 11 | 28 | — |
| Illutalik | 19 | 16 | 32 | 19 |
| Kipisa | 33 | 37 | 46 | 35 |
| Keemee | — | — | — | 25 |
| Opinivik | 21 | 12 | — | — |
| Kingmiksoo | 48 | 9 | 52 | 50 |
| Totals | 434 | 477 | 551 | 604 |

they represented helped to maintain the independence of the Cumberland Sound Inuit well into the 1960s.

## THE APPROACH OF MODERN TIMES, 1962 TO 1970

For decades the combined effects of official policy and Inuit tradition served to prevent the formation of a modern community in Pangnirtung. With the help of high prices for seal skins,[93] the Cumberland Sound Inuit entered the 1960s with their economy and community structures relatively unchanged from those of forty years earlier (Goldring 1989: 51). And when distemper decimated the dog population in the Sound over the winter of 1961–62, Cumberland Sound's Inuit still retained 'the rapidly disappearing virtue of being a self-reliant people'.[94]

Although a few camps refused to be evacuated, most people were transported to Pangnirtung in the spring of 1962. Here, relief was issued and a dog breeding program was begun by government officials. Toward the end of the year, however, most Inuit returned to their settlements. With the skin of the common jar seal fetching over $12, most camps prospered over the next three years (Table 10).[95] Even though the annual summer caribou hunt had been discontinued (Haller et al. 1966: 97)—canvas tents had replaced skin ones, and

TABLE 10. Pangnirtung seal, whale, fox, and other returns. After Haller et al. (1966: 186).

| YEAR | JAR SEAL | SILVER JAR | FOX | BELUGA | OTHER |
|---|---|---|---|---|---|
| 1958–59 | 2405 | 696 (+63 whitecoat) | 600 | 62 | 11 bear |
| 1959–60 | 1893 | 2798 (+28 whitecoat) | 98 | 153 | 20 bear |
| 1960–61 | 3750 | 2749 | 964 | 155 | 19 bear |
| 1961–62 | 2771 | 1751 | 545 | 60 | 9 bear, 90 narwhal |
| 1962–63 | 4880 | 2553 | 4 | — | 15 bear, 33 narwhal |
| 1963–64 | 6020 | 3809 | 46 | — | 8 bear |
| 1964–65 | >12,490< | | 691 | — | 15 bear |
| 1965–66 | >11,002< | | 67 | — | 21 bear, 295 harp |

store-bought clothing was being purchased in ever increasing quantities—the average Cumberland Sound family still lived on the land and continued to depend principally upon the ringed seal for its livelihood. While sealing had developed primarily into a commercial undertaking and a year-round occupation by the early 1960s,[96] the seal still provided many useful and traditional products. For example, seal meat was the principal food for both Inuit and their dogs, seal blubber was the major source of heat and light (most dwellings owned four soapstone lamps), and seal skins were used for dog traces and footwear (Haller et al. 1966: 85).

By 1965 the dog population had risen to pre-1962 levels and the population of the region had grown to 867.[97] With the establishment of government services in Pangnirtung, welfare, social assistance, and wage labour became increasingly important sources of income. Even so, as late as 1966 native products accounted for over half the annual income, while household income averaged $1737 (Haller et al. 1966: 197). Compared with other regions of Baffin Island,[98] a continuing emphasis on the seal allowed the Cumberland Sound Inuit to maintain their economic independence and social structures well into the mid-1960s. However, the introduction of the snowmobile, government housing, and education, health, welfare, and other govenment services would soon change all this.

The advantages of the snowmobile were realized soon after its introduction in 1964; the hunter could now cover more ground in less time, which allowed travel to traditional hunting grounds and back within a day. By 1967, 60 snowmobiles had been purchased and only a few individuals owned dog teams. With the adoption of snowmobiles, outboard motors, and freighter canoes, the hunting economy of the Cumberland Sound Inuit reached a critical stage whereby income from the sale of native products was no longer sufficient to cover operating and depreciation costs in the mechanization of the hunt (Haller et al. 1966: 197). The high price of fuel for snowmobiles and outboard motors, in particular, often resulted in a financial loss to the hunter. Largely because of the mechanization of the hunt, the Sound's economic base changed from the sale of native products to wage labour employment in the late 1960s

(Haller et al. 1966: 89–90). By 1967–68, 73 houses had been built in Pangnirtung and its population rose to 531.[99] Moreover, only two camps (Kipisa and Seegatok) remained occupied, and only three hunters still relied on dogs for their livelihood.[100]

The attraction of government housing, health, and other services on one hand, and the mobility offered by the snowmobile on the other, encouraged most Inuit families to settle in Pangnirtung. However, in exchange for these benefits and services, they had to agree to the formal schooling of their children. No longer could they live out on the land where various agencies found it difficult to attend to their needs. The federal government operated under the belief that traditional Inuit culture was doomed to extinction and that the best solution for all concerned would be to integrate them as quickly as possible into the Canadian 'mainstream' by creating a healthier, better educated work force for future economic development (Mayes 1978).[101] These factors, combined with a drop in the price of seal skins during the late 1960s and the attraction of wage labour positions in construction, forced the Cumberland Sound Inuit to participate in, and be affected by, the rapid and uneven changes that were sweeping across Arctic Canada.

# Chapter 4

## Culture Change and Continuity, 1840-1970

The Cumberland Sound Inuit have an association with western civilization that is as rich and complex as any in Arctic Canada. But how did 130 years of interaction with Qallunaat affect Inuit culture and society in Cumberland Sound? Specifically, how did participation in commercial whaling, the adoption of Christianity, population decimation by foreign diseases, and other influences alter traditional Cumberland Sound Inuit social organization? Until this issue is addressed we cannot realize our goals of determining if local group composition in Cumberland Sound between 1920 and 1970 is representative of precontact social organization, or whether these data might possibly inform the structure of Central Inuit social organization.

This study examines cultural change and continuity among the Inuit of Cumberland Sound by employing a form of analysis known as 'mode of production' theory. All societies, past and present, must reproduce themselves from one year to the next by maintaining their population and replenishing their physical stocks (Kay 1976: 13). They do so by engaging in the process of material production which, in a real sense, is the starting point of society. This process, however, involves not only a material dimension, but a social one; human beings must enter into social relationships with each other in the process of material production. While the former is described as the 'means of production' for a society, the latter are known as 'relations of production'. The combination of the two aspects of production—production as a material process on the one hand, and as a social process on the other—constitutes a 'mode of production' (Kay 1976: 22).

'Mode of production' theory offers distinct advantages over ecological-evolutionary based analyses insofar as it rejects the notions that social institutions are structured solely on the basis of techno-environmental factors, and that the process of change operates completely outside the consciousness of human beings (Asch 1979: 88). While ecological-evolutionary approaches may be able to classify forms, they do not illuminate the processes by which social formations maintain themselves or change. In short, in their claim that the techno-environment determines both the level of complexity a society can develop and the shape of social institutions intimately connected with the productive process (Asch 1979: 86), such approaches fail to consider the formative role of social relationships in shaping the process of the material reproduction of society.

Fundamental to mode-of-production analysis is the view that social change is the result of dynamic, dialectical historical processes; no society can be known through its phenomenal surface alone (O'Laughlin 1975). Such perspectives hold that underlying the operation of all social systems are relations which embody incompatibilities, and that social change occurs through the synthesis of these contradictions. In this regard, some would view the tension between the forces of production and the relations of production as the central one in society. Forces of production include the raw materials, tools and technical knowledge, as well as the organization of the labour force in the productive process. In contrast, relations of production structure the economic rationality of the material process of production, i.e., the use to be made of resources, who shall or shall not work, who controls surplus, and how the product of labour is to be appropriated (Friedman 1974: 446). In effect, relations of production are essential for reproducing social relationships that motivate a particular mode of production (Asch 1979). One interpretation of the systemic relationships of these elements, following Friedman (1974: 445), is illustrated in Figure 15.

Some social anthropologists (e.g., Friedman 1974, Godelier 1966) maintain that there are two fundamental contradictions in society, (1) 'intrasystemic contradiction', where a dominant class controls the means of production of another, while controlling and appropriating surplus, and (2) 'intersystemic contradiction', where the structure of the relations of production conflict with the forces of production.[1] Because the structure of material reproduction incorporates a social as well as a technological component, and because a mode of production for a given society may be said to be the resolution of productive forces and relations, we are primarily concerned here with intersystemic contradiction. More specifically, we want to know if during the historic period fundamental contradictions between forces of production and relations of production arose in Cumberland Sound Inuit society such that a significant transformation in social organization occurred.

New relations of production may result from external as well as internal forces. Under the former process, a situation might arise whereby the adoption of new economic pursuits or technologies might engender a reorganization of production which ultimately transcends the logical structure of the existing system. Alternatively, existing relations of production might demand ever-increasing production (Asch 1979), as would be expected among most Inuit societies where productivity and prestige are intimately related. Nonetheless, forces of production (e.g., the environment) fix a finite limit on the productive potential of societies such as that of the Inuit. Although disparities between

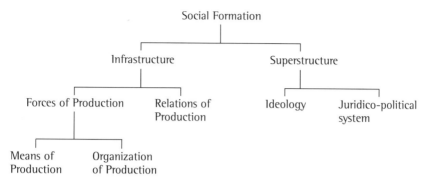

FIGURE 15. Constituent components of a social formation and their systemic relationships.

productive forces and relations might not be apparent at first, eventually antagonisms between the two could become so great as to cause a collapse in the entire mode of production, thus setting the stage for a structural transformation (Asch 1979: 92).

In light of these considerations, it is apparent that the Cumberland Sound Inuit were particularly predisposed, both internally and externally, to structural change. Whether or not a structural transformation in social organization took place is the subject of what follows. Our analysis begins by documenting changes in the means of production and how they may have affected the organization of production in Cumberland Sound. We then examine if and how changes in the forces of production affected relations of production. Finally, we explore whether alterations in religious ideology (superstructure) influenced productive relations and the organization of production.

## FORCES OF PRODUCTION

As early as the 1850s Euroamerican observers remarked that the Cumberland Sound Inuit were fast losing their aboriginal hunting skills through trade and intercourse with the whalers: 'It (employment in the whale fishery) was very detrimental to the habits of the poor things, as their children were not then trained in the use of the bow and arrow or canoe (*sic*), but trusted to the ships coming' (Barron 1895: 44). Some time later, Kumlien (1879: 14) noted reprovingly that 'the Cumberland Eskimo of today, with his breech-loading rifle, steel knives, cotton jacket, and all the various trinkets he succeeds in procuring from the ships, is worse clad, lives poorer, and gets less to eat than did his forefathers, who had never seen or heard of a white man.' Successive visitors to Cumberland Sound painted progressively bleaker pictures (e.g., Wakeham 1898: 75; Low 1906: 10).

## Changes in Technology and Subsistence

Like so many observations of the era, those above were clouded by distorted perceptions of Inuit life and the impact of Euroamericans on Inuit culture. The fact is that the concern revealed in these quotes was misplaced. The new technology was never forced on the Inuit and they always exercised choice in their adoption of Euroamerican ways. For instance, in 1857 Warmow met an old but skilful and industrious man who 'shunned Europeans, though the whalers had promised him much should he only come to work for them' (Harper 1981: 47). In another example, the trading opportunities afforded by the presence of a whaling ship at Anarnitung in August of 1841 did not prevent the entire population of this settlement from repairing to its annual fishing grounds (Wareham 1843: 24–5).

To be sure, the rifle quickly rendered the use of the composite bow obsolete. But not everyone in aboriginal society was familiar with the bow's use or manufacture; both appear to have been reserved for the specialist (e.g., M'Donald 1841, Kumlien 1879: 37). And while the whaleboat assumed many functions of the *umiaq* and kayak, the latter was still used at the *sina* and in open water up to the 1950s, when rowboats began to replace kayaks. Similarly, the hand harpoon continued to be used extensively for sealing at breathing holes and at the ice edge (Boas 1964, 1883–84: 41–42B) well into the 1950s. Other aboriginal weapons, such as the seal skin float (*avatak*) and whaling harpoon (*sakurpang*, Boas 1964: 92), found no superior or consistently available replacements in the white man's technology, and were used for decades after the coming of the whalers (e.g., Kumlien 1879: 35). Steel and wood supplanted the use of stone and whalebone in the manufacture of the woman's knife (*ulu*), the man's knife (*savik*), the *qamutiik* (sled), the sealing harpoon, and a host of other items, but their form and function remained unchanged.[2] Various elements of Euroamerican technology were subjected to experimentation, such as the use of metal for *qudlit* (seal oil lamps), trap-rifles for *mauliqtuq* sealing, and certain articles of clothing, but these too were found to be inferior to traditional counterparts, and were soon abandoned (Etuangat Aksayuk, personal communication, 1983). Other items of technology, such as seal and fish nets, introduced by the HBC, found no traditional equivalents and were quickly adopted. While aboriginal hunting implements such as bows and arrows were still prized for competitive recreational uses well into the twentieth century (Ross 1985c: 238), the Cumberland Sound Inuit had seized the opportunity to hunt more efficiently and live more securely than before, even if it meant sacrificing certain traditions and periodic abundances: 'they prefer restricted but assured gains to those which are perhaps larger, but uncertain in amount' (Hantzsch 1977: 40). Relying on indigenous hunting skills and knowledge passed down through the generations, the Inuit thoroughly integrated imported weapons into the hunting economy by the turn of the century (Goldring 1986: 163).

TABLE 11. Inventory of hunting equipment, Cumberland Sound, summer 1966. After Haller et al. (1966: 160).

| Settlement | Dogs | Sleds | Snow-mobiles | Whaleboats/Canoes | Row-boats | Out-boards | Rifles | Nets |
|---|---|---|---|---|---|---|---|---|
| Pangnirtung | 375 | 54 | 27 | 8wb, 7fc | 41 | 44 | 143 | 10s, 29f |
| Tuapait | 50 | 9 | 5 | 2wb, 4fc | 1 | 5 | 12 | 2f |
| Idlungajung | 74 | 9 | 0 | 1wb, 1fc | 6 | 10 | 17 | 2s, 4f |
| Sauniqtuajuq | 105 | 12 | 0 | 2wb, 2fc | 7 | 7 | 21 | 6s |
| Iqalulik | 41 | 3 | 0 | 2wb | 3 | 3 | 7 | 3s |
| Keemee | 83 | 6 | 0 | 2wb, 1fc | 3 | 4 | 12 | 5s, 2f |
| Illutalik | 42 | 6 | 0 | 1wb | 1 | 3 | 8 | 2s, 2f |
| Kipisa | 66 | 8 | 2 | 4wb, 1fc | 3 | 3 | 17 | 7s |
| Kingmiksoo | 104 | 11 | 2 | 2wb | 6 | 6 | 26 | 4s |

Note: wb=whaleboat, with or without Acadia engine; fc=freighter canoe; s=seal net; f=fish net.

With the incorporation of commercial whaling into the annual round, a number of changes in seasonal subsistence patterns followed. For example, the autumn caribou and spring basking seal hunting seasons became more abbreviated, as hunters found various forms of employment in the fall and spring whale fishery. Yet, the acquisition of the rifle extended the caribou hunt into other seasons,[3] and seals continued to be hunted throughout the winter to feed both Inuit and wintering whalers. During the 1850s and 1860s hunting intensified as the bowhead whale became almost the exclusive pursuit of Inuit employed in the fishery.[4] However, after 1870 commercial hunting diversified as seals and other marine mammals were incorporated into the annual hunts. The acquisition of whaleboats also extended the open-water hunting of seals into the fall and spring. Unlike the *umiaq* and kayak, the whaleboat could venture forth in stormy or ice-congested seas. Trapping never supplanted the pursuit of sea mammals in Cumberland Sound, and its incorporation into the annual round had little impact on the traditional economy, though fox was more actively pursued in some years than others. The commercialization of the beluga whale hunt likely intensified the hunting of this animal, particularly after 1920. But the beluga whale was apparently always hunted regularly at Milurialik and opportunistically elsewhere.

Mechanization of the hunt and the incorporation of commercial whaling into the local economy altered some traditional hunting patterns, while reinforcing others. And although the Cumberland Sound Inuit had come to depend on international markets to maintain their traditional hunting economy, there appears to have been no significant transformation in aboriginal economy from precontact times to the mid-1960s. Figure 16 presents the seasonal round of most hunters attached to Kekerten during the second decade of this century, while Table 11 indicates that as late as 1966 virtually all camps in Cumberland Sound were still outfitted and organized for the hunt. The absence of steel traps in the 1966 inventory, though perhaps more apparent than real, is instructive. It would take the wholesale adoption of snowmobiles, the abandonment of most camps in the Sound, and the shift from an economy based on native products to one based on wage labour, welfare, and social assistance in the late 1960s before the traditional economy was replaced by a mixed economy. Although Inuit in Pangnirtung still rely on animals and the land for a substantial part of their nutritional, social, material, and emotional needs, they have come to depend on wage labour and government assistance to keep up the hunt.

**Changes in Organization of Production**
Discussion of change in the organization of production necessarily involves consideration of relations of production; in Inuit society the two are intimately related. In fact, it is simply impossible to talk about economic organization without discussing social organization. Although the social cannot be divorced from the economic in Inuit society, we will discuss changes in technology and subsistence and how they affected the size and organization of productive units, before examining whether changes in the forces of production and religious ideology influenced the structure of social relationships in Cumberland Sound.

Mechanization of the hunt and the incorporation of commercial whaling and trapping into the annual round of the Cumberland Sound Inuit appear to have had little impact on the traditional economy. Even so, it is possible that the introduction of the new technology altered the traditional character of social relationships between local groups. With many indigenous items finding far superior counterparts in the whiteman's technology, the social alliances and trading partnerships that formerly facilitated the acquisition and exchange of raw materials, capital equipment, and economic information in precontact times may have begun to break down. At the very least, the new technology may have created a greater dependence on Qallunaat. However, whether the erosion of traditional concepts of reciprocity accompanied the incorporation of the whiteman and his technology into the culture of the Cumberland Sound Inuit, and whether non-kin relationships became more important in strategies of affiliation, as they did in northwest Alaska (e.g., Burch 1975), is unlikely. This proposition finds support in the allegiance given to Inuit leaders such as Angmarlik and Kanaaka by their kinsmen during the 1920s, and the fact that sharing customs remained virtually unchanged throughout the historic period.

Two items of Euroamerican technology, the rifle and the whaleboat, warrant special consideration here, as more than any other introduced artifacts they may have affected the organization of production in Cumberland Sound.

Wherever the rifle was introduced throughout the Arctic, it appears to have produced a marked individualization in hunting patterns and a reduction in co-operative and sharing practices (e.g., Balikci 1960, 1964; Burch 1975). As one of Balikci's (1960: 144) Netsilingmiut informants remarked when asked

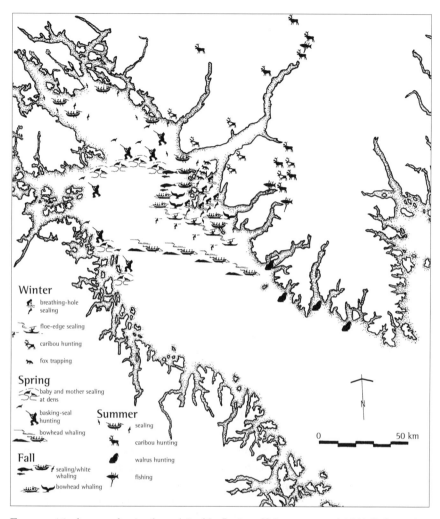

FIGURE 16. Areas and animals exploited by Inuit at Kekerten around 1918. Information supplied by Etuangat Aksayuk and Qatsu Eevic (1983, 1988).

why he didn't share his food as much as he had previously, 'Now everybody has a rifle and can go out and get food for himself, there is no need for much sharing.' This individualization of hunting patterns resulted in a reduction of the importance of the extended family and a concomitant emergence of the nuclear family as the basic socioeconomic unit (Balikci 1960). Since a hunter with a rifle no longer needed to depend on his father, brothers, or other kinsmen to 'put food on the table', the economic rationale of the extended family began to crumble. While the size of families became smaller, the size of dog teams grew considerably, increasing mobility and reinforcing the individualization of the hunt (Balikci 1964: 48–9).

As among hunters elsewhere (e.g., Balikci 1964, Burch 1975), the rifle appears to have had a similar atomizing effect on the organization of the hunt in Cumberland Sound. We know that caribou hunting among the Cumberland Sound Inuit was formerly a co-operative activity involving numerous hunters and a division of labour (Boas 1964: 93–4, Hantzsch 1977: 196). We also know that the introduction of the rifle served to individualize the caribou hunt. While family groups continued to travel to Nettilling Lake in the summer for caribou after the introduction of the repeating rifle, caribou were more commonly taken throughout the year within a few miles of the coast by Inuit hunting alone or in pairs (Etuangat Aksayuk, Qatsu Eevic; personal communications, 1983, 1988). Caribou were still procured in the late summer primarily for their skins, but this animal had also become an alternate source of meat during the rest of the year.

The rifle had a similar impact on seal hunting. Breathing-hole sealing in the Sound was formerly a communal activity involving several people (Boas 1964). While this type of hunting was practised up to the 1960s, the rifle individualized winter sealing by shifting emphasis away from *mauliqtuq* sealing towards the more solitary hunting of seals at *sarbut* and the *sina*, though the latter still required co-operation between adult hunters, particularly if rowboats were used. The adoption of the rifle in Cumberland Sound, then, appears to have resulted in a reduction in both the complexity of the division of labour and the degree of collaboration in the autumn caribou and winter seal hunts. Nonetheless, although the rifle resulted in a decrease in sharing and the importance of the extended family as the basic productive unit in other areas of the Central Arctic (Balikci 1960, 1964), such appears not to have been the case in Cumberland Sound.

Dwellings in Cumberland Sound began to be dominated by nuclear family houses in the mid-nineteenth century. However, no fundamental change in the structure of local groups appears to have attended this shift, as the residential group remained the primary unit of production and consumption in most camps (see Chapter 5). As Guemple (1972a: 83) has observed, all that seems to occur in the reduction of household size is that constituent nuclear elements construct separate dwellings close by their nearest relatives, with the structure of household placement continuing to mirror the structure of social relationships within the group.[5] Thus, *irniriik* or *nukariik* cores remained the socioeconomic basis of most local groups in Cumberland Sound, and sharing practices continued to be dominated by *nekaishutu*, the community-wide distribution of food and blubber. The question that arises is this: Why didn't the nuclear family emerge as the basic socioeconomic unit in Cumberland Sound, as it did in other places?

Perhaps Guemple (1972a: 96) was correct when he asserted that firearms were not nearly so effective in augmenting food production among Inuit subsisting predominantly on marine mammals. For one thing, since the

Cumberland Sound Inuit were primarily sealers and whalers who, until recently, hunted caribou almost exclusively for clothing, the rifle may not have affected social organization to the extent that it did among groups having a greater nutritional dependence upon the latter species (Balikci 1964). Alternatively, whereas the organizational autonomy of the individual in the hunt, initiated by the acquisition of the rifle, appears to have been reinforced in the Pelly Bay area by the adoption of trapping and fish nets—both require little or minimal co-operative activity—the latter may have had less impact on the economy of the Cumberland Sound Inuit. *Iqaluk* appears to have been nothing more than a seasonal diversion in an otherwise steady diet of marine mammals in Cumberland Sound as Inuit inhabiting its shores rarely built up reserves of fish, except for dog food. More importantly, trapping was never embraced by the Cumberland Sound Inuit, even during the mid-1920s and again in the mid-1940s when the high price of fox stimulated increases in trapping. Rather, they remained dependent upon seals, whales, and other marine mammals for their livelihood well into the 1960s.

Although the adoption of seal nets and rowboats, which required some cooperative effort, may have helped to maintain the economic foundation of the extended family, the whaleboat, perhaps more than any other single material introduction, assisted in the preservation of extended family structure in Cumberland Sound. Whaleboats had been acquired by the Cumberland Sound Inuit for almost a century before the HBC stopped shipping them to the eastern Arctic some time around 1950 (Nichols 1954). However, whaleboats were always a scarce commodity as few individuals had the skills, acumen, or resources to obtain these craft. Yet those who did continued to attract relatives, thus preserving the structure of the extended family. This fact was not lost on contemporary observers: 'The whalers left the camps very well supplied with boats . . . usually the headman of the camp owns the boat (and) having one kind of holds the camp together. . . .'[6] The whaleboat also served to increase the authority of the eldest productive hunter or *angajuqqaq* in economic and ultimately social matters. In contrast to the rifle, which undermined the authority of the *angajuqqaq* by individualizing the hunt and placing less importance on the organization of production, the whaleboat reinforced the authority of its owner by encouraging organized activity and collaboration in both commercial and domestic whale and seal hunts.

Information supplied by Boas (1964) indicates that there were two methods of hunting whales aboriginally, one involving the use of an *umiaq* and a specialized technology, the other a large number of kayaks with less specialized weaponry. The *umiaq* whaling crew consisted of a harpooner (*sivutik*), boat-steerer (*aggutik*), and paddlers. The duty of lancing the whale, the most dangerous task of all, was reserved for the *aggutik*, who was usually the boat owner or *umialiqtak*. The adoption of the whaleboat in the context of commercial whaling increased the division of labour among aboriginal whaling crews

TABLE 12. Heads of whaleboat crews participating in HBC white whale drive prior to 1928 and between 1930 and 1932, as listed in the Pangnirtung Post Journal Diaries (HBCA B455/a/9–12).

| Whaleboat owners/operators before 1928 | Whaleboat owners/operators, 1930–32 | |
|---|---|---|
| 1) Angmarlik (2 whaleboats, 1 motorized) | 1) Adjalik | 2) Akpalialuk |
| 2) Aksayuk | 3) Angajuqaaqjuaq | 4) Angmarlik |
| 3) Attaguyuk | 5) Aksayuk | 6) Attaguyuk |
| 4) Eevic (in charge of Angmarlik's boat?) | 7) Eevic | 8) Keenainak |
| 5) Joanasie | 9) Kakka | 10) Kingu |
| 6) Keenainak | 11) Kukkik | 12) Maniapik |
| 7) Maniapik | 13) Ooneasagak | 14) Ooshutapik |
| 8) Mike (in charge of son's boat?) | 15) Nakashuk | 16) Nowlalik |
| 9) Veevee | 17) Nowyook | 18) Nukeeruaq |
|  | 19) Padlu | 20) Peterosie |
|  | 21) Veevee | 22) Tiosaqjuak (Ittusarjuaq?) |

since the additional technology (sail rigs, dart-guns, explosive harpoons, tubs of rope, etc.) required more specialized positions (e.g., bowman, tub oarsman). Remarkably, however, the relationship between the *sivutik* and *aggutik* was identical to that of the commercial whalers (Ansel 1983), and thus remained unchanged throughout the commercial whaling period (Etuangat Aksayuk, personal communication, 1983). However, seal hunting from whaleboats with rifles, whether for trading or domestic purposes, was new to the Cumberland Sound Inuit; seals were traditionally pursued in open water with harpoons from kayaks. Nonetheless, sealing from whaleboats required a greater degree of organization than sealing from kayaks, as whaleboat crews usually consisted of a boatsteerer, oarsmen, and a rifleman, who was normally the *umialiqtak* (Etuangat Aksayuk, personal communication, 1983). Whaleboat crews were similarly organized for the white whale drive, and the addition of one-cylinder engines in whaleboats after the late 1930s did little to alter this arrangement, although it likely added another rifleman or two to each boat crew.

In precontact times, social groups integrated by various kinship ties, collective activities, and reciprocal relationships probably characterized most Cumberland Sound camps. During this period, Inuit headmen were needed in the organization of summer whale, winter seal, and autumn caribou hunts. With the adoption of rifles, nets, and trapping, conditions were set for the emergence of the nuclear family as the basic socioeconomic unit in all seasons (e.g., Balikci 1964: 77). The acquisition of the whaleboat counteracted this tendency by helping to preserve economic ties among extended family members as well as the authority of the *angajuqqaq*.[7] Even so, the whaleboat appears to have had a positive role in promoting economic co-operation and mutual sharing within the group only insofar as it remained attainable by a minority of hunters. As the supply of whaleboats increased during the late 1920s (Table 12), so too did the number of camps. At the same time, settlement size decreased. In 1923 the average size of 11 settlements in Cumberland Sound was 29.5, whereas ten years later the number of settlements had grown to 16 with an average population of just over 21 in each (Table 8).

It would be misleading to suggest that the effects of the rifle and whaleboat on the organization of society in Cumberland Sound cancelled each other. Such a perception misconstrues the facts that the adoption of the former increased the individualization of some economic pursuits, while the acquisition of the latter increased the organization of others. Both affected the organization of production within the local group in different ways. However, the net effect was that the acquisition of the whaleboat reinforced group solidarity, sharing practices, and authority patterns within the local group at a time when the organizational autonomy engendered by the adoption of the rifle threatened to undermine the socioeconomic basis of the extended family.

## RELATIONS OF PRODUCTION

### Interregional Group Relations

Perhaps the most obvious impact that the whalers had on Inuit social organization in Cumberland Sound was in the arena of 'intertribal' relations. Boas (1964: 271) noted that the seven tribes of Baffin Island differed widely from each other. For example, the distinctiveness of the Nugumiut was apparent in their hair styles, tattooing, and aggressive tendencies (Barron 1895: 98). However, by 1863 this regional group was living side by side with the Talirpingmiut at Naujateling (Barron 1895: 162). The intermixing of groups within the context of commercial whaling, which began in Cumberland Sound in the late 1840s and on Davis Strait at least a decade earlier, continued throughout the nineteenth century (e.g., Howgate 1879: 23) and well into the twentieth century. As Hantzsch (1977: 38) observed, 'since the coming of the whalers ... so close a bond has been created between different districts that a general and very pronounced mixing of the inhabitants has taken place.' In fact, regional mixing seems to have increased during the early twentieth century as new whaling and trading stations were established throughout the eastern Arctic.

Wintering ships, followed by the establishment of shore-based whaling stations and, later, trading posts led to an increase in contact among many local groups. While 'intertribal' mixing seems to have been a regular, albeit limited, occurrence in precontact times (Boas 1964: 54–7), its increase in the mid-nineteenth century almost certainly resulted in a breakdown in social barriers and hostilities between local groups both inside and outside the Sound. This mixing undoubtedly also served to increase the size and composite organization of settlements (Guemple 1972a), at least temporarily. However, whether this 'compositization' initiated a change in the relations of production seems unlikely. Even though individual personality and productivity were highly valued within the context of trade and employment in order to obtain articles of value, the larger composite structure of seasonal aggregations did not shift emphasis away from traditional kin-based obligations in the formation of productive units. The fact that the latter still played an important role in the selection of whaleboat crews and the formation of hunting partnerships well into the 1920s is apparent in the strong bonds of affection that existed between Inuit leaders and their same-generation kinsmen. It is also evident in Angmarlik's decision around 1919 not only to place his sister's son's step-son (Etuangat Aksayuk) on a whaleboat crew at the tender age of 13, but to train him as a harpooner's apprentice (personal communication, 1983). Individual ability figured largely in the appointment of *aggutik* for the fall and spring whale hunts; only the most capable and knowledgeable were entrusted with boats belonging to the whaling stations. Individual ability also played an important role in the selection of whaling crews at Kekerten. In particular, Angmarlik would not let anyone participate in the fishery whose lack of experience or ability could jeopardize the hunt or the safety of the crew (Etuangat Aksayuk, personal communication, 1983). However, beyond these criteria, the selection of whaleboat crews operated primarily within the context of kinship relations as members of individual boat crews were more closely related to each other than to crew members serving on other boats.[8]

Despite the mixing of groups from both inside and outside the Sound, as well as the increased size and 'compositization' of settlements, some regional divisions of Cumberland Sound Inuit appear to have retained their individual identity, although perhaps not autonomy. Specifically, the annual pattern of dispersion and nucleation that characterized Inuit settlement and subsistence in the Sound during the late nineteenth and early twentieth centuries allowed most groups to maintain connections to their traditional camps, hunting grounds, and patterns of living. Even after most families became permanently attached to Kekerten or Umanaqjuaq, they frequently returned to their former camps and hunting grounds. Thus, as commercial bowhead whaling declined, most Cumberland Inuit simply repaired to the same locations where their ancestors had lived for centuries and where they themselves had hunted for decades.

Although the Kinguamiut and Kingnaimiut appear to have lost any cultural differences that formerly existed between them as they settled at Kekerten, the Talirpingmiut seem to have retained their distinctiveness. Tentative evidence was provided in Chapter 2 indicating organizational differences between the former two regional divisions. However, the fact that they seem to have forged a single identity in the context of interaction suggests that, former hostilities notwithstanding, their organizational differences were not as great as previously assumed. While these two regional divisions may not have always got along, they may have been similarly structured, as their tendencies towards well developed leadership would seem to indicate. It would appear appropriate, then, to regard the Cumberland Sound Inuit during the twentieth century as being composed of two major regional divisions, the Qikirtarmiut (formerly, the Kinguamiut and Kingnaimiut) and the Umanaqjuarmiut (formerly, the Talirpingmiut). In 1936, one authority differentiated between these two divisions on the basis of economic and geographical factors:

> The Pangnirtung (read Qikirtarmiut) group have more dealings with the white population. They are employed by the Hudson's Bay Co., during the whaling. They have more advantage to our medical service. They are the ones employed as servants. On the other hand, the Blacklead group can devote all their time to hunting. They do not gather in large numbers as they do here in Pangnirtung. They have not the feeling that they are living close to the whiteman.[9]

However, the possibility that differences between these divisions are more deeply rooted in social and cultural factors than economic and geographical ones is evident in the fact that Inuit in Pangnirtung still distinguish themselves today on this basis of whether they are, or are descended from, Qikirtarmiut or Umanaqjuarmiut.[10]

**Leaders and Followers**
Well-defined leadership appears to have been a common feature of aboriginal Cumberland Sound Inuit society, especially, it seems, at the head of the Sound, where, if Boas' (1964: 57) and M'Donald's (1841) observations are correct, Inuit attained considerable influence in the context of advanced age. We have seen how the whaleboat helped to preserve traditional leadership roles and decision-making relationships within the extended family at a time when the rifle and other external forces threatened to undermine the authority of the *angajuqqaq*. However, other factors such as the superior procurement technology of the white man, participation in commercial whaling, and the diversification of the resource base may have also resulted in a greater emphasis on individual leadership.

The introduction of rifles, whaleboats, telescopes, and steel for knives and harpoons increased overall hunting effectiveness by reducing search, pursuit, and handling costs, while buffering against the occasional threat of periodic food shortages. The social advantages of the new technology were also not lost on the Cumberland Sound Inuit; prestige is closely linked to productivity in Inuit society. In order to maximize access to this technology, individual competition may have been minimized in favour of allowing a few select individuals to act as brokers for other Inuit. Individual competition for the new technology would not only have been a breach of traditional patterns of authority and respect, but it would have placed control of the exchange relationship in the hands of the whalers. In order to maximize access to the rarer, more attractive goods in possession of the whalers (i.e., rifles, whaleboats, telescopes), prominent individuals assumed responsibility for their acquisition and distribution. In this connection, it was probably more the rule than the exception in 1859 when Tesuwin bargained with the masters of the *Emma* and *Sophia* at Kekerten on behalf of 20 or more hunters (i.e., three or four boat crews).

Individuals such as Tesuwin may have attained even greater influence after they came to represent the interests of Qallunaat in negotiations with other Inuit. By the late 1870s, and perhaps much earlier, Inuit middlemen were trading with other natives on behalf of white interests. For example, in 1877 Tesuwin traded with Inuit at Naujateling for Captain G. Tyson, who was wintering over 100 km away at Anarnitung (Howgate 1879: 30). In the same year, the noted hunter and whaler Nepekin of Imigen traded with Molly-Kater, i.e., Malukaitok (or Nettilling) Fiord natives on behalf of Tyson, who, incidentally, remarked that 'of course his boat's crew will do as he tells them' (Howgate 1879: 28). Okaitok of Anarnitung and Kekerten also appears to have represented Tyson in negotiations with Malukaitok Fiord Inuit (Howgate 1879: 71). This trend seems to have continued into the mid-1920s, and perhaps later, as Angmarlik occasionally travelled to Kivitoo to trade on behalf of the HBC trader in Pangnirtung.[11]

Participation in commercial whaling resulted in a greater diversification and specialization of economic activities as individuals assumed a wider variety of functions and duties in the procurement, transportation, and processing of whales. In turn, certain individuals may have accrued more authority and influence as there would have been an increased need to organize, co-ordinate, and rationalize the greater complexity and division of labour. This would have been particularly so among those task groups operating independently of the whalers (e.g., Tesuwin's). Employment in the fishery also resulted in a delay between production and consumption. With the exception of times of anticipated resource stress (i.e., the late fall), when food was stockpiled, production for immediate consumption characterized Inuit economy in the Sound. While Inuit had immediate access to whale carcasses after the blubber and baleen had been removed, wages were paid on a weekly and semi-annual basis. As the relationship between production and consumption became more structured and delayed, it had to be rationalized and legitimized, a role that undoubtedly fell to those already in positions of power and influence.

The gradual change in economy from commercial whaling to general trading in furs, skins, blubber, and ivory further reinforced the authority of some individuals by providing opportunities for them to assume larger roles as middlemen and organizers of various hunts. After the diversification of the resource base and the return of most Inuit to their original settlements, trading companies usually entered into agreements with the most productive hunter of each camp whereby the latter co-ordinated the activities of resident hunters while representing the interests of both the trader and his local group in economic transactions:

The system adopted has been to leave specific trade goods with an intelligent native and expect him to make the best possible returns of oil, furs, skins, etc. (He) will outfit and supply himself with the goods, but other renumeration is nebulous. Their duty as they see it is to obtain what their employer requires in exchange for his goods.[12]

This system seems to have been a significant departure from earlier times wherein individuals trading on behalf of white interests did so primarily with Inuit living in other camps, i.e., unrelated Inuit or distant kinsmen (e.g., Howgate 1879: 28, 30, 71). Yet, with the institution of the 'camp boss', the leading hunter came to represent the interests of both the trader and his immediate kinsmen, solidifying his position as middleman in transactions between the two. In order to satisfy the needs and wants of both parties, the camp boss had to walk a fine line between obligations to his employer and to his *ilagit*.[13] Subordinating the interests of either would have jeopardized his position as he risked losing either his followers or his ability to provide for them. As long as the camp boss was successful at satisfying and manipulating the needs and wants of both his kinsmen and the trader, his authority and the welfare of the camp prospered. While the need to co-ordinate and organize activities in the whaling industry enhanced the authority and influence of certain individuals, the emergence of the middleman within the context of general trading placed even more control in the hands of prominent Inuit.

In this connection, the origin of the term *angajuqqaq* merits examination. The use of this term was not recorded by Boas; he states that village leaders were known as *issumautang*, 'the one who thinks', or *pimain* (1964: 179). However, my informants have little knowledge of the latter in reference to leadership (although *pimaji* is still used in northern Quebec) and the former was used only rarely, if at all, in the Sound during the twentieth century. Rather, *angajuqqaq* came to denote secular leadership on South Baffin Island and in adjacent regions (e.g., northern Quebec, Hudson Strait, etc.). As *angajuqqaq* was also used to refer to Euroamerican whaling masters and traders, it is possible that the word initially arose within the context of Inuit-white interaction. However, the root of this word is derived from *angak* (mother's brother) and forms the basis of other terms denoting positions of influence and power (e.g., older brother/sister = *angajuk*, shaman = *angaquk*, etc.), and translates roughly as bigger, more substantial, etc. In turn, one is expected to obey and follow the instructions of one's *angajuqqaq* be it his/her parents, older sibling, or whaleboat captain. Among the Central Inuit, the word *angaquk* was reserved for sacred leadership, while on the west coast of Alaska it simultaneously designated 'chief' and 'uncle on the mother's side' (Fainberg 1967: 247), a fact which led Thalbitzer (1941: 629–31) to conclude that the 'Eskimo' family was formerly matrilineal. While it is tempting to speculate that *angajuqqaq* was coined to designate authority figures in the commercialization of the hunt, its use among the Unalit of Alaska to refer to members of the community who had earned powerful status owing to their wealth (Fainberg 1967: 248) suggests that the word has considerable antiquity and its use probably predates contact in Cumberland Sound. Perhaps Lantis (1987: 191) and Kellerman (1984: 71) come closest to distinguishing conceptually between *isumataq* and *angajuqqaq* when they differentiated 'headship' from 'leadership':

> A head's (read *angajuqqaq*) authority and relationship to subordinates is maintained by an organized system . . . whereas the leader (read *isumataq*) is accorded his authority by group members who follow because they *want* to rather than because they *must* (original emphasis, Kellerman's).

Thus, while the term *angajuqqaq* may not have been entirely an invention of Inuit participation in a foreign economic system, as leadership roles expanded in Cumberland Sound it appears to have supplanted the use of the term *isumataq*. In other words, as the position of leader became more structured and vested with authority in the commercialization of the hunt, the more appropriate term, *angajuqqaq*, was adopted. However, in order to maintain his position, the *angajuqqaq* could not neglect his role as *isumataq* nor abuse his authority. To do so would have resulted in a loss of respect and possibly one's influence, kinsmen, and power base:

> The boss of the (Kivitoo) camp . . . is an old man, an ex-whaler type, and is considered too bossy by the majority of the Eskimos. He is not a good leader and it appears that he often uses his position of boss just to show the Natives he is the boss, rather than to direct and lead them in sensible plans that would benefit the community.[14]

Whereas the above individual may have been an *angajuqqaq*, he was clearly not an *isumataq*. Such men, however, were probably more the exception than the rule.

Over the years numerous individuals attained positions of considerable power and influence in the contexts of commercial whaling and general trading in Cumberland Sound. The well known Inuk whaler, Tesuwin, was perhaps one of the first. Another may have been Pakaq, the oldest and most influential man at Kekerten during Boas' stay there in 1883–84. A year after Boas left Cumberland Sound, Pakaq acted as an executioner of a murderer as well as of an old woman (Boas 1964: 260–1).[15] The latter case is noteworthy as, even though the woman was well provided for and apparently unrelated to Pakaq, 'he deemed it right that she should die'. Pakaq appears to have carried out his intentions with little interference from the old woman's relatives.

In 1883–84, Pakaq, Kanaaka, Okaitok, Metiq, and Nepekin appear to have been the most influential Inuit in the Sound (Boas 1883–84). While the former three lived at Kekerten, Metiq, the oldest man in the Sound at 80, and Nepekin lived at Anarnitung and Imigen, respectively. At the latter camp Boas ran into opposition from Nepekin after the anthropologist was blamed for bringing sickness and death upon the inhabitants of this village. Even though the villagers would have been rewarded handsomely, Nepekin declared that no one was to allow Boas into their dwellings, lend him their dogs, or have anything to do with him (Boas 1883–84). Fifty years later, Angmarlik and Kanaaka exercised the same control over their followers. By the early twentieth century, Okaitok, Kekerten's chief *angaquk* (Greenshield 1914: 13), and the younger Angmarlik were the most influential men at Kekerten. Across the Sound at Umanaqjuaq, Pawla served as the pilot on the whale hunts and bargained for furs on behalf of the station manager at Kekerten, while the older Ittirq was another 'sort of a foreman in the service of the trading-station' (Hantzsch 1977: 53, 94). At the same time, Tooloogakjuaq appears to have also shared in the direction of Umanaqjuaq's whaling fleet.[16] During the second decade of this century, Kanaaka, 'once the leading Angokok (*sic*) of Blacklead Island, who had extraordinary influence over his people and was a great opponent of Christianity' (Greenshield 1914: 13), ran two trading stations near the mouth of the Sound for the Sabellum Company (Parmi n.d.: 22). No hunter in Cumberland Sound during the 20th century acquired more prestige, power, and influence than Angmarlik. Apparently, if a younger or unrelated individual wished to speak to Angmarlik, they normally approached a third party who was usually more closely related in age, social status, and/or blood tie (Isa Papatsie, personal communication, 1988).

The correlation between leadership and camp prosperity was also not fortuitous:

> Camps that have the leadership of a good headman seem to get by very well indeed. . . . This was especially noticeable at Bon Accord Harbour (Idlungajung) and Imigen (Sauniqtuajuq), where Angmalik and Johanasee, respectively, are leaders. These camps are in good shape. Other Native camps in the Gulf do not come up to the standard of the above mentioned.[17]

In 1927 Keenainak, Angmarlik, and Attaguyuk were the 'camp bosses' of Nunaata, Idlungajung, and Sauniqtuajuq, respectively. While these settlements struck one police officer as 'fairly healthy, contented and prosperous camps', Opinivik and Kingmiksoo 'seemed the reverse'.[18] Although Idlungajung was by no means the largest camp in the Sound, it was always known as the most prosperous so long as it was under Angmarlik's leadership.[19] Another camp where leadership and prosperity were strongly correlated was Padloping Island or Padli, which was under the control of Kingudlik, a well known native whaler, trader, and catechist.[20]

The institution of prominent men as 'camp bosses' served to strengthen individual claims to leadership, a tendency recognized by both Inuit and Qallunaat alike:

> I believe that if we are going to make any progress with these ... people, we must throw more responsibility on these headmen. I did this when it came to issuing destitute rations of ammunition. There is no use giving ammunition to a poor hunter or rather one that will not get seal with it. Letting these head men (*sic*) handle the situation tended to increase their authority.[21]

The fact that these 'camp bosses' were issued ammunition not only by the trader(s) but by various other contact agents also increased their status relative to others. While the selection of prominent Inuit to serve as mediators between various Qallunaat institutions and the rest of the Inuit population in Cumberland Sound did little to enhance the socioeconomic standing of the average hunter, he benefited by it, at least materially:

> (Ammunition is issued) to the natural leaders of the camps. When the good fellows have plenty, we do not have to worry about the less fortunate. What actually happens when we issue ammunition to the good hunters is that they supply the destitute with game.[22]

The emergence of middlemen and the need to organize and co-ordinate various activities within the context of commercial whaling and general trading placed a greater emphasis on authority and leadership roles. While this allowed already influential Inuit to accumulate even more wealth, power, and prestige, it also led to a greater socioeconomic differentiation of individuals within the community. In 1910 Hantzsch (1977: 79) described the less fortunate Inuit at Umanaqjuaq as a 'proletariat, which through awkwardness, laziness, indifference or poor health never thrives, but remains in inferior station'. Although Hantzsch observed that 'will power and ambition to improve their lot is wanting', he also conceded that 'it is so difficult to make a start ... when one has only an antiquated gun, no dogs, no sledge, no kayak or (whale)boat, with a wife who is a poor manager and a flock of hungry children, it is just as hard to mend his lot as it is for such like folk in our civilized environment.' The consequences of being materially impoverished were obvious: 'In winter the people who had dog teams would leave for a better place to hunt, but the families who didn't have dogs stayed where they were, and sometimes in the winter, they would go hungry because they had to go on foot to hunt' (Pitsualak 1976: 19). Conversely, Umanaqjuaq's well-off could travel to *sarbut* near Nettilling Fiord where seals were plentiful. However, access to such sealing grounds was reserved only for the wealthiest families (i.e., those headed by Pawla and Tooloogakjuaq) as few hunters possessed enough ammunition to stop for long at such a place (Hantzsch 1977: 94). The disenfranchised also appear to have been more prone to accidental death, particularly in the fall when, lacking kayaks, they were forced to use ice cakes, at the floe edge.[23] Inequality within the context of commercial whaling seems to have been a major source of contention:

> For working so hard on the whales, we got a new pair of pants, shirt, smoking pipe, and tobacco. The person who had shot the whale would get a boat and a rifle. We never got what we wanted ... even though we worked so hard on the whales. We, the whalers, didn't get what we deserved to get. Now that I think about it, we were all fooled. (Pitsualak 1976: 24–5)

Resentment of the well-off was apparently not uncommon. For example, Hantzsch (1977: 195, 205, 350) observed that two of his hired hunters, Ittusarjuaq and Aggakdjuk, were envious of the more productive and skilful Ittirq. Not unexpectedly, the materially disenfranchised often shared a general feeling of helplessness and languor:

> One often becomes indifferent to higher aims and wishes in life, loses the ambition to belong to the class of the capable and industrious, and sinks into destitution and squalor and becomes the universally disliked 'sponger' who barely scrapes a living by his own naïve impudence and the good of others. (Hantzsch 1977: 116)

Acquisition of western technology, participation in commercial whaling, and the institution of general trading appear to have resulted in greater social and economic differentiation than was the case aboriginally. But these were not the only external forces that may have engendered changes in relations of production during the historic period. Before examining whether a structural transformation occurred in social organization, the effects of epidemic disease and the adoption of Christianity must be considered.

### Epidemics and Social Change?

The historic Cumberland Sound Inuit, living on the margins of a global economic system, were always susceptible to diseases introduced by contact agents. Every fall after the supply ship left, many Inuit in the Sound came down with 'ship's flu'. In most years, few deaths occurred, despite recurrent food shortages during the late fall. Yet, in some winters, such as 1899–1900 and again in 1941–42, a significant portion of the Sound's population perished. However, no years were worse than the mid-1850s, when the frequency and duration of interaction with commercial whalers increased by orders of magnitude. While the exact number of Inuit who died during this period may not be as great as that implied by Warmow's 1857 population estimate, there can be little question that scores, if not hundreds, of Inuit lost their lives to foreign diseases soon after the onset of over-wintering by the whalers. The issue that remains to be addressed is whether the loss of so many people resulted in a significant transformation in relations of production. More specifically, did population decimation help to initiate a fundamental change in leadership patterns and the emergence of class structure?

We cannot answer these questions with the precision we would like. But we can rationalize that, following Guemple (1972a), there is little theoretical basis from which to argue that epidemic disease during the mid-1850s facilitated the emergence of new social formations in Cumberland Sound. For one thing, decreases in group size mean only that some parts of the social apparatus fall into disuse, so that while some 'boxes' in the structure collapse, it does not always mean that the structure itself collapses (Guemple 1972a). Moreover, the process of creating new social formations is often a difficult process necessitating new organizational principles and the institution of concomitant infrastructures to make them work. Conversely, social change might be expected under conditions of expansion whereby new 'boxes' in the social apparatus require the adoption of new organizational principles and structures (Guemple 1972a: 106). Perhaps more importantly, transcending a mode of production requires conscious knowledge of an alternative method of material reproduction, which in itself entails the ability to 'conceptualize the negation of revealed realities' (Asch 1979: 93). Out of this dialectical process, alternative conceptions can arise through the recombination of existing elements and their negations. As there was no socioeconomic basis from which to formulate alternative social structures, Inuit might have chosen to respond to population decimation in Cumberland Sound through more simple and traditional means, such as migration or the suspension of customary laws and practices, rather than restructuring their social system or its ideological foundations.

Certainly, the loss of scores of people during the mid-1850s might have resulted temporarily in the relaxation of marriage and residence rules. Local groups simply may have found it difficult to remain economically viable by adhering rigidly to customary law. Yet such circumstances usually do not call for experimentation with new social formations. On the contrary, increased emphasis on social conservatism, i.e., reliance on the old social order, might be anticipated as a means of reproducing the social and material conditions with which people were familiar and trusted. In this regard, it is likely that a greater degree of mutual co-operation and bonding among surviving relatives may have occurred as they attempted to cope with the losses of their recently departed kinsmen. While increases in the size and 'compositization' of settlements may not have hampered the operation of the old social order, a greater emphasis on secular and sacred leadership might be expected in order to rationalize population loss and the impacts of other external forces.[24] Under such conditions, then, it is difficult to see how fundamentally different leadership patterns, organizational structures, or new social formations could emerge. Put simply, while the Cumberland Sound Inuit may have been decimated by foreign diseases during the mid-1850s, their social system, albeit under considerable strain, did not collapse to the point where a structurally different social formation arose from its ashes.

## CHRISTIANITY AND LEADERSHIP: A NEW ORDER?

Sacred and secular leadership appear to have always been correlated strongly in Cumberland Sound; a village's *angajuqqaq* and *angakuq* were usually one and the same. For example, in 1840 the elderly Anniapik appears to have been both the principal political and religious authority at Anarnitung. Nearly half a century later, Pakaq, the source of much of Boas' information on Oqomiut religious ideology, was the most influential man at Kekerten. By the late nineteenth century, Kanaaka and Okaitok were the principal shamans and foremen of whaling operations at Umanaqjuaq and Kekerten, respectively. Several years later, Angmarlik and Tooloogakjuaq assumed positions of considerable secular and sacred importance at these settlements. While the latter individuals appear to have been minor *angaqut* of some influence around 1900, both came to accept Christianity, Tooloogakjuaq much sooner and more thoroughly than Angmarlik.

Bilby (1923: 136) and Cardno (Ross 1985c: 234) indicate that *angaqut* held the first place in public esteem and common council, after which the village was ruled by the elders and successful hunters. However, as a leader's productivity declined with age, it is likely that his wisdom and experience, and thus influence and control over spiritual and everyday matters, remained undiminished. In other words, there appears to have been no clear separation of secular and sacred leadership in Cumberland Sound. Just as the social cannot be divorced from the economic in Inuit society, neither can the secular be separated from the sacred; they, too, are intimately linked with control over natural and cultural forces. The question that remains is this: Did the adoption of Christianity alter the traditional relationship between secular and sacred leadership, and ultimately superstructure and infrastructure, in Cumberland Sound?

Christianity threatened to undermine the power and influence of various leaders in Cumberland Sound by usurping their control over spiritual and secular matters:

> The Eskimo, being pagan, were under the authority of their religious leaders, who were unprincipled and crafty men, shrewd enough to appreciate the fact that if (missionaries) were successful in their efforts they would destroy the power of the pagan leaders. (Fleming 1932: 41)

It comes as no surprise, then, that prominent individuals such as Angmarlik and Kanaaka rigorously opposed the teachings of the Anglican missionaries. Nonetheless, some leading figures, such as Tooloogakjuaq, adopted the new religious ideology much more readily than others. Women also accepted Christianity much sooner than men, apparently because it represented a better way of life for them:

> This morning Timukka, one of our oldest Christians and one of the first baptized by Peck, gave birth to a little son. While I was visiting she commented on how much the position of women has improved, staying in the comparative warmth of her tupik instead of being banished to an individual snow hovel, there to remain unattended and alone in her distress, considered as one unclean and unfit to approach for some time afterwards.[25]

Thus, the new religious order eradicated many of the taboos and ritual injunctions to which women were subject and forced to adhere to under the old religious ideology. Tooloogakjuaq's motivations for adopting Christianity, however, may have been different. Despite the latter's hunting prowess and leadership capabilities (Hantzsch 1977: 39–40), Pawla (the native son of Paul Roche) and Kanaaka (who was originally from the other side of the Sound) appear to have attained positions of greater influence, though perhaps not higher esteem, at Umanaqjuaq. In other words, Tooloogakjuaq may have adopted Christianity initially as means of improving his socioeconomic standing, particularly since the mission served as an alternative source of ammunition and relief in times of need. Across the Sound at Kekerten, Ooneasagak, 'a keen young hunter, thoroughly able and reliable, known by the traders as one of their best workers' and this settlement's first lay preacher (Greenshield 1914: 13), may have adopted Christianity for the same reasons. Thus, it seems that well-established leaders initially opposed Christianity, while those with less influence and stature appear to have embraced the new ideology much sooner.[26]

Eventually, however, Kanaaka, Okaitok, and Angmarlik came not only to accept Christianity (Greenshield 1914), but to preach it (Munn 1932). This leads us to consider whether many of the roles of shamanistic leadership continued to be played out within the context of Christianity. Furthermore, while many of the old shamanistic beliefs were replaced by Christian concepts, the structure of religious ideology appears not to have been as radically or quickly altered as the missionaries believed. Put another way, while many traditional beliefs were discarded, the structure of the belief system, and the role of the supreme deity, in particular, remained virtually unchanged. That the Christian God simply assumed the sea goddess's benevolent/malevolent role as giver/taker of life without a change in the overall structure of the belief system is apparent in Kingudlik's teachings that 'game (came) in answer to prayer, and bad accidents (were) punishment for sin'.[27]

Most Inuit, and even some lay preachers, failed to grasp throughly many of the basic concepts of Christianity. For example, Angmarlik at Kekerten in 1902 preached a religious doctrine that fused many elements of the old religious order with the new ideology. While Peck felt that he had destroyed Angmarlik's peculiar brand of syncretism, the latter won him many followers twenty years later at Pond Inlet:

> (Angmarlik) had become an ardent convert to Christianity. He had done some active proselytizing in Pond's Inlet, and made many converts, though his explanations of the Christian tenets were vague and crude. (His) explanations only enabled them to tack the new belief on to their older one, and I was sometimes called on to explain difficult theological problems. (Munn 1932: 245–7)

The facts that Angmarlik and his wife (Ashivak) believed that a young girl in Pond Inlet had conceived by immaculate conception, and that Munn (1932: 248–9) failed to convince them otherwise, provides a clue as to the level of comprehension of the new religious doctrine. Similarly instructive is the belief among the Qivitormiut of Kivitoo that when the 'holy spirit' entered their bodies 'their feet would leave the floor . . . until they would be standing in the air a few feet off the floor'.[28]

The possibility that some headmen employed Christianity in much the same way as shamanism, i.e., as a means of control over both natural and sociocultural forces, is evident in the case of the Home Bay Murders. Although Niaqutsiaq, the *angajuqqaq* of Kivitoo, received incomplete instruction from the missionaries, he undertook to convert his followers.[29] On various occasions, Niaqutsiaq commanded acts of incest—perhaps the greatest crime of all under customary law—and abstinence from food, sleep, and sexual relations in order to test the faith of his people. To make individuals confess their belief in Jesus and to cleanse their souls, Niaqutsiaq physically assaulted his followers and threatened their lives at knife-point. In a notably symbolic act he ordered the sacrifice of three dogs belonging to each man. Eventually, Niaqutsiaq declared that two men should die. His instructions were followed, and Niaqutsiaq was in the process of taking a third life when he himself was killed. The possibility that Niaqutsiaq may have been mentally unstable is suggested by the fact that he claimed to be Christ. However, many of the events in this case are reminiscent of the type of control exhibited and sanctioned under shamanism. It was not until Niaqutsiaq began to exhibit overtly bizarre and unpredictable behaviours—screaming and running around like a madman—that he was murdered.

An additional element of interest about the above case concerns the treatment of Niaqutsiaq's body. After his body was washed a hunter threw three handfuls of 'bloody water' in the face of each grown person and one handful in the face of each child. Under the old religious ideology, human blood and death had a tremendously contaminating effect on the hunter and the hunted. Thus, bodies were not touched directly, let alone washed.[30] The acts of washing a body and then throwing the water on the faces of the living represent an explicit negation of this religious belief, while resembling elements of Christian practice.

Despite the casting off of many former religious beliefs, some continued on in the same or a slightly altered form. In 1910 Hantzsch (1977: 119) remarked that the 'conceptions (of the Umanaqjuarmiut) still have a strong tincture of heathendom, (though) this heathendom is modest and of more pleasing form than before.' That the missionaries failed to destroy immediately all or even most shamanistic beliefs and practices is evident in Hantzsch's (1977: 107) observation that 'Sedna' was still remembered and thanked in 1910. Over a dozen years later, baptized Inuit from Cumberland Sound were still practising polygyny and ritual spousal exchange,[31] and enlisting the services of *angaqut* in times of sickness (Munn 1932: 220). Even by 1934:

> [T]ime has not yet erased from their memory the magic performed by their shamanistic healers: (Neither) has the association of sickness with taboos, and superstitions etc., been replaced by anything that the white man has up to this time brought them.[32]

The same astute observer also recognized that 'the process of erasing traditional gods from their minds must be calculated in terms of generations'.

Some fundamental pre-Christian concepts, in fact, continue to this day. Paramount among these is the belief that the souls of the deceased are reconstituted in the newly born. While most Pangnirtarmiut embrace the Christian belief that the soul goes to its reward after death, they also believe that a part of the soul, or the 'name soul', continues to live on through a newly born child if it receives the name of a recently departed kinsman. This is the reason why so many people of the same age in Pangnirtung have the same name.[33] So strongly is this belief held that many Pangnirtarmiut still continue to apply the kin term appropriate to the deceased. For example, one of my elderly informants calls her nephew *anik*, as he is named after her older brother, whereas he refers to his aunt as *nayak* (sister). The fact that the belief in a second soul weathered the wrath of the missionaries underscores the fundamental role this custom still plays in reproducing relations of production from one generation to the next in Cumberland Sound Inuit society.

Similarly, although most people no longer believe that individual animals have souls that need to be placated, there is a general feeling that seals, for example, have a collective soul or consciousness that obviates mistreatment of the animal after death. Thus, children are still instructed in the proper use and handling of animals, e.g., one should not throw caribou bones into the ocean, or mix products of the land and sea.

The fact that many old beliefs did not die off until the old leaders passed away is apparent in archaeological investigations undertaken by Doug Stenton (1989). Stenton found a number of historic Inuit campsites on the shores of Nettilling Lake, which he differentiated on the basis of age and mode of refuse disposal. Specifically, in the earlier group of sites, caribou bones, even after they had been broken for marrow and grease extraction, were collected and deposited selectively in stone caches. This practice reflects the traditional belief that for fear of offending the spirits of the caribou, caribou bones should not be allowed to lie strewn about campsites where dogs could gnaw on them. However, in sites dating to the mid-1930s and after, this mode of disposal was discontinued as caribou bone was found scattered randomly across campsites. In this regard, I believe that it is not fortuitous that many of the old leaders such as Angmarlik, Tooloogakjuaq, Keenainak, Attaguyuk, Maniapik, and others became infirm or passed away about this time. In other words, as the influence of leaders raised under the old religious ideology faded, so too did the primary means by which many pre-Christian beliefs and customs were socially sanctioned and enforced.

One concept introduced by the missionaries that may have affected social relationships and authority patterns was Christian charity. Prior to the Christian era, sharing was practised predominantly, albeit not exclusively, along kin lines; members of local groups were almost always related in some way. However, with the acceptance of Christian charity, the needy, whether relatives or not, were provided for by the more fortunate. While this served to equalize the distribution of food throughout the community and between communities, it also tended to enhance the status of food-givers by engendering a sense of indebtedness and inferiority among the food-takers. While women appear to have been at the heart of the distribution of food to the needy at both Umanaqjuaq and Kekerten, in the latter settlement the destitute also freely helped themselves to Angmarlik's cache of food which he placed in a dozen or so wooden casks directly in front of his *qammaq*. Interestingly, even though the needy had free access to Angmarlik's food supply, the sharing of capital equipment such as rifles and whaleboats was restricted only to kinsmen (Etuangat Aksayuk, personal communication, 1988). Had no importance been placed on productivity or no stigma attached to the acceptance of charity, the adoption of this institution would have served only to smooth out imbalances in the food supply. However, as it was, the adoption of Christian charity increased the status of the well-to-do relative to the less fortunate. Eventually, however, the adoption of Christianity may have served to undermine traditional authority patterns as white missionaries came to assume many of the non-secular roles and functions of aboriginal leadership. While this was mitigated somewhat by the instruction of native catechists, white missionaries remained the ultimate dispensers and authority of religious ideology.

The adoption of Christianity did not significantly alter, at least not initially, the relationship between secular and sacred leadership in Cumberland Sound. Many beliefs were modified and eventually abandoned, to be sure. Moreover, the adoption of Christian charity may have temporarily increased socioeconomic differentiation. Nonetheless, the structure of traditional religious ideology appears to have remained virtually unchanged for decades. It would take the passing of the old leaders before religious authority came to be vested entirely in the hands of the white missionaries. Even so, a number of fundamental traditional beliefs and customs remain to this day. Although the religious conversion of the Cumberland Sound Inuit warrants far more extensive treatment than has been provided here, it is obvious that this process was characterized by a greater degree of syncretism than has heretofore been recognized.

## Enduring Features of Cumberland Sound Inuit Social Organization

No one specific external mechanism, be it participation in commercial whaling, population decimation by foreign diseases, or the adoption of Christianity, appears to have provided the conditions necessary to transform the structural basis of Cumberland Sound Inuit social organization. This may be traceable ultimately to the fact that the traditional economy of the Cumberland Sound Inuit was never subordinated or destroyed by external forces. Indeed, some outside influences (e.g., the acquisition of whaleboats) appear to have reinforced the traditional mode of production in Cumberland Sound. In other words, relations of production and the necessities required to reproduce social relations that motivated the system in precontact times continued to work well within the context of Inuit-white interaction. Yet there appears to have developed a greater social and economic separation between individuals than may have been the case aboriginally. Did the many external forces to which the Sound's Inuit were subject during the historic period place increased demands on leadership so that there was a fundamental change in relations of production?

### Social Stratification: the Emergence of Class?

A number of powerful individuals appear to have emerged during the historic period within the context of Inuit-Qallunaat interaction, perhaps none more so than Tesuwin and Angmarlik. This is not to say that Inuit did not attain positions of similar socioeconomic standing in precontact times. Certainly, if Mary-Rousselière's research into the Qitlarssuaq migration and Boas' observations on aboriginal leadership and feuding are considered, individuals in Cumberland

Sound rose to positions of substantial influence long before the arrival of Qallunaat. If the dynamic, richly-textured stage of Inuit-Qallunaat interaction did engender social change, it was in the direction of greater socioeconomic differentiation, not an overall transformation in the structure of social relations. As Asch (1979: 92) has noted, modes of production often have flexible structures that can accommodate many variations and changes without changing their fundamental form. In this connection, all that appears to have happened in Cumberland Sound was that the 'rich got richer and the poor got poorer'. At the same time, a particular mode of production is not infinitely malleable, for contradictions between relations and forces of production might eventually become so great as to warrant a structural transformation. While no apparent structural change in leadership occurred in Cumberland Sound, the stage appears to have been set for the emergence of different socioeconomic classes.

One corollary of the emergence of class structure is the occurrence of marriage between groups or families of similar socioeconomic standing, and in this regard there appears to have been a well-developed tendency in Cumberland Sound for the well-to-do to marry amongst themselves. For example, Angmarlik was married to Ashivak, the sister of H.T. Munn's headman at Pond Inlet (Munn 1932: 245), while his sisters, Kowna and Peeka, were married to Niaqutsiaq (of Kivitoo) and Ooneasagak (of Kekerten), respectively. In turn, Angmarlik's daughter, Qatsu, was married to Eevic, the son of Keenainak, the *angajuqqaq* of Nunnata. Angmarlik, it seems, was especially intent on seeing that his children and grandchildren took spouses of high standing and ability:[34]

> The headman had but one daughter and no sons so he adopted four boys and a girl. They are all now grown and with families of their own. His adopted sons are all excellent men trained by their adopted father and his two 'hand-picked' sons in law (sic) are of the same caliber. This old man is far above the average Eskimo of this area in every way. He is also determined he is going to pick husbands and wives for his grandchildren who will meet his standard.

In another example, Kingudlik, the camp boss of Padloping Island, travelled more than 300 km to Idlungajung and back to arrange a marriage between his daughter and a young productive hunter named Etuangat, the step-son of Aksayuk, a prominent hunter from Kekerten. While more evidence of this well-developed tendency will be provided in the following chapter, the important point to consider here is whether a true class system emerged in the wake of this propensity.

The tendency to marry up, or hypergamy, is to be expected in a society where prestige is so closely linked to productivity. Thus, we might anticipate hypergamy to have been a common feature of precontact society in Cumberland

Sound. Recall Eenoolooapik's persistent overtures to marry Anniapik's daughter in 1840, despite the fact that they belonged to different groups. Although the payment of bride-price was less a material exchange between two families than a symbolic gesture establishing and acknowledging certain reciprocal rights and obligations (e.g., access to surplus in times of need), not all families could exact or pay the same bride-price. Nor could families of disproportionate or inferior socioeconomic standing expect to meet or fulfil all the duties and obligations of the contractual arrangement instituted by their marriage alliance. Simply stated, while there may have been a tendency towards hypergamy, individuals of different productive capabilities might be expected to have married generally into families of similar socioeconomic status.

But the very conditions which laid the groundwork for the emergence of endogamous class structure appear to have prevented it. Individuals born into privileged families were certainly predisposed to success in the productive process. However, as elsewhere in the world of the Central Inuit, social standing in Cumberland Sound during the historic period ultimately depended on individual initiative and productivity. As Hantzsch (1977: 83) noted in 1910:

> Relatively quickly can the bright youngster become independent, prosperous, respected and successful. Unknown are the advantages derived from high birth, personal beauty or gaudy possessions. The right of the stronger, of the more intelligent, and of the more ambitious, is the only right acknowledged, whether of man or woman.

Moreover, because one's social position was determined more on individual ability than birth right, there was no basis for antagonisms based on class differences to emerge. In contrast to the Kachin of highland Burma (Friedman 1974), for example, the well-to-do did not become a non-producing class that controlled and appropriated more surplus while producing less and less. On the contrary, families of high socioeconomic standing continued to enhance their positions relative to others by producing more and more surplus. Those individuals and families unwilling to assume their position in the social hierarchy had the universal option, so common among hunter-gatherers worldwide, of 'voting with their feet', i.e., aligning themselves with kinsmen in other camps.

Terray (1984) identified several types of 'class' societies based upon the relationship between producers and non-producers in the use and control of the means of production. Only when groups of producers are separated from their means of production, do true class societies develop. Exploitive relationships existed in Cumberland Sound, to be sure. Boas (1964: 173), for example, observed that adult men without relatives or other means of support were sometimes incorporated into households as servants. Moreover, sons almost always served their fathers in the process of production until they acquired the necessary experience, skills, and resources to strike out on their own. Influential men may have secured surplus from the labour of others, but it was for the good of the community and they could not dictate their conditions to the exploited nor the amount of surplus to be appropriated (cf. Terray 1984: 88). Thus, whereas social differentiation was an accepted fact of life in Cumberland Sound, there appears to have been no structural basis for the emergence of a true class system.

In support of this proposition, there seem to have been no significant changes in sharing patterns such as might be expected to accompany a transformation in social structure. *Nekaishutu*, the village-wide sharing of food whereby the fortunate hunter or his wife invited the rest of the villagers to share in his success, remained the predominant sharing custom well into the 1960s. Individuals responsible for the division and distribution of game sometimes varied—Kumlien (1879: 21) states that the younger men assumed this responsibility, while my elderly informants assert that the successful hunter, and often the *angajuqqaq* or his wife, performed this task. Nonetheless, *nekaishutu* continued to fulfil its main functions of (a) acknowledging and celebrating the productivity of the successful hunter, thus elevating his social standing, and (b) smoothing out imbalances in the food supply. While *nekaishutu* instilled a sense of solidarity among the community, it also had the potential to engender a feeling of indebtedness and social differentiation if the roles of producer and non-producer were not reversed, at least periodically.

Although it might have proven advantageous to do so, most Cumberland Sound Inuit refused to alter their sharing arrangements if economic gains resulted ultimately in social losses:

> One would expect that the introduction of our economic standards by the trading companies would quickly alter theirs. But the department is extremely fortunate in that we are dealing with a . . . race that seems to have a strong passive resistance to any alteration. We know that they still share. In good times the good hunters purchase new rifles. Their old ones are handed on to the less fortunate and I do not think that they are bartered. I know when it came to supplying me with seal for which he would be paid . . . or whether Ahnmahle (Angmarlik) would give it to Newyillia who needed it, but could not pay, Newyillia was given the seal.[35]

This anecdote reveals that traditional relations of production still largely influenced sharing practices; an individual was only as successful and influential as the number of people he could depend on in times of want or need. Traditional relations of production also played a role in the rejection of credit. While most Cumberland Sound Inuit were indebted in some way to their kinsmen, they were not willing to go into debt to Qallunaat for any length of time:

(They reject HBC) attempts to outfit themselves in good years and going into a large debt, despite that the present manager has done everything he possibly can in that direction. The natives pay him the compliment of listening very carefully but immediately trade for tobacco or flour. They do the same with all of us and our progress in overcoming this has only been to the extent of cutting relief to zero for the past year. They will not voluntarily set aside a credit for years when the fur return is only (sporadic).[36]

The rejection of credit, which continues to this day, particularly among the elderly, represents an explicit resistance to enter into socioeconomic arrangements that would alter or jeopardize the structure of traditional productive relationships. Just as the Cumberland Sound Inuit determined which resources ultimately became the focus of production in the commercialization of the hunt, so too did they maintain control over those relations of production that obtained in the reproduction of society. In other words, credit, like trapping, was resisted simply because its use undermined the traditional mode of production.

The possibility that the Cumberland Sound Inuit 'mode of production' did not undergo a significant structural transformation, despite the complexity and dynamism of their interaction with Euroamerican society, is further supported by analyses of marriage practices, residence patterns, and the kinship system. Boas' (1964) 'salvage ethnography' indicates that the Cumberland Sound Inuit traditionally practised kin and local group exogamy, and such appears to have been the case throughout the historic period. Instances are known where close relatives married. For example, in 1930 a native of Saumia 'cast off his wife in order to take his brother's daughter, which (was) against the native code'.[37] But such marriages were extremely rare; 'even when remote camps permit a close marriage, the people are naturally concerned to avoid them, and don't need to be discouraged.'[38] Even today, marriages between close relatives to the second degree of collaterality are frowned upon and the topic of much gossip. In this connection, a young Pangnirtarmiut couple, who were considered related, recently married against their parent's wishes. In order to avoid ridicule, this couple moved elsewhere.

Cardno (Ross 1985c: 235) observed that mothers and grandmothers formerly arranged marriages as 'a woman with a marriageable daughter is fully alive to the advantage of seeing a good hunter attach himself to the domestic circle.' In a society where the death rate among adult males was high, as it was in Cumberland Sound well into the twentieth century, particularly among the Talirpingmiut (see Chapter 6), women might be expected to be responsible for such decisions. However, my informants suggest that both parents participated in arranging marriages for their children. Yet, in some instances, men such as Angmarlik and Kingudlik appear to have been primarily responsible for finding spouses for their children. The high socioeconomic standing of these individuals and their great degree of mobility undoubtedly favoured them in this regard. Child betrothal seems to have been common in the Sound up to the 1960s when people began to move into Pangnirtung. Nowadays, young people exercise more choice in the selection of their marriage partners, as evidenced in the above example.

Patrilocality, preceded by an extended period of bride-service, has always appeared to have been the preferred residence arrangement in Cumberland Sound (Etuangat Aksayuk, personal communication, 1988). Although Valentine (1952: 159) states that the Oqomiut, in contrast to all other 'Eskimo' groups he examined, exhibited a matrilocal residence pattern, it seems that he has confused bride-service/uxorilocality with intitutionalized rules of matrilocality. In contrast to the Iglulingmiut, who normally engaged in one year of bride-service (Damas 1963), there was no specified period for this type of residence in Cumberland Sound. As such, a man's obligations to his *sakkiik* could encompass one season or an entire generation, depending on a multiplicity of factors. However, after living with his wife's parents, a man was expected to return to the camp of his birth, if his parents were still alive, for this was where his primary kinship connections and societal obligations lay.[39] Indeed, parents often refused to allow their sons to join their parents-in-law after marriage (Boas 1964: 171). Patrilocal residence following a period of bride-service was the stated ideal during the twentieth century (Etuangat Aksayuk, personal communication, 1988) and, on the basis of Sutherland's (1856: 210–12) 1846 census of Kingmiksoo, it appears that it was the predominant residence arrangement in early contact times as well (Figure 14).

There seems to have been relatively little alteration in marriage or residence patterns between 1840 and 1963. Nonetheless, the kinship terminology may have undergone some changes during this period (e.g., compare Figures 13 and 17). As kinship structures may function as both relations of production and ideologies on which mythologies are constructed (Friedman 1974: 445, Levi-Strauss 1969), any change in terminology may be significant potentially.

The most apparent difference between the terminologies recorded by Morgan (1870) and the one in use today in Pangnirtung is that the term for MBW, formerly *aiyug(ga)* (my aunt or mother's sister), has been replaced by *ukuaq* (in-marrying female). The only other Central Inuit group to extend *ukuaq*—a term usually reserved for in-marrying females in Ego's and all descending generations—to in-marrying females in the first ascending generation is the Iglulingmiut (Damas 1975c). However, while they do this for female Ego only, among the Pangnirtarmiut *ukuaq* is used by both female and male Egos. *Ukuaq* is also used by Inuit in Cumberland Sound for FBW, just as it was in the 1860s when Morgan recorded their terminology, although he spelled the term as *ukuu(nga)* (my in-marrying aunt) (Figures 13 and 17).

Morgan recorded the term *iei(nga)* for MZH (Figures 2 and 13). We may assume that the root for this term is the more conventionally-spelled *ai*, or

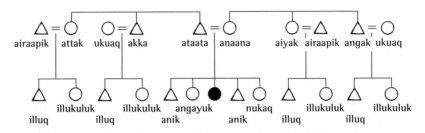

FIGURE 17. Present-day Cumberland Sound Inuit kinship terminology, Ego's and first ascending generations (female Ego). Note that the spellings of most terms differ from those recorded by Morgan (1870) and reflect more conventional usage (e.g., Damas 1964, 1975c).

in-marrying relative of the opposite sex in Ego's and the descending generations. However, the Cumberland Sound and Iglulik Inuit also extend *ai* to the first ascending generation. Yet, whereas only males use the term *ai* to refer to in-marrying females in the first generation among the Iglulingmiut, this term in Cumberland Sound was applied to MZH by both male and female Egos. Today, the same structure prevails, although the term for both MZH and FZH is *ai(raapik)* (Figure 17). Assuming that there was some logical continuity in the use of the term *ai* between the mid-nineteenth century and today, we may speculate that the missing term for FZH in Morgan's terminology is, in fact, *iei(nga)*.

Equally problematic is Morgan's recording of *atchu(nga)* for FZ. Today, the term for FZ is *attak*, which has a universal application across the central and eastern Arctic.[40] Although *atchu(nga)* (i.e., my aunt) is uncommon among the Central Inuit, *attak* can also be pronounced as *achak* or *atsak*, depending on the dialect spoken. In this light, the confusion is lessened and we may assume that Morgan's *atchun(ga)* equates with today's *attak*.

By way of summary, the first ascending generation kinship terminology of the aboriginal Cumberland Sound Inuit appears to have placed an emphasis on the matri-line whereby MBW was equated with MZ. Conversely, in-marrying females (and presumably, in-marrying males) on the father's side, as well as in-marrying males on the mother's side, were classified as subordinate relatives and lumped together with affinal relations in Ego's and the descending generations. What this seems to suggest, in contrast to previous speculations (see Chapter 2), is that aboriginal social relationships may have been skewed towards the matri-line, wherein in-marrying females on the mother's side were merged with consanguines, and treated accordingly. Societies which place an emphasis on cross-sex sibling relationships and, to a lesser extent, female sibling relations would be the most likely candidates to employ such a terminology. However, it is apparent that the Cumberland Sound Inuit no longer accord special status to in-marrying females in the first ascending generation—both MBW and FBW are *ukuaq* (Figure 17).

I have speculated, based on a variety of sources, that the Cumberland Sound Inuit traditionally may have demonstrated a male bias on a bilateral structure (see Chapter 2). At the same time, however, there appears to be little evidence for significant change in other aspects of Cumberland Sound Inuit society. How, then, does one reconcile these conflicting interpretations? First of all, it must be assumed that, because of their extensive distribution among regional groups with markedly different characteristics (e.g., the Netsilingmiut and Oqomiut), affinal-including aunt-uncle terms are ancestral to Central Inuit kinship systems.[41] Thus, their occurrence among the aboriginal Cumberland Sound Inuit probably has an ancient history. Yet, affinal-including aunt-uncle terms no longer exist today in Cumberland Sound. Given (a) the lack of evidence for significant change in other facets of Cumberland Sound Inuit society subsequent to contact, and (b) the existence of different kin terms in the past, the most parsimonious explanation is that Morgan's terminology was not representative of the Cumberland Sound Inuit at all. Rather, it was only illustrative of the regional division to which Morgan's informants belonged, i.e., the Talirpingmiut. As the Qikitarmiut in the late nineteenth and twentieth centuries were superior to the Umanaqjuarmiut in number, wealth, and influence, the possibility that their kinship usage simply swamped that of the Talirpingmiut (Umanaqjuarmiut), in the context of Inuit-Qallunaat interaction and greater interregional mixing, seems likely. Under this scenario, no significant alteration in kinship terminology appears to have occurred in Cumberland Sound during the historic period. Rather, certain regional divisions originally may have possessed different systems—a subject that will be addressed in the following chapters.

The kinship terminology of the Cumberland Sound Inuit underwent some additional changes during the historic period.[42] However, one factor which suggests that there was no significant transformation in the structure of relations of production as encoded in the kinship system is the emergence of matriarchs. At first glance, the ascendancy of women to positions of leadership would appear to run counter to the kinship directives. Yet a number of women appear to have assumed positions of considerable authority and influence during the nineteenth and twentieth centuries. For example, in 1868 the mate of the New Bedford barque, *Milwood*, told many yarns, some seemingly improbable, about an old woman named 'Molly-Kater' (Malukaitok), after which the fiord was named, who reputedly maintained 'despotic matriarchal power over a large settlement'.[43] Over half a century later, two women, Nuneeaguh and Kowna, ran small trading operations at Mingoakto (in Frobisher Bay) and Kivitoo, respectively, for the Sabellum Company. Malukaitok's heritage or marital status are not known. Nonetheless, it is instructive that both Nuneeaguh and Kowna were widows of prominent headmen and former traders, Godiliak and Niaqutsiaq.[44] In regard to Kowna, it was observed that 'she . . . has been a good influence on the small settlement (Kivitoo), for she has far more initiative than the men, both of whom seem to obey her instructions in all matters.'[45] Oqomiut women appear to have always held positions of considerable influence and authority in the domestic sphere. One need only refer to Hantzsch's (1977) ongoing tirade about the domestic tyranny of Sikirnik, Ittusarjuaq's wife, to appreciate this fact. However, the ascendancy of Kowna and Nuneeaguh to local group leadership could only have been sanctioned if all males in their respective camps were subordinate to them, a situation made possible, barring unusual circumstances (e.g., infirmity of resident males), if they were immature and/or in-marrying males (e.g., *ningaut*). While individual abilities were as important as ever within the context of Inuit-Qallunaat relations, the structural basis of social relationships appears to have remained virtually unchanged as dominance-subordinate relations were still based largely on some combination of gender, age, and blood ties with respect to the primacy of the resident (extended) family. In the cases cited above, the latter two factors may have conspired against the former to permit matriarchs to ascend to positions of considerable influence.

The structure of the extended family in Cumberland Sound was subject to many external pressures during the historic period, perhaps none more so than decimation by foreign diseases. In this connection, there appears to have been some subtle alterations in the terminology recorded by Morgan and, by extension, possibly associated behaviours. Perhaps I may be attaching too much importance to Morgan's terminology. After all, he did not record the terms for FZH, and his terminology was obtained from only two Kingmiksormiut. Consequently, the latter may simply reflect subregional variation, and may not be representative of kinship terms in use among other regional divisions in Cumberland Sound—a distinct possibility suggested by the evidence presented above and below. However, for the sake of argument, let us assume that differences between the old and new kinship terminologies are the result of change over time across the Sound, whereby affines in the first ascending generation assumed less consanguineal status and same-sex cousin relationships intensified (i.e., terms of endearment—*illukuluk* for female cousins and *illuajuk* for male cousins—were appended to the cousin term *illuq*). Attendant with these developments, social directives implicit within same-sex and consanguineal relationships may have become more pronounced in the formation and maintenance of productive relationships.[46] But this does not necessarily mean that there was a structural change in the basis of the extended family. Rather, it merely reflects a slight adjustment in emphasis—in this case, a greater reliance on blood and gender relationships in order to maintain the viability of local groups. Perhaps this is what most Arctic anthropologists mean when, in reference to Inuit culture, they employ that odious, over-used, and poorly understood concept, 'flexibility'.

## Conclusion

The foregoing analysis leads inescapably to the conclusion that there was no significant structural transformation in Cumberland Sound Inuit social organization during the historic period. At no time was the traditional mode of production, and especially the aboriginal relations of production that underpinned Inuit society and economy in Cumberland Sound, subordinated by the capitalist mode of production of the whalers and traders. In fact, it can be argued that Inuit participation in the capitalist mode of production reinforced the traditional mode, and that only recently has the latter begun to articulate with the former. Rey (1971, cited in Foster-Carter 1978: 55) has distinguished three stages in the articulation of pre-capitalist with capitalist modes of production: (1) an initial link in the sphere of exchange, where interaction with capitalism reinforces the pre-capitalist mode, (2) a stage in which capitalism takes root, subordinating the traditional mode of production, but still making use of it, and (3) the eventual disappearance of the pre-capitalist mode. The traditional mode of production remained dominant in Cumberland Sound throughout the historic period, in spite of the fact that new means of production were introduced, because no new relations of production accompanied Inuit participation in capitalist economy, and the exchange of labour and produce in the context of commercial whaling and general trading reinforced traditional relations of production, particularly authority relationships. Thus, for 130 years capitalism depended exclusively upon traditional relations of production in Cumberland Sound for providing its goods (cf. Foster-Carter 1978: 59).

Bradby (1975: 147) has suggested that 'the process of capitalist reproduction only implies the expansion of capitalist relations if it is taking place in a social formation where capitalism is already dominant.' That participation in commercial whaling and general trading served only to underpin the existing modes of production is evident in the fact that the Cumberland Sound Inuit explicitly rejected capitalistic relations, such as those which might have been instituted by the adoption of credit.

Even though the Cumberland Sound Inuit eventually came to depend upon international markets to maintain their traditional mode of production, only since the mid-1960s has this mode begun to articulate with the capitalist mode. This, in large measure, is the result of the demise in the market for seal skin products and the adoption of cash as a medium for exchange. The processes of articulation and eventual subordination of traditional modes of production in pre-capitalist societies, however, are extremely complex and poorly understood. As Foster-Carter (1978: 60) notes, how can capitalism take root in social formations 'in which capitalism (itself) is not born from the self-destruction of previous relations of production'? Some take the view that capitalism can only become implanted through transitional forms of production which develop in the womb of colonialism. Whatever the case, one should not view traditional Inuit economy as the passive and formless victim in the process of articulation with capitalist economy that many Arctic anthropologists have heretofore assumed; the resultant social formation reflects the dynamics of both, as well as their articulation, by which, indeed, it is constituted.

Today, in Pangnirtung, it is gratifying to learn that the recent development of a winter turbot fishery and potential markets for seal products in the Orient, and to a lesser extent, a summer shrimp/scallop fishery, has provided a stage for the reconstitution of traditional productive relations not seen since the demise of the seal skin market in the late 1970s and early 1980s. However potentially rewarding and worthy of investigation, an analysis of the articulation of traditional and capitalist modes of production over the last two decades in Pangnirtung remains outside the scope of this study. While the reader is referred to Mayes (1978) for a comprehensive study of government-enforced change in Pangnirtung during the late 1960s and early 1970s, we must be content presently to acknowledge that the Cumberland Sound Inuit did not undergo a significant transformation in social organization prior to this period. And this despite the fact that few other regional groups in the eastern Arctic experienced as long or as intense an association with Qallunaat culture and institutions.

What this means in terms of the broader objectives of this study is that data on local group composition presented in the following chapter may be considered to be representative of precontact social organization in Cumberland Sound. This is not to suggest that camps inhabited between 1920 and 1970 were identical in size or composition to those occupied before 1840. Such an assumption would ignore the many events and processes that took place during the historic period. Rather, because the Cumberland Sound Inuit did not experience a significant transformation in social organization, we can expect that the same structural principles which underpinned society in precontact times continued to do so throughout the historic period. This being the case, we would anticipate that an analysis of local group composition might illuminate not only precontact social organization in Cumberland Sound, but possibly also the broader structural principles of Central Inuit social oganization.

# Part Three

# Cumberland Sound Inuit
# Social Structure

# Chapter 5

## Cumberland Sound Inuit Kin and Local Groups, 1920–1970

The Cumberland Sound Inuit appear not to have undergone a significant transformation in social organization as a consequence of their contact with Qallunaat. The size and composition of social groups in Cumberland Sound during the historic period may not be identical to those during the precontact period. However, there are no apparent grounds for dismissing the proposition that the same principles upon which local groups were constructed in late prehistoric times continued to provide the basis for the formation of productive activity and social relationships well into the twentieth century.

In this chapter the local group composition of major settlements occupied in Cumberland Sound between 1920 and 1970 are described. This period, known as the 'contact-traditional' period (Damas 1988, Helm and Damas 1963), is understood to be that stage in the meeting of Euroamerican and native cultures during which the traditional economy continues to hold sway, although its technological base may have been altered markedly (Usher 1965: 49). The process began in Cumberland Sound after 1850, but was soon followed by a transitional stage where face-to-face encounters with Qallunaat were greatly reduced. The contact-traditional period in Cumberland Sound is considered to have begun with the demise of commercial whaling and subsequent dispersion of the population around 1920, and to have ended with the centralization of the population at Pangnirtung in the mid-1960s (Goldring 1986). Inuit experienced sustained contact with various contact agents during this period, but until the hegemony of the HBC, Anglican missionaries, and the RCMP was broken in the mid-1960s by the establishment of government services (Goldring 1989, Mayes 1978), life still retained a predominantly traditional character.

The following detailed examination of local group organization during the contact-traditional period is intended to flesh out the structural features of Cumberland Sound Inuit social organization. If no significant changes in social organization followed in the wake of the Qallunaat, then the data in this chapter may shed light on the broader structural features of Cumberland Sound and Central Inuit social organization.

METHODOLOGY AND PRESENTATION

The information on local group composition provided below was obtained primarily from 21 Pangnirtarmiut, most of them elderly, and secondarily from federal government records and other archival sources and historical materials. The problems inherent in attempting to reconstruct data derived largely from informant memory are obvious; time sometimes plays tricks on the mind. Yet, despite this limitation, such data also contain certain strengths. It is evident that information on local group organization obtained after the fact from informants advanced in years differs from that which an ethnographer would observe first hand. Such data are abstractions, or models if you will, of social groups that lived at specific times and places that exist primarily in the mind of the participant, not the outside observer. However, this is seen as less a problem than an advantage, for such models reflect the cultural biases and perceived realities of the informant rather than those of the anthropologist, though the latter should be held entirely responsible for their interpretation. Moreover, because two or more profiles of local group composition from different time periods were usually obtained for each settlement, the presentation below traces the social history and evolutionary trajectory of kin groups at specific locations.

Nevertheless, the fact that the following data were obtained decades after most camps were occupied cannot readily be dismissed. Understandably, whereas many informants were unequivocal in their reconstructions of the past local groups to which they belonged,[1] the recollections of some were less precise. Although most informants had little trouble recalling the heads of specific families at certain times and locations, some occasionally experienced difficulty remembering the exact number and/or gender of children attached to various households. While family censuses taken by the RCMP and other contact agents during the 1920s and 1930s helped to resolve this problem, the number and, to a lesser extent, the gender of children provided in the following kinship diagrams are not as precise as would have been the case had an anthropologist been present to record this information. Just as importantly, whereas many elders could recall who lived in their camps during a particular period of time, others were less certain as to what temporal period they were describing; 'they did not keep track of time in those days.' Moreover, even if some informants believed that a person belonged to a particular group and place, a few were not sure as to whether all individuals associated with a specific occupation lived together there at the same time. Whereas archival research and direct questioning helped to overcome these deficiencies, most profiles of local group composition probably describe not one moment in time, but a period spanning one to several seasons. In this sense, the following data truly constitute emic reconstructions of social reality, while combining the synchronic approach of ethnography and the diachronic perspective of history.

In regard to kinship connections between household heads, I have attempted to overcome variability in informant memory by asking two or more informants to provide group profiles for each specific time period represented. Thus, most kinship diagrams are composites based on information derived from two or more informants. While this procedure sometimes introduced new problems, it served generally to clarify the composition of local groups at specific locations. Still, because of the nature of informant recall, local group reconstructions vary in detail from one camp to another. In addition, for the sake of clarity, only the most direct or proximate kinship connections between individuals are described.

Interviews with elders were carried out systematically during the summer of 1989, as well as in a more *ad hoc* manner over a period of several years prior to this date. All interviews in 1989 were conducted in Inuktitut, with the assistance of a translator. Notes were taken during each session and an attempt was made to tape all interviews in both English and Inuktitut. Where the use of a tape recorder was not feasible owing either to weather or to the preference of the interviewee, notes were the only source of documentation. Interviews were conducted individually as well as in small groups. Both methods offered advantages and disadvantages. For example, while consensus-derived descriptions helped to alleviate the informant memory problem, the 'younger' elders in the interview group, because of the nature of Cumberland Sound Inuit social organization, often deferred to the oldest, even though they may have been more an authority on the camp in question. Whenever feasible, historic settlements were visited and recorded with one or more elder(s) present. Where elders were unable to accompany the writer to their former camps, they were asked to produce maps identifying various households and major features. From these exercises it soon became clear that there was a positive correlation between kin relatedness and the locations of houses, i.e., between social distance and spatial distance. The same fit has been observed among other Inuit groups (e.g., Burch 1975, Graburn 1964), and is to be expected in a society where kinship has always played a predominant role in shaping social interaction. Guemple's insistence on the ascendancy of locality over kinship notwithstanding, even he has observed this correlation among the Belcher Islanders (1972c: 74).

Description of contact-traditional local group composition in Cumberland Sound begins with the earliest occupied Qikirtarmiut camps at the head of the Sound. While all local groups were mobile to some degree or another from late spring to early fall, each was attached to a permanent location where it spent most of the rest of the year. Not all settlements occupied between 1920 and 1970 in Cumberland Sound are described, however. A comprehensive description of most contact-traditional settlements is provided in the thesis upon which this study is based (Stevenson 1993), and the reader is referred to this work for information on local groups of brief duration. For the sake of brevity, only the most continuously occupied camps are described in detail. Other camps are not described owing to insufficient information stemming from either a lack of informants or a lack of recall. For example, as virtually no information was obtained on settlements inhabited by the descendants of the Saumingmiut, their camps (e.g., Saumia and Aukadliving) are excluded from the present discussion. Also omitted are temporary camps formed in the interests of the HBC (i.e., Kingua and Kaneetookjuak). Nor are detailed descriptions of local group composition at Umanaqjuaq, Kekerten, and Pangnirtung provided. Whereas Pangnirtung existed primarily as a white settlement to serve Inuit needs, its Inuit population was constantly in a state of flux owing to the nature of the various Euroamerican institutions represented there. Moreover, Pangnirtung's half dozen or so permanent Inuit families were engaged in the service of the white population. Similarly, because of the presence of whites and Euroamerican amenities at Kekerten and Umanaqjuaq, and because these centres were not formed primarily out of aboriginal interest, they cannot be considered traditional camps in the normal sense of the term. Although most settlements at the head of the Sound were occupied by the Qikirtarmiut, this was not exclusively so; Inuit from both Kekerten and Umanaqjuaq overlapped at certain times and places. Nonetheless, in order to facilitate analyses in the following chapter, camps are grouped according to whether they were occupied predominantly by Qikirtarmiut or Umanaqjuarmiut.

Notwithstanding the exclusion of the above settlements, we are left with a data base that includes most major Qikitarmiut and Umanaqjuarmiut contact-traditional period camps—locations which have been inhabited more or less continuously for hundreds of years (Figure 18). Thus, in spite of certain limitations, the following descriptions hold the potential to illuminate the dynamics of group formation in Cumberland Sound and possibly the structural principles of Central Inuit social organization.

## QIKIRTARMIUT SETTLEMENTS

### Nunaata

Shortly after the closure of the HBC post at Kingua in 1924, a small group of Qikirtarmiut settled on Nunaata Island near the entrances to Issortuqjuaq (Clearwater) and Shark Fiords. Nunaata was occupied continuously for the next four decades before being permanently abandoned in the 1960s. The location of this settlement was determined by a combination of factors. While surrounding waters provide good winter habitat for ringed seals, strong tidal rips at entrances to these two fiords keep small bodies of water ice-free throughout much of the winter. Nunaata also offers excellent access to large concentrations of beluga, which congregate annually each summer at Milurialik. In addition, caribou and char can be taken at the heads of numerous small inlets and bays nearby. Fox may have also been a determinant in the selection of this

146    Part Three: Cumberland Sound Inuit Social Structure

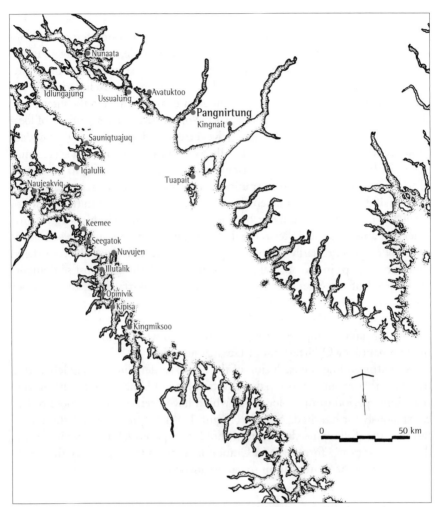

FIGURE 18. Locations of contact-traditional period settlements referred to in text.

site. As noted previously, the HBC enticed the better hunters in the Sound to work for the Company and to relocate their winter camps to more productive fox trapping areas by promising them rifles and often whaleboats—agreements that did not come without strings attached.[2] In return for a new whaleboat, Keenainak, a prominent whaler from Kekerten, moved permanently to Nunaata in the fall of 1924 and was trapping for the HBC by winter.

*1925–1927*
Residential solidarity at Nunaata during the mid-1920s was based largely on male sibling ties between Keenainak (1) and his two younger brothers, Koodlooalik (2) and Ishulutaq (3) (Figure 19).[3] While Keenainak's father (4) was

the head of this extended family unit at Kekerten, by 1920 he was no longer capable of group leadership and had become a dependent of 2. In accordance with behavioural directives, 4 relinquished leadership to his eldest son (1), transforming the structural basis of this extended family from an *irniriik* core to a potentially less stable *nukariik* core. Keenainak's ascendancy to group leadership, however, was not challenged. Not only was he a catechist, but he owned the only whaleboat in the settlement,[4] and was regarded as an excellent hunter.

Keenainak's abilities as a hunter and leader, in turn, appear to have attracted a large following of his wife's (5) relatives including the latter's mother (6), sister (7), niece (8), and cousin (9), together with their respective husbands (10, 11, 12, and 13) and families. Indeed, the size of this affinal group is surprising given the generally acknowledged dominance and superiority of the consanguineal extended family in Central Inuit society. Although *nukariik* bonds between 5 (Avingaq) and 7, as well as *panniriik* ties to their mother (6), undoubtedly gave stability to this affinal unit, 11, 12, and 13 might have found more viable socioeconomic relationships with members of their own kin groups. While 12 may have been performing an extended period of brideservice for his father-in-law (11), the latter individual, who came from Ussualung, may have not had the option of residing with his immediate kin as he was from Frobisher Bay. With the exception of Keenainak's father (4), and possibly his father-in-law (10), all male affines were terminologically subordinate to Keenainak. While Keenainak may have had co-operative sibling-like relationships with 11 and 13, *naalaqtuq* directives explicit in their affinal linkage would have overshadowed most bonds of affection and closeness. Under such circumstances we might anticipate the eventual disintegration of this affinal group, and this is precisely what happened.

During the late 1920s, 11, 12, and 13 left Nunaata for 'greener pastures'. Their brief occupation at this location as well as the placement of their houses (E, F, and G, respectively) on the periphery of the camp,[5] attest to their relatively subordinate positions in Nunaata's social hierarchy (Figure 20). At the same time, features A, D, E, and F form a chain of affinally linked residences parallel to the shoreline which is in perfect harmony with kinship directives. Here, 5's mother (6), sister (7), and niece (8) have located their households (D, E, and F, respectively) progressively further away from Keenainak's residence (A).

Social solidarity at Nunaata was expressed not only in the layout of the camp, but also in game-sharing practices. As noted previously, the village-wide sharing of game (*nekaishutu*) and the more restricted distribution of food between two families (*piutuq*) dominated sharing practices in Cumberland Sound. The butchering and distribution of game during *nekaishutu* would normally fall to the man most experienced at butchering and fair at distributing the catch. In times of scarcity, this duty normally fell to the camp leader. Keenainak almost always performed *nekaishutu* at Nunaata, distributing meat and blubber according to need and tradition (e.g., larger families received

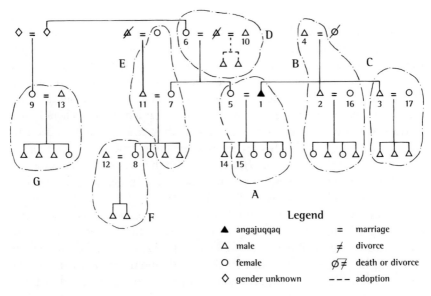

FIGURE 19. Social composition of Nunaata camp during the mid-1920s.

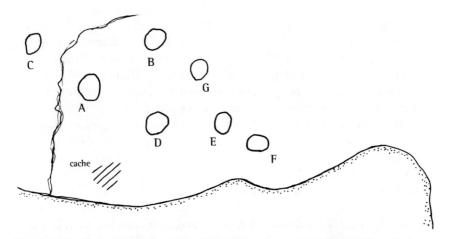

FIGURE 20. Map of Nunaata, mid-1920s. Redrawn from original by Elija Keenainak.

larger portions of food, while parents of the hunter usually received the choicest parts of the seal, i.e., shoulders and front-side flippers). Although the frequencies with which *piutuq* and *nekaishutu* were practised were dependent upon the availability of game as well as other factors, *nekaishutu* was always the most common form of food sharing at Nunaata, a fact which appears to have strengthened the solidarity of the residential group as well as Keenainak's social position.

Co-operative hunting among household heads occurred only during the open water season. Throughout the remainder of the year, men preferred to hunt with their eldest available son(s). The Pangnirtung (HBC) Post Journals support my informants on this issue; Keenainak is almost always recorded as travelling with his sons as opposed to his brothers.[6] So great was this tendency that, until 2's sons were old enough to accompany him, he hunted alone. Keenainak's brothers and male affines forged co-operative hunting partnerships only as members of his whaleboat crew during the early summer beluga whale, mid-summer seal, and early fall caribou hunts. Accordingly, during the open water hunting season all meat and blubber was stored in a large central cache owned and controlled by Keenainak. For the rest of the year each household maintained its own food cache in the porch of its dwelling.

Membership in Keenainak's whaleboat crew undoubtedly contributed to residential stability, but it also served to strengthen *naalaqtuq* directives implicit in his relationship with his affinal relations. Whereas individuals 11, 12, and 13 benefited economically from their membership in Keenainak's crew, they paid a price through increased deference to Keenainak. Given the subordination of these individuals in economic production as well as the social system, it is surprising that they lived at Nunaata at all. While membership on a whaleboat crew at a time when such boats were scarce may have motivated 11, 12, and 13 to settle at Nunaata, their social and economic subordination undoubtedly provided the incentive to leave. A contributing factor in the dissolution of this affinal group may have been that Keenainak's oldest sons, 14 and 15 were fast approaching the age—c. 17 and 15, respectively, in 1927— where they could hunt and assume positions on their father's whaleboat.

Leadership at Nunaata remained well-developed and uncontested under Keenainak's tenure, and the size and style of construction of his *qammaq* and cache reflected his social position; they were the largest wooden structures in the settlement. Keenainak's material possessions, hunting prowess, and other capabilities surely solidified his position as *angajuqqaq*, as did his role as Nunaata's chief trader. As noted, the HBC normally engaged one hunter from each camp to collect and transport furs and blubber skins to Pangnirtung, and to distribute trade goods to the community upon his return. The acquisition, allocation, and distribution of trade goods, like that of game, undoubtedly served to bolster Keenainak's authority and social position in the community.

The ascendancy of patrilocality and bride-service over other forms of residence at Nunaata during the mid-1920s is obscured somewhat by the presence of Keenainak's affinal relations. Although individual 12 may have been performing an extended period of bride-service for 11, the latter as well as 10 and 13 have no blood relatives in the same or ascending generations at Nunaata. The prevalence of these forms of post-nuptial living arrangements over others becomes apparent after the departure of Keenainak's affinal relations. Keenainak's brothers' wives (16 and 17) had no kin relations at Nunaata, whereas his eldest son (14) and youngest brother (Eevic) appear to have been performing bride-service elsewhere. Although the former returned to Nunaata around 1927, the latter remained permanently at Idlungajung, the camp of his father-in-law, Angmarlik. Over the next two decades, 2 and 3 lived periodically at Idlungajung, where they presumably benefited by their younger brother's (Eevic) association with Angmarlik.

*1940–1942*

Throughout the 1920s and early 1930s social integration at Nunaata was still largely, though weakly, based on *nukariik* ties. However, by 1935 a shift in structure occurred (Figure 21). This transformation began with the maturation of Keenainak's eldest sons, Ashuluk (14) and Lazalusie (15), in the late 1920s, but was complete by the late 1930s. By 1940, Keenainak's younger brothers (2 and 3) no longer resided at Nunaata, his father (4) had died, and residential stability was accomplished through Keenainak's relationships with his married and unmarried children. Other changes are also apparent. Avingaq (5) has died and Keenainak has taken another wife (18) from Padloping Island. Keenainak's eldest married daughter (19) has moved back to Nunaata with her husband (20) and his mother (21). The death of 20's father, Maniapik, at Iqalulik apparently motivated this move.

Despite the fact that Keenainak remarried, his first wife's mother (6) continued to reside at Nunaata, providing kinship connections to the heads of Nunaata's remaining families (22 and 23). While individual 22 was apparently 6's *inngutaq* (grandchild) whom she raised in her household, 23 is the husband of 22's wife's cross-cousin (25). Individual 24 is also related to Keenainak, as she is 2's daughter, but from a different man. Although Keenainak has multiple, though somewhat distant, ties to 22 and 24, the latter's union would appear to constitute a local group endogamous marriage since both were raised at Nunaata.

Matrilocality appears to be more prevalent than patrilocality at Nunaata in the early 1940s. Individual 26 is married to Keenainak's daughter (27) and is performing bride-service in the household of his *sakiik*. Within a year or two, this couple moved back to 26's parents' place of residence, Pangnirtung. As noted above, the death of 20's father encouraged him to settle permanently in the camp of his wife's father. Although Mike (23), who died shortly after he took up residence here, might have found closer relatives to reside with had he been an Oqomiut, he was a Netsilingmiut/Aivilingmiut from Repulse Bay via Coral Harbour with no direct relations in the Sound other than his son, Akpalialuk at Ussualung. Individual 22, who grew up in the household of 6, exhibits the ideal marriage arrangement as he remained in his adopted parent's camp with his wife joining him from elsewhere. That patrilocality was not well-developed at Nunaata during the early 1940s appears, in part, to be related to the fact that, while Keenainak's remaining sons and step-son were of

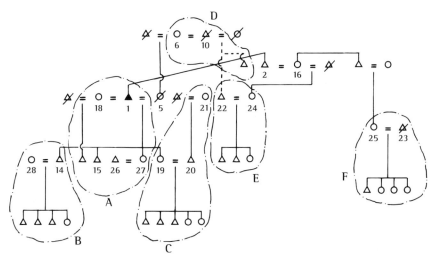

FIGURE 21. Social composition of Nunaata during the early 1940s.

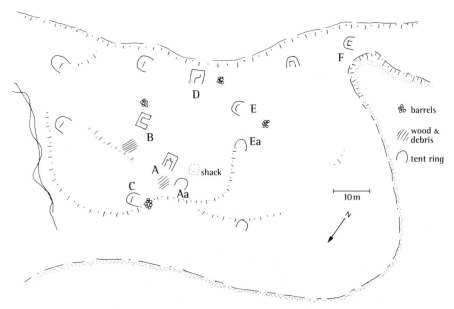

FIGURE 22. Plan of Nunaata, 1940–1942, drawn by author. Unlettered dwellings unoccupied.

marriageable age, they were not yet married. Two instances of hypergamy are evident in the marriages of Keenainak's eldest son to 28 and eldest daughter to 20. Whereas Karpik (20) was the son of Maniapik, the *angajuqqaq* of Iqalulik, Kilabuk (28) was the daughter of Veevee who had been adopted by Attaguyuk, the leaders, respectively, of Naujeakviq and Sauniqtuajuq.

An almost perfect correlation is seen to exist between social distance prescribed by the kinship system and the location of households (Figure 22). Keenainak's *qammaq* (A) is flanked by those of his eldest son (B), which is one of the most substantial house foundations at Nunaata (personal observation), and his daughter (C). Outside this core live his deceased wife's mother (D), and a couple with which he has *qangiariik* relations (E). Finally, his brother's wife's niece and her husband, in agreement with being on the periphery of Keenainak's kinship network, live on the edge of the camp in household F. It is noteworthy that this household, which a son of Keenainak's (29) later joined after 23 died, was occupied for only two years before it was abandoned. Considered together these data suggest that not only were social and spatial distance at Nunaata positively correlated, but households of those more closely related to Keenainak appear to have been more substantial and occupied for longer periods of time than those of individuals who were more distant socially. Similarly, two of the three summer tent rings recorded, Aa and Ea, were used each summer by the female heads of two of the most permanently occupied dwellings at Nunaata (households A and E). The remaining tent site was inhabited by a number of families over the years.

## 1960–1962

Throughout the 1940s and early 1950s Nunaata's social composition did not change significantly. Some individuals left, while others shifted residences or built new ones as deaths and marriages warranted the formation of new social relationships and economic arrangements. For example, after the death of Keenainak's daughter (19), 20 remarried and moved into a vacant *qammaq*. However, Keenainak and his married sons still formed the basis of residential stability. Nonetheless, sometime during the mid-1950s Keenainak died, leaving his whaleboat not to his eldest resident son (15)—14 had predeceased his father—but to his third eldest son (29) (Figure 21). It is not certain why 15 did not inherit his father's boat, as convention would dictate. Individual 29 apparently had an investment in this craft, which raises the question of why 15 did not have a greater interest in the boat.

Whatever the case, by 1960 the basis of residential solidarity at Nunaata had reverted back to a male sibling core. In contrast to earlier times, group leadership was weakly developed. It is plausible that the contravention of *naalaqtuq* directives in 15 and 29's relationship undermined the authority and power base of the former. Then again, perhaps *ungayuq* behaviours between 29 and 15, who were close in age, overshadowed *naalaqtuq* directives, thus suppressing the emergence of strong individual leadership. Alternatively, many of the traditional roles and functions of aboriginal leadership may have become obsolete by 1960.

The most striking feature of Nunaata's social composition during the early 1960s is the presence of two village endogamous marriages (Figure 23). After

# Cumberland Sound Inuit Kin and Local Groups, 1920–1970

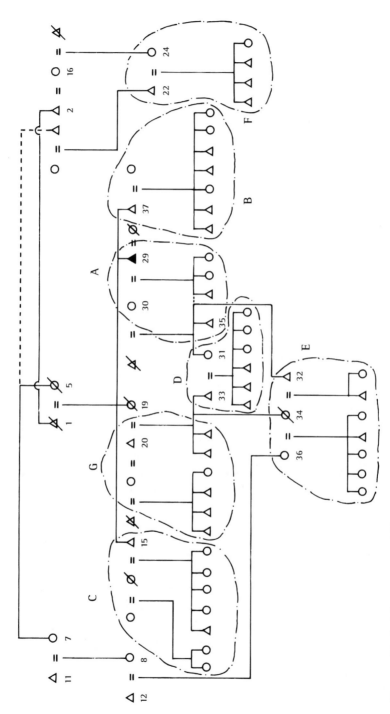

Figure 23. Social composition at Nunaata during the early 1960s.

29's first wife died, he took a second wife (30) with several children. The two eldest of these (31 and 32) married 19's children (33 and 34). Both 31 and 32 appear to have lived in 29's household for at least a decade, and considered 29 as an adopted father (*ataatasaq*), though 32 took his natural father's name (Veevee) as his own. Guemple (1979) has clearly demonstrated that, among most Central Inuit, the age of a child at adoption determines directly the extent of affectional bonding and the use of kinship terms between adoptive parents and children. As both *ungayuq* and *naalaqtuq* directives are more strongly developed the younger a child is brought into the household, the preferred arrangement among most Central Inuit is to adopt children at birth. Thus, while 29 called 31 *paniksaq* and 32 *irniqsaq*, he referred to 35 as his *irniq* since the latter was only an infant when he remarried. While the marriages in question were arranged by 19 and 30, the death of the former may have helped eventually to sanction these unions in the eyes of the community. Even though my principal informant, 29, felt that such close marriages were a violation of traditional customs—he observed that parent-child relationships were dominated more by respect and obedience behaviours in the past—he was powerless to prevent these unions from taking place. Apparently, some relatives thought that, because 31 and 32 were not blood-related to 33 and 34, it was permissible for them to marry. As will be noted throughout this chapter, the statuses of adopted and foster children appear to be considerably more negotiable than those of actual blood relatives.

While kin endogamy may be obviated in this case, local group endogamy is not. However, throughout the late 19th and early 20th centuries, local group endogamy may have become less a violation of marriage customs than an accepted fact of life. In other words, with much of the population of the Sound concentrated at Kekerten and Umanaqjuaq, locality may have become relatively unimportant as a criterion of marriage. After the death of 34, 32 married 34's second cousin (36). This was not a village endogamous union, nor, for reasons just cited, was it regarded as a kin endogamous marriage. Patrilocality appears to be particularly well-pronounced at Nunaata during the early 1960s as all seven married women live in their husbands' camps, although households D and E might also be interpreted as ambilocal or bilocal arrangements.

Once again, a positive, albeit weak, correlation between social distance and the placement of households was apparent in this occupation. Individual 29's *qammaq* was flanked by those of his younger (37) and older (15) brothers. But, also located adjacent to 29's dwelling were the houses of more distant relations, including his second wife's son and daughter, and his deceased sister's children, or *uyuruk*. On the periphery of the site 29's former *ningauk* (20), took up residence.

Nunaata was one of the first settlements to be abandoned after the events of 1963. Given the proximity of this camp to Pangnirtung and the wholesale

FIGURE 24. Angmarlik (left) and Jim Kilabuk (right) standing in front of the last bowhead taken by Angmarlik, Kingua Fiord, August 1945 (HBCA W-46, 80/124).

adoption of snowmobiles, and the nature of the relationship between 29 and 15, especially the lack of strong leadership directives, Nunaata was never reoccupied.

### Idlungajung

Idlungajung, or Bon Accord Harbour, was occupied intermittently throughout the commercial whaling era, and probably much earlier; Thule Inuit houses are said to have once been plentiful here.[7] In fact, Idlungajung and the traditional village of Noodlook (M'Donald 1841: 89) may be one and the same. Together with the nearby village of Anarnitung, which is located across the harbour from Idlungajung, Bon Accord Harbour formed the principal seat of settlement in the upper half of the Sound prior to contact. However, Idlungajung remained sporadically occupied until about 1917, when a small group of Qikirtarmiut under the direction of Angmarlik took up residence here. Over the next half dozen years Angmarlik moved often between Idlungajung and Kekerten, where he conducted whaling and trading campaigns for the Cumberland Gulf Trading Company. With the purchase of Kinnes' holdings by the HBC in 1923, Angmarlik moved to Idlungajung permanently, the camp of his birth and formative years (Figure 24).

Idlungajung and Nunaata are situated much further away from the *sina* than other contact-traditional camps in Cumberland Sound (Figure 18). However, strong tidal rips, which keep small bodies of open water ice-free year round, are found near each site. These *sarbut* are favourite hunting areas as they attract ringed seal and other wildlife throughout much of the winter.

A large, well known *sarbuk* occurs in a group of small islands just southwest of Idlungajung at the mouth of Kangiloo Fiord and may have been a major factor in the location of this camp. Ringed and harp seals also congregate in the same location during the summer and fall (Haller et al. 1966). Bon Accord Harbour, which forms part of the strait between Idlungajung and Anarnitung Island, is itself a *sarbuk* which funnels ringed seals through its waters during winter tidal changes. In earlier times, bowhead whales were apparently common in the waters around Bon Accord, and the people of Anarnitung were known as expert whalers. In 1840 Penny found several whale carcasses on a nearby beach, one of them only ten days old. Two small islands in the harbour also provide excellent vantage points for sighting beluga whales during their annual migration to the head of Kingua Fiord. Today, Idlungajung remains a favourite summer camp for seal and beluga hunters.

## *1922–1924*

During the early 1920s Idlungajung was composed of a single kin group and several related families centred around Angmarlik.[8] Angmarlik's hunting prowess, material wealth, and leadership capabilities provided the basis for this aggregation as he was regarded as the most influential man and best hunter in Cumberland Sound. He owned two boats, and normally filled both with game when he went hunting. He also possessed many other items of foreign manufacture including the largest alarm clock collection in the Arctic (M. Haycock, personal communication, 1986)—a fact which apparently awed many Inuit as it suggested that he could control time (Isha Papatsie, personal communication, 1989). While Angmarlik was the 'wealthiest native in the district', he was apparently in 'no way spoiled by his contact with the white race'.[9] Under Angmarlik's leadership, Idlungajung was always known as the most prosperous camp in the Sound.

Although Angmarlik's (1) *ningaugiik* relationship with 7 (Eevic), who was Keenainak's youngest brother, contributed to group solidarity, Angmarlik's relationships with his younger brother (2) and sister (3) appear to have been the major unifying ingredient in this aggregation (Figure 25). Angmarlik also attracted another sister's son (4) and daughter (5) and their families, while his oldest daughter, 6 (Qatsu), resided permanently at Idlungajung with Eevic (7). Eevic, in turn, attracted his older brother (8) from Nunaata.

Four of the seven married couples at Idlungajung demonstrate patrilocal tendencies, since these men live in the same camp as other blood relatives while their wives don't. Still, matrilocality appears common. Individual 9 (Ooneasagaq) lives with Angmarlik's sister (3), while 7 and 10 reside, respectively, with Angmarlik's daughter (6) and niece (5). Some of these marriages, however, may represent bride-service. Eevic (7), in particular, appears to have been performing an extended period of bride-service for Angmarlik, a duty he had been performing since he left Kekerten around 1918. Although Eevic could have

Cumberland Sound Inuit Kin and Local Groups, 1920–1970    157

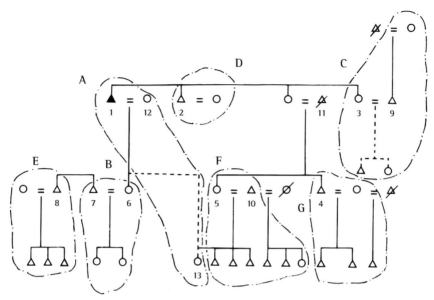

FIGURE 25. Social composition of Idlungajung during the early 1920s.

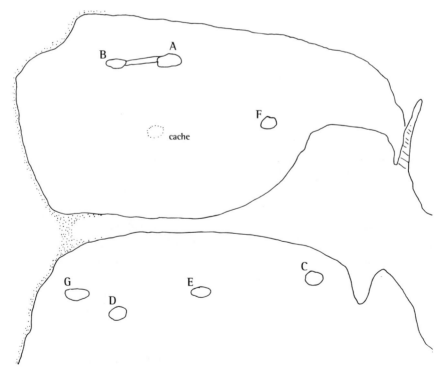

FIGURE 26. Plan of Idlungajung, 1922–1924. Redrawn from original by Pauloosie Angmarlik.

rejoined his older brothers at Nunaata, where he may have benefited socially and economically from his relationship with his eldest brother, Keenainak, he chose to remain at Idlungajung. Eevic's decision to reside permanently at Idlungajung was undoubtedly influenced by (a) his relatively junior position among his own kin group, (b) Angmarlik's considerable material, social, economic, and political advantage, (c) the fact that Angmarlik had no sons, and (d) the very close bonds of affection that existed between 6 (Qatsu) and her parents—Qatsu was Angmarlik and Ashivak's (12) only natural daughter.

In lieu of the death of 11, 10 may have also been fulfilling a period of bride-service for Angmarlik, his wife's *angak*. Similarly, Ooneasagaq (9) may have been engaged in bride-service for Angmarlik, his brother-in-law or *sakiaq*, as Angmarlik's father had passed away. Angmarlik's prowess notwithstanding, Ooneasagaq moved to Tesseralik over the winter of 1924–25 where he headed his own small extended family unit. Ooneasagaq was an accomplished hunter and whaler, as was 4 (Aksayuk). Although Angmarlik and Aksayuk hunted and travelled extensively together throughout the 1920s, the latter being considered Angmarlik's 'helper', Aksayuk appears to have moved to Koangoon in Nettilling Fiord sometime prior to 1927. The HBC encouraged both Ooneasagaq and Aksayuk to relocate to better fox trapping areas. However, the fact that both men remained in subordinate social positions to Angmarlik at Idlungajung for as long as they did is mute testament to Angmarlik's substance and leadership.

Adoption appears to have been a common practice at Idlungajung, especially by the wealthiest families, i.e., those headed by Angmarlik and Ooneasagaq. Although it is not known from which families Ooneasagaq and his wife obtained their adopted son and daughter, equal representation of each sex appears to be a desired goal of most childless couples in Cumberland Sound. Indeed, as Ashivak (12) gave birth to only one daughter (Qatsu), she and Angmarlik adopted another daughter (13) around 1918, and two sons in the mid-1920s. However, only the former (13) appears to have been related to Angmarlik as she was his sister's *inngutaq*, and thus terminologically his grandchild as well. It is interesting that the decision to give 13 up for adoption was a unilateral one made not by 5, but her husband (10), suggesting that this adoption may have served to solidify his relationship with Angmarlik.

Idlungajung's inhabitants were separated physically by a tidal basin, which provided excellent shelter for whaleboats, but divided the community at high tide (Figure 26). Whereas Angmarlik lived on one side of the camp, Ooneasagaq settled on the other. Angmarlik inhabited a large rectangular wood dwelling with two large, facing rooms joined by a common passage/entry. Whereas Angmarlik, his wife, and adopted daughter lived in one room (A), his daughter (6) and Eevic (7), and their children occupied the other (B). Eevic's relationship with Angmarlik apparently went beyond that of a normal *ningaugiik* tie to one that resembled an *irniriik* bond. Indeed, my informants, as well as historic records,[10] indicate that Angmarlik and Eevic travelled extensively together during the 1920s. By way of contrast, Angmarlik's brother (2) and Eevic's brother (8), respectively, lived in houses (D and E) on the opposite side of the site.

*Nekaishutu* was the most common form of food sharing at Idlungajung and, not surprisingly, Angmarlik controlled the allocation and distribution of meat and blubber. While Angmarlik also engaged in *piutuq*, he never favoured kin over non-kin nor differentiated between the wealthy and the poor. Everyone was treated fairly. Unlike most camps in Cumberland Sound, there was only one large food cache at Idlungajung, which was owned and controlled by Angmarlik. This cache of a dozen or more wooden casks and iron rendering vats, was located due east of Angmarlik's dwelling. As time passed, these containers were replaced by a large wood shack. While Angmarlik was either directly or indirectly involved in obtaining the bulk of the meat and blubber for the community, hunters who obtained their own game deposited it in Angmarlik's central store, regardless of the season. Only when Angmarlik's influence and productivity begin to decline in the late 1940s did families begin to manage their own individual food caches.

As noted above, Angmarlik owned two whaleboats, one of which was equipped with a motor by 1928. While Angmarlik operated the latter craft, he normally selected Aksayuk (4) and a small crew of hunters to man his second boat, which he would tow to various hunting grounds. Eevic also owned a whaleboat during the 1920s, which he obtained from Angmarlik in exchange for hauling blubber from Idlungajung to Kekerten during the late 1910s. The selection of Idlungajung's whaleboat crews seems to have been based more on availability than anything else, and as local group membership changed so too did crew membership. It should be observed that the *aggutik* of the summer hunt was always the eldest in the crew, not necessarily the one who owned the boat. As at Nunaata, whaleboats provided the basis of co-operative hunting partnerships during the summer. Yet, while the heads of most families preferred to hunt with their sons during the winter, some also occasionally hunted with Angmarlik, especially the less productive hunters. As a matter of note, most of the less accomplished hunters in the Sound normally hunted seal one day at a time during the winter. Lacking ammunition, these hunters were usually relegated to hunting seals at breathing-holes. Alternatively, the more prosperous hunters, possessing larger dog teams, better *qamutiit*, more ammunition, better rifles, etc., often hunted at the *sina* and various *sarbut* for several days before returning home. This created a significant imbalance in mobility and hunting success, not to mention social status—a fact born out by the Pangnirtung Post Journals, whereby numerous entries record the constant arrivals and departures of prominent Inuit for trading purposes.[11]

*1934–1936*

The basis of residential solidarity at Idlungajung does not appear to have changed substantially during the 1920s. However, by 1930, the departures or deaths of Angmarlik's younger brother (2), sister (3), and brother-in-law (9), as well as the adoption of two boys (14 and 15) changed the structural foundation of this camp from one based on sibling ties to one founded on bonds between Angmarlik and his children (Figure 27).

Other changes in local group composition and physical site structure also occurred. Like Ooneasagaq (9), Aksayuk (4) left to head a small settlement of his own, while Eevic (7) moved into Ooneasagaq's vacant *qammaq* opposite Angmarlik (Figure 28). Although Aksayuk and Ooneasagaq settled elsewhere, they still maintained co-operative economic relationships with Angmarlik. For example, Angmarlik appears to have travelled often to and from Aksayuk's camp in Nettilling Fiord during the late 1920s.[12] The individuals who gave 14 and 15 up for adoption to Angmarlik (Kopalee and Kookootok, respectively) were prominent *aggutiit* in the bowhead fishery. Upon the demise of commercial whaling in the Sound, they moved to the Pangnirtung and the Saumia areas.

Too many dogs on the west side of the camp has been cited as the reason for Eevic's shift in residences. However, by the mid-1920s Eevic had begun to accumulate considerable wealth and status, and was beginning to head a family of his own. Given these factors, Eevic's move may have been less a logistical exercise to relieve congestion, than a symbolic attempt to break away from Angmarlik's dominance and to establish his own support group. Alternatively, Angmarlik's adopted sons were approaching the age where they could hunt and travel with their adopted father. Indeed, the Pangnirtung Post Journals record that Angmarlik almost always travelled alone with his two sons after 1927; rarely was he accompanied by Eevic after this date, except for the annual beluga whale drive.[13]

By the mid-1930s Idlungajung's prosperity attracted several families with which Angmarlik appears to have had no apparent kinship connection (Figure 27). Individual 16 may have been included in Angmarlik's terminological network; he was Angmarlik's sister's step-son (*uyuruksaq*?). However, their relationship was probably not very close as 16 never lived in Angmarlik's sister's household. The male heads of two of the three remaining families (17 and 18) form a sub-*ilagit* integrated on the basis of *irniriik* directives. The third family, headed by 19, has no kin relations at Idlungajung. During the 1920s and 1930s this latter family resided at at least three different camps in the Sound including, in order of occupation, Ussualung, Idlungajung, and Kingmiksoo. Individual 19 apparently had few immediate relations in the Sound other than his maternal uncle (*angak*), Tooloogakjuaq, the well-known lay-preacher and hunter from Kingmiksoo. As might be predicted, 19 eventually moved to Kingmiksoo where he remained longer than at any other camp in the Sound.

# Cumberland Sound Inuit Kin and Local Groups, 1920–1970 161

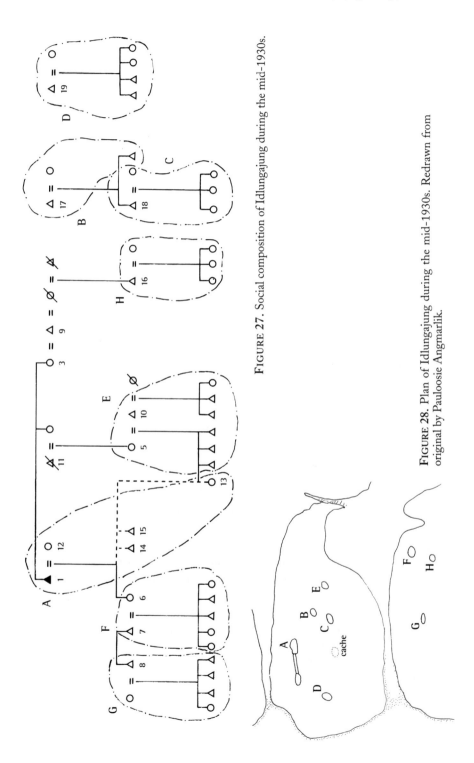

FIGURE 27. Social composition of Idlungajung during the mid-1930s.

FIGURE 28. Plan of Idlungajung during the mid-1930s. Redrawn from original by Pauloosie Angmarlik.

Contrary to expectations, geographical distance and social distance as determined by kinship, do not exhibit a strong correlation at Idlungajung during the 1930s beyond the level of *irniriik*-like relationships (Figure 28). The nearest detached households to that of Angmarlik's (A) are occupied by three families that trace no apparent kinship linkages to Angmarlik (i.e., B, C, and D). The location of household D is particularly difficult to explain as its inhabitants were considered 'outsiders'. It is possible that 19 was exercising some distant kin or other social tie with household A; both 19's *angak* and Angmarlik's wife (12) were originally from Umanaqjuaq. Even though Angmarlik adopted a child from his niece (5), the location of her household (E) is somewhat curious given Angmarlik's closer kinship connection to his daughter (6) on the opposite side of the tidal basin. That families most distant to Angmarlik in terms of the kinship system would reside nearest to him, while those with closer kinship ties would locate their dwellings on the opposite side of the settlement is, at first glance, puzzling.

It is possible that Angmarlik emphasized *naalaqtuq* directives at the expense of *ungayuq* behaviours in his relationships with his siblings and children. More plausibly, however, is that Angmarlik's material wealth and that of his closest kin may have simply required the construction of *qammat* on both sides of the basin. In other words, there was not enough room for everyone to settle permanently on one side of the camp. In particular, Angmarlik's dogs, and to a lesser extent Eevic's, may have discouraged the construction of too many permanent dwellings on Angmarlik's side of the settlement. The likelihood that Angmarlik needed considerably more space than the average hunter for his dogs, belongings, and activities is apparent in statements made by my informants.[14] Finally, Eevic's growing family, not to mention social status, like those of Ooneasagaq's before him, may have simply motivated him to move to a different part of the camp.

Families that did locate beside Angmarlik's may have done so with an explicit understanding that it was on a temporary basis—a possibility supported by the transient nature of the inhabitants of households B and D. The occupants of C also shifted residences frequently as they lived at Ussualung, Sauniqtuajuq, and Iqalulik during the 1920s before settling in Kingnait Fiord. A similar situation might have occurred on Eevic's side of the camp. Here, 16 has placed his house (H) beside Eevic's *qammaq* (F). In fact, this dwelling is situated considerably closer to F, than is Eevic's to his brother's household (G). Such alliances may have been accompanied by specific understandings and agreements whereby immigrant families served in the capacity of indentured servants providing labour for resident families in exchange for certain material benefits—a scenario strengthened by the fact that there were more whaleboats at Idlungajung during the 1930s than there were people to fill them.

## 1945–1947

A virulent strain of flu claimed as many as ten lives at Idlungajung, most of them elderly, over the winter of 1941–42. Even so, by the mid-1940s Idlungajung had replaced Kingmiksoo as the largest settlement in the Sound. And while Eevic's extended family expanded in size and influence throughout the late 1930s and early 1940s, residential solidarity at Idlungajung was still largely based on Angmarlik's ties with his now married children. Only after the late 1940s when Angmarlik's productivity declined did his influence begin to wane.

Figure 29 illustrates the social composition of Idlungajung during the mid-1940s. This aggregation is considerably more complex than those which characterized the settlement during previous decades. Beginning at the top generation only one of Angmarlik's siblings, his older widowed sister 20 (Kowna, the former *angajuqqaq* of Kivitoo) appears to have resided at Idlungajung. While Angmarlik and Ashivak's (12) children no longer live with them, they have adopted one *inngutaq* from each of their four children (21, 22, 23, and 24). As this situation represents the highest incidence of grandchild adoption recorded among any Inuit family in Cumberland Sound, it warrants further examination.

Guemple (1979) identified two basic types of Inuit adoption. One involves adoption as a form of exchange based on the reciprocation of gifts for children, which is more symbolic or social rather than economically motivated. The other appears to be a kind of 'caretakership' between relatives where, if gifts are given, they are for the support of the child. Adoptions of the latter variety often form the basis of alliance between families even if they are close kinsmen (Guemple 1979: 33).

However, when Angmarlik and Ashivak adopted 14 and 15, it was not from immediate relatives, but from genealogically and/or geographically distant individuals. Although the natural parents of 14 and 15 may have lived at Kekerten when Angmarlik contracted their adoptions, they soon moved, respectively, to Pangnirtung and the Saumia area. The pattern of adoption prevalent in Angmarlik's household during the 1940s differs from that in the 1920s. While only one of this household's three early adoptions could be considered as an exchange between kinsmen of different generations, all four adoptions in the 1940s were intergenerational or grandparental adoptions. The giving of a grandchild to Angmarlik and Ashivak by each of their children undoubtedly strengthened bonds of affectional solidarity between parent and child, while simultaneously acknowledging the *naalaqtuq* directives implicit within their relationship. In this regard, it is noteworthy that Guemple (1979) has observed that children who are adopted often give to their parents in their later years a child out of a sense of respect, duty, and allegiance.

All four of Angmarlik and Ashivak's children have remained at Idlungajung with their spouses, giving a bilateral appearance to the aggregation. Angmarlik's oldest adopted son, Kingu (15), has attracted his older natural brother

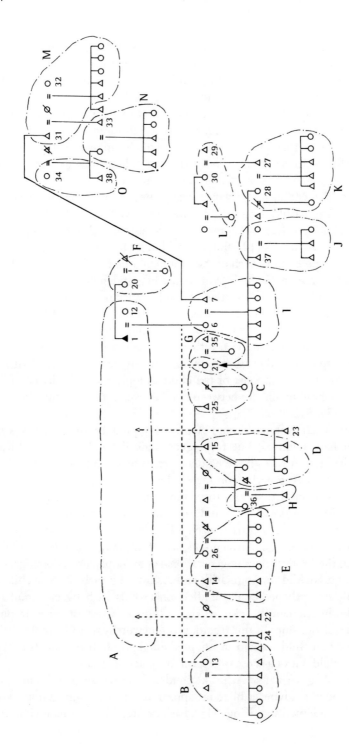

FIGURE 29. Social composition of Idlungajung during the mid-1940s.

Cumberland Sound Inuit Kin and Local Groups, 1920–1970 165

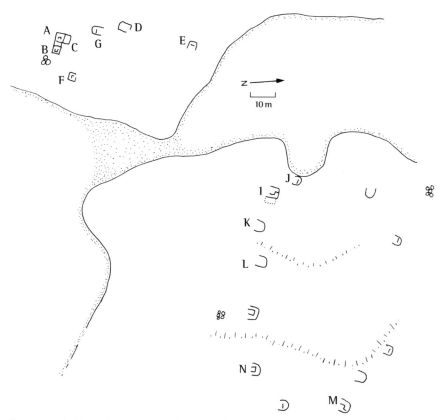

FIGURE 30. Plan of Idlungajung during the mid-1940s. Sketched by author. Unidentified *qammat* not occupied during this occupation.

(25) and younger sister (26), both of whom have married into the group. While 25 married Angmarlik's adopted daughter (21), 26 married Angmarlik's youngest adopted son (14) soon after the death of her second husband (Mike from Nunaata). The former marriage does not represent a kin endogamous union, as, even though 25 was Kingu's brother, he was not considered Angmarlik's relation. It is not known what gifts, if any, were given by Angmarlik to Kingu's natural parents in his adoption. Nevertheless, what appears to have been implicit in this arrangement is the option for Kingu's siblings to reside in the camp of the most productive and influential hunter in the Sound. Individual 25 may have also been performing a period of bride-service, as he resided at Idlungajung with 21 in Angmarlik's storage shack for a year before

he left for Kekerten, whereupon 21 remarried. Individual 25's bride-service was transferred to Angmarlik as a particularly close bond of affection existed between her and her grandparents. Besides, another man (27) was apparently already performing bride-service for Eevic. All of Eevic and Qatsu's married children have remained at Idlungajung, again giving a bilateral appearance to the next descending generation, with their eldest daughter (28) attracting her *sakkiik*, 29 and 30. Similarly, Eevic has attracted his older brother (31) and his wife (32) (i.e., 2 and 14 from Nunaata). A son from 31's first marriage (33), in turn, resides at Idlungajung along with his wife's widowed mother (34).

Of the 11 recorded couples at Idlungajung during the mid-1940s, six exhibit patrilocal tendencies, three demonstrate matrilocal arrangements, while another two are bilocal, insofar as one or both parents of each married individual were co-resident at Idlungajung. Although 27 and 33 live in the same camp as their fathers (29 and 31), the latter appear to have been exploiting the benefits, respectively, of their son's and brother's matrilocal living arrangements. At least one case of matrilocality in this group represents a case of bride-service.

In contrast to the 1920s, there appears to be a strong correlation between kin relatedness and the spatial arrangement of houses during the mid-1940s (Figure 30). The division of the site into areas occupied by kin groups headed by Angmarlik and Eevic is evident. No longer do unrelated families reside beside Angmarlik (A); his married children and grandchildren now occupy these dwellings and presumably economic roles. Angmarlik's daughter (13) and her husband live in Angmarlik's *qammaq* in household B, while his adopted granddaughter (21) and her first husband (25) briefly inhabited his storage shed (C). Angmarlik's two sons (14 and 15) and older sister (20) also lived nearby in households D, E, and F, respectively. However, the closest dwelling (G) to Angmarlik's during the 1940s belonged to 21 and her second husband (35). The remaining house on this side of the settlement was inhabited by Angmarlik's adopted son's wife's sister (36) and her son. In agreement with 36's relatively distant kinship connection to Angmarlik, her *qammaq* (H) was located further inland at a different elevation.

On the east side of the camp Eevic and Qatsu's *qammaq* (I) is flanked by those (J and K) of their eldest son (37) and daughter (28), respectively. Adjacent to the latter household is the house of 28's husband's parents (L). Located at a higher elevation some distance away is a more dispersed group of three dwellings (M, N, and O) occupied, respectively, by Eevic's older brother (31), the latter's eldest son (33), and 33's wife's mother (34) and brother (38). The distance between Eevic's household and that of his brother's betrays the closeness one might expect male siblings to exhibit in a camp where both were *ningaut*. However, a fundamental conflict between kinship and economic statuses may have contributed to this situation. While Eevic was subordinate to 31 in terms of the kinship system, he was superior to the latter in the economic system. In fact, whereas 31 never owned a boat, Eevic apparently owned two whaleboats, one motorized, which he owned and operated with his eldest son (37). Not surprisingly, by 1950, 31 had moved back to Nunaata. Although 31 may have had access to one of Eevic's boats, they apparently never formed a co-operative hunting partnership. Rather, until their sons were old enough, Eevic hunted with Angmarlik or his oldest daughter (28), while, as will be recalled, 31 hunted alone.

If one considers the spatial arrangement of dwellings as well as the structure of hunting partnerships at both Idlungajung and Nunaata an emphasis on parent-child ties or vertical bonds at the expense of sibling or horizontal bonds is apparent. Only during the last occupation of Nunaata do *nukariik* ties appear to have approached the stability of previous *irniriik* relationships. Perhaps not coincidentally, leadership was weakly developed and multiple ties between key individuals were not uncommon.

Around 1950 Angmarlik's health deteriorated and he moved to Pangnirtung. Three years later he died. While 14 assumed the position of *angajuqqaq*, his role seems to have been more symbolic than anything else as Eevic (7) was now superior in both the social and economic spheres—by 1950, Eevic's kin group had surpassed that of Angmarlik's in size. However, Eevic too soon died. Idlungajung was occupied principally by Eevic's and Angmarlik's children and grandchildren prior to being abandoned permanently in 1966.

## Avatuktoo

The settlement of Avatuktoo is located at the head of a small inlet 15 km east of Ussualung on the northeast side of the Sound (Figure 18). Although a late prehistoric/early historic Inuit site is found nearby, the first recorded occupation of Avatuktoo during this century occurred in the late 1920s when a small group of Inuit settled here. The availability of ringed seals, which were plentiful in the inlet year-round, particularly the fall, was the initial reason for the selection of the site. However, the number of resident seals during the winter was soon depleted, and the inhabitants of Avatuktoo were forced to live on food reserves built up during the fall and on Arctic char, which inhabit a series of small lakes behind the site. This latter resource, good fall sealing, and Avatuktoo's proximity to Pangnirtung were apparently the major reasons why people chose to remain at this camp.[15]

### *1928–1932*

Avatuktoo appears to have been occupied briefly sometime during the last half of the 1920s by two families, one headed by individual 10 (Aulaqeak) from Idlungajung. These men were married to two sisters whose close relationship provided the basis of residential solidarity at Avatuktoo during this period. Agalik (1) was generally regarded to be the leader of the camp, despite the fact that he had no boat and was married to the younger of the two sisters (3), the sister of the trader's (William Duval) wife at Ussualung. Although it is uncertain whether 4 was similarly recognized, she was thought to have been related distantly to Agalik. The union of 1 and 3 was probably not regarded as kin endogamy, however. This couple had a son and a daughter, while 2 and 4 had at least one son—the number, gender, and names of other children were not remembered. The latter family apparently lived here only intermittently during the 1920s and 1930s as it is known to have also lived at Idlungajung and Kivitoo. Sometime around 1930, these two families were joined by two others when Nukinga (6), a son of Aulaqeak and individual 12 from Nunaata, accompanied by his wife's parents (7 and 11 from Nunaata), settled here.

Interestingly, the arrangement of dwellings at Avatuktoo during the early 1930s does not reflect the degree of closeness expected by kinship connections among this group. While the households of 1 and 2 are located on the same bench at the west end of the site, with 5's (Sukulak) *qammaq* situated nearby, Nukinga (6) built his house at the opposite end of the camp. What prompted Nukinga to locate his household so far from those of his father (2) and father-in-law (5) is not known. Nukinga's period of bride-service may have been over and he may have had the only food cache (and boat?) at Avatuktoo,[16] but this does not explain the spatial, and presumably social, distance between him and his father, uncle, and father-in-law. Perhaps, Aulaqeak's (2) close relationship with Agalik (1) may have undermined the reaffirmation and maintenance of *ungayuq* behaviours between Nukinga and his father. Whatever the case, within a few years 1 and 2 left Avatuktoo, leaving Nukinga behind with his wife, 7 (Soudlu), and family. This move changed the structural basis of residential solidarity from one based on female sibling ties to one founded on parent-child ties.

## 1941–1943

After 1 and 2 left Avatuktoo, 5 (Sukulak) assumed the role of camp leader, even though 6 (Nukinga) owned the only whaleboat here. At the same time, 5's only son (8) and two daughters (9 and 10) had taken spouses (11 and 12) and were living at Avatuktoo (Figure 31). While the latter individuals initially may have been performing bride-service for 5 and 13, both chose to stay at Avatuktoo permanently, as 11's parents lived across the Sound at Kipisa, and 12's parents were 'nomads' with few other relations in the Sound.

Sukulak (5) and his wife (13) occupied the same house (C) they built around 1930, while their children (8, 9, and 10) resided near them in households A, E, and B, respectively (Figure 32). The spatial separation noted previously between 5 and 6 has been maintained; the latter remained in *qammaq* D throughout the early 1940s until his mysterious suicide in 1943.[17]

## 1958–1960

After Sukulak's (5) death around 1950, his son, 8 (Angutitaluk), took over the position of *angajuqqaq*. At about the same time, Sukulak's daughters (9 and

10) and their families left Avatuktoo for other camps. Throughout the early 1950s only two families headed by 7 and 8 appear to have resided at Avatuktoo. However, several deaths and marriages occurred during this decade that once again changed the social composition and structure of the camp. Most notably, Angutitaluk (8) died sometime during the early 1950s, and Soudlu (7) assumed the role of *angajuqqaq* (Figure 31).

Female leaders, while rare, were not unknown in Oqomiut society (see previous chapter), and the emergence of a matriarch at Avatuktoo may be the consequence of a number of factors. During the previous decade, three of Nukinga (6) and Soudlu's (7) daughters (14, 15, and 16) took husbands (17, 18, and 19). Although the latter individuals may have moved to Avatuktoo initially to perform bride-service, each, like the preceding generation, decided to settle permanently here. By the late 1950s all three couples were raising large families of their own. Yet, since 17, 18, and 19 were in-marrying males, they were subordinate to Soudlu (7). As Soudlu had been raised as a hunter—she was her 'father's assistant'—and as she had lived at Avatuktoo for the last two decades, she assumed the role of leadership and decision-making for the camp. Even though 18 owned his own motorized whaleboat, which he obtained from his father (Etuangat Aksayuk) in Pangnirtung, Soudlu held the position of camp leader until she was too old to carry out effectively the duties of this office (including *nekaishutu*), at which time 17 and 18 assumed joint leadership of the camp.

Individuals 17, 18, and 19 might have enjoyed greater social status elsewhere. Yet, they remained in subordinate positions to Soudlu (7). While Avatuktoo's proximity to Pangnirtung might have encouraged 17, 18, and 19 to stay at Avatuktoo, their options were limited. Individual 18's father was employed by the mission hospital in Pangnirtung—a poorer hunting area than Avatuktoo's. Alternatively, 19 was the youngest son of a relatively destitute family in Padloping Island. Interestingly, 19's father (20) soon joined him at Avatuktoo. Finally, 17 was Veevee's adopted grandson from Naujeakviq. With Veevee's death in the late 1950s, the latter apparently had no other alternative but to remain at Avatuktoo, as his father, Veevee's youngest son, was apparently not an accomplished hunter—a consideration which may have influenced the latter's decision to give his two sons in adoption to Veevee (see below). With the exception of 18, my informants pointed out that in-marrying males of the last two generations at Avatuktoo were generally younger and poorer-off than their siblings in other camps, and they remained at Avatuktoo because they thought they could make a better life for themselves than in their camps of origin.

The possibility that Inuit kinship directives have the potential to create, under certain circumstances, disenfranchised individuals seems apparent as birth order appears to be a major factor in determining the hierarchy of social and material advantage within the family. Personal attributes aside, elder sons generally fare better than younger sons because both *ungayuq* and *naalaqtuq*

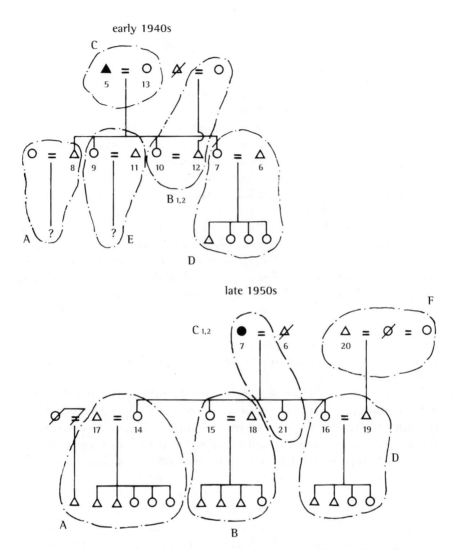

FIGURE 31. Social composition of Avatuktoo during the early 1940s and late 1950s.

directives within *irniriik* relationships are stronger and better developed. Additional sons usually do not have the opportunity to forge comparable bonds with their fathers. Although the Cumberland Sound Inuit have always had a high birth rate, the low survival rate of infants traditionally keep terminological-behaviourally disenfranchised individuals to a minimum. However, with the establishment of medical services at Pangnirtung during the late 1920s, infant survival rates appear to have increased. In turn, this may set the stage for the creation of more disenfranchised Inuit, and thus more anomalous residential arrangements.

Cumberland Sound Inuit Kin and Local Groups, 1920–1970    171

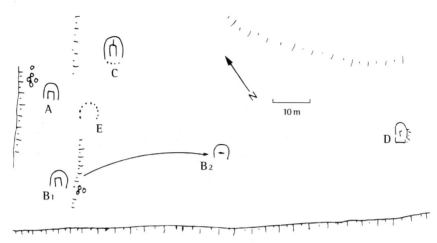

FIGURE 32. Plan of Avatuktoo during the early 1940s. Sketched by author.

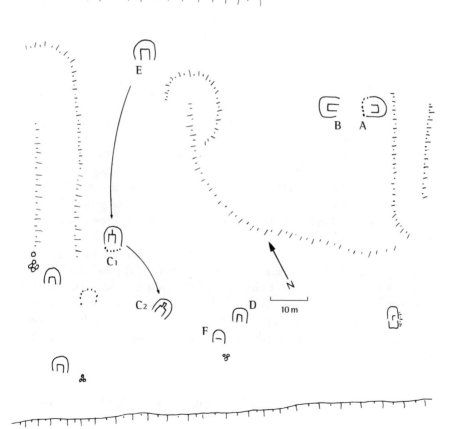

FIGURE 33. Plan of Avatuktoo during the late 1950s. Sketched by author.

*Panniriik* relationships formed the basis of residential solidarity at Avatuktoo throughout the 1950s and 1960s. However, the arrangement of dwellings suggests a greater degree of heterogeneity than otherwise expected (Figure 33). Soudlu's two eldest daughters (14 and 15) and their husbands live at the extreme northeast end of the site. Here, they built their houses (A and B) facing each other to create a common work space. While 17 and 18 appear to have had a close relationship, *nukariik* ties between their wives were clearly the *raison d'être* and stabilizing influence among this subgroup. Conversely, Soudlu and her remaining two daughters (16 and 21) lived at the opposite end of the camp in households C and D, where changes in residence were common, especially following the deaths of Sukulak and later his widow (Emanapik, 13). A small house foundation (F) beside 16 and 19's *qammaq* (D) may have belonged to the latter's father (20).

Avatuktoo was temporarily abandoned during the dog epidemic of 1961–62. The following year only the occupants of households A, B, and C returned. While Soudlu and her co-resident daughters had a new house constructed, a young couple moved to Avatuktoo and took up residence in Soudlu's original house (C1). Although the camp was abandoned permanently a year or two later, Avatuktoo remains a popular summer fishing site to this day.

**Tuapait**

The recent historic site of Tuapait is located in a small bay on the south shore of one of the most northerly islands in the Kikistan Island chain at the mouth of Kingnait Fiord (Figure 18). Two families of Qikirtarmiut moved permanently to Tuapait sometime around 1950. However, they were not the first; old sod houses were apparently once abundant at this location, though none remain today. A popular anchorage for wintering whalers during the late 1850s and early 1860s, Tuapait Harbour contains the remains of the *Hannibal*, an American whaler wrecked here in 1861. Though this wreck provided lead for the manufacture of bullets well into the twentieth century, ringed seal was the primary reason for the location of this settlement. The Kekerten area has always been known as an excellent year-round sealing ground. Throughout the winter, seals can be taken at *aglu* between Kekerten and the mainland, while in early spring denning sites are plentiful off the mouth of Kingnait Fiord. In addition, the *sina*, which often forms along the west coast of these islands, provides access to various species of marine mammals, including bowhead whales, as the name of the nearby island, Akviqsurapiq ('the place to look for whales'), suggests.[18]

*1954–1956*

The families that originally settled at Tuapait were headed by two unrelated individuals, Kisa (1) and Koonooloosie (2). Both appear to have moved to this location from Pangnirtung, where they had worked, respectively, for the

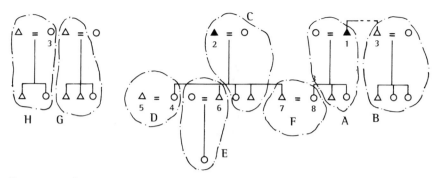

FIGURE 34. Social composition of Tuapait during the early 1960s.

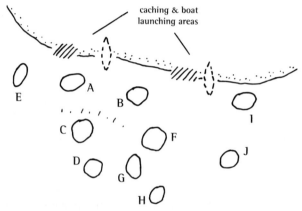

FIGURE 35. Plan of Tuapait during the early 1960s. Redrawn from original by Michael Kisa.

Anglican Church and the RCMP, although Koonooloosie also lived at Kekerten for a few years. Kisa also attracted his adoptive brother (3), while Koonooloosie's eldest daughter (4) lived at Tuapait with her husband (5), who may have been performing bride-service. Though Kisa was older than Koonooloosie, and possessed the only whaleboat, which was equipped with a motor, there was no overall leader at Tuapait; each was regarded as the *angajuqqaq* of his respective *ilagit*. There also appears to have been little sharing between these two kin groups. Each *ilagit* located its dwellings at different elevations, while each had its own cache and boat launching area.

## 1960–1962

Over the next half dozen years or so Tuapait expanded in size as Kisa and Koonooloosie's children began to marry (Figure 34). Although 7 and 8's union represents the first instance of local group endogamy and intermarriage between the two groups, the relative spatial autonomy of these groups was maintained (Figure 35). A number of apparently unrelated families also took up residence at Tuapait, including that headed by 9 (18 from Avatuktoo). Accordingly, the dwellings of these families (G and H) were located on the periphery of the site, as were houses intermittently occupied by Kisa and Koonooloosie's children residing in Pangnirtung (I and J).

Sometime before the mid-1960s Koonooloosie and his married children (4 and 6) moved to Broughton Island (Kekertakjuaq), while Kisa (1) and his adopted brother (3) passed away. Subsequently, 7 assumed the role of *angajuqqaq*. Tuapait remained one of the wealthiest and most productive seal hunting settlements in the Sound (see Haller 1967) until the winter of 1966–67, when it was abandoned permanently. Tuapait continues to this day to be an important summer sealing camp for Inuit living in Pangnirtung.

### Sauniqtuajuq

Sauniqtuajuq is located on the south end of a small island just off the northwest coast of Sauniq or Imigen Island (Figure 18). Sauniqtuajuq was originally settled in 1923–24 by Inuit from Kekerten and Umanaqjuaq. As one of the principal settlements of the Kinguamiut (Boas 1964: 27), however, Imigen Island has been occupied for hundreds of years. Although caribou and char are not abundant around Sauniqtuajuq in the summer, the *sina* is often located within several kilometres of Imigen during the winter. Numerous *sarbut* in the vicinity of the Drum Islands, 15 kilometres northeast of the settlement, provide alternative prospects for winter sealing. Also, ringed seal can be taken in the vicinity of Sauniqtuajuq throughout the winter at breathing holes and during the spring at denning sites. In addition to Sauniqtuajuq's favourable maritime hunting location—Imigen Island was known as one of the best winter hunting grounds in Cumberland Sound (Boas 1964: 27)—fox also played a role in the selection of this site. In 1923 the HBC provisioned two prominent hunters, Maniapik and Attaguyuk, to relocate here in order to trap fox and trade with the area's inhabitants. Maniapik, in fact, was given a whaleboat to move to Sauniqtuajuq. Attaguyuk, on the other hand, owned an old American whaleboat which he obtained at Kekerten during the early twentieth century. In exchange for a new boat, which apparently took the HBC five years to deliver, Attaguyuk agreed to move to Sauniqtuajuq. Unlike other camps established in the interest of the fur trade, and subsequently abandoned within a year or two, Sauniqtuajuq was occupied continuously for the next four decades. Sauniqtuajuq's favourable hunting location appears to have permitted an intensity and duration of occupation not found at other HBC-sponsored settlements.[19]

### 1923–1925

The most interesting feature about Sauniqtuajuq's composition during the early to mid-1920s is that it was composed predominantly of two different *ilagiit* (Figure 36), one headed by Attaguyuk (1), the other by Maniapik (2). It

Cumberland Sound Inuit Kin and Local Groups, 1920–1970    175

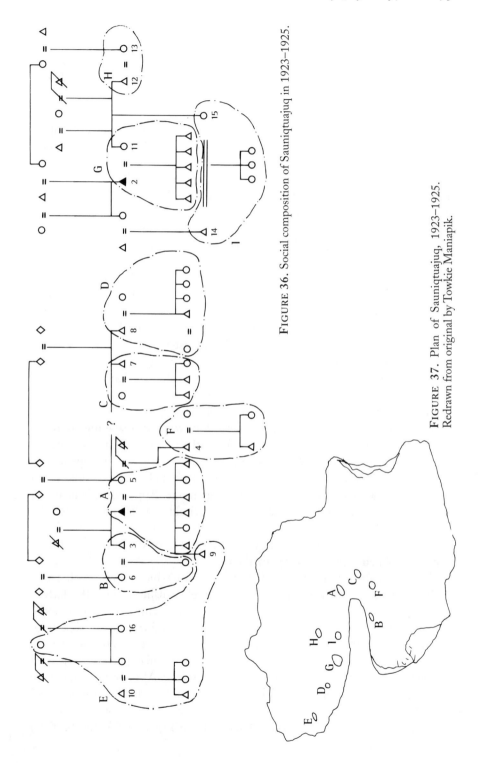

FIGURE 36. Social composition of Sauniqtuajuq in 1923–1925.

FIGURE 37. Plan of Sauniqtuajuq, 1923–1925. Redrawn from original by Towkie Maniapik.

is possible that these two groups were related in some way, but my informants could not agree how. Although there is some question as to Maniapik's place of origin, he had extensive kinship ties with people from Umanaqjuaq and was considered to have been an Umanaqjuarmiut. Attaguyuk, on the other hand, came from Kekerten where he was a prominent *aggutik* in the bowhead whale fishery.

Maniapik and Attaguyuk had a *mangnariik*, or rough joking relationship in which each tried to outdo the other. They called each other *avik*, which translates as 'to come apart', and were recognized to be *avikgiit*, 'those who have separated'. Accordingly, the houses of their kinsmen were located in different areas of the site (Figure 37). In fact, Maniapik's and Attaguyuk's kinsmen originally settled on opposite sides of a small ravine, where each group placed its *qammat* facing the other. Also living at Sauniqtuajuq during this time was a local family that appears to have been unrelated to either Maniapik or Attaguyuk.

In terms of size, the dominant *ilagit* at Sauniqtuajuq during the early to mid-1920s was headed by Attaguyuk (1). *Nukariik* and *irniriik* relationships between Attaguyuk and his older brother (3) and step-son (4), respectively, appear to have been the main unifying ingredients in this aggregation, although the fact that Attaguyuk's wife (5) and sister-in-law (6) were cousins also undoubtedly added an element of stability to this aggregation (Figure 36). Attaguyuk's ties to 3 and 4 suffered from certain inherent deficiencies, however. Whereas 3 (Koseaq) was superior to Attaguyuk in the terminological-behavioural system, he was subordinate to his younger brother in both the socioeconomic and political spheres, a fact which did not escape the attention of the authorities.[20] That they originally placed their *qammat* (A and B) directly across the ravine from each other suggests a certain degree of independence between these households (Figure 37). Alternatively, individual 4 (Nukeeruaq) was Attaguyuk's wife's son from a previous marriage. Apparently, Nukeeruaq lived for only a short time in Attaguyuk's household as a dependent before he married. Nevertheless, Attaguyuk appears to have maintained an *irniriik*-like relationship with Nukeeruaq as they travelled and hunted together until their sons were old enough to accompany them.

Attaguyuk also attracted two other families with which he was related affinally. These included the families of his wife's cousin (7) and the latter's brother (8). The exact nature of 5, 7, and 8's relationships are sketchy as the primary kinship ties between their parents, who originated from Kekerten, could not be recalled by my informants. It is clear that individuals 7 and 8 were natural brothers. However, while 5 (Makitoq) and 8 (Tujarapik) were regarded as cousins, 7 (Akatoogaq) was considered 5's brother. Although 5 and 6 were apparently cousins, their parents' relationships could not be determined. As 7 was adopted into 3 and 6's household prior to the early-1920s from 6's cousin or sibling, he would appear to be both Attaguyuk's *sakiaqsaq* (wife's step-brother) and *qaniaksaq* (older brother's adopted son). While Akatoogaq (7) and Tujarapik (8) formed a sub-*ilagit* integrated on the basis of *nukariik* directives, the bond between them could not have been very strong as they built their houses, C and D, respectively, at opposite ends of the camp (Figure 37). By 1927 Tujarapik (8) had moved away to Pangnirtung.

The adoption of Attaguyuk's eldest son (9) into the household of 10 undoubtedly intensified this family's socioeconomic ties with Attaguyuk. However, this adoption took place not in infancy, but when 9 (Ekaliq) was a young adult, which raises the question of who was adopting whom. In other words, Ekaliq may have lived in the household of 10 primarily in a productive or apprenticeship capacity. While overcrowding in Attaguyuk's household may have motivated this arrangement, Ekaliq still maintained strong attachments to his father and siblings. Even though Ekaliq's adoption may have served to integrate 10, who was originally from Umanaqjuaq, and his extended family into the community, the location of his household (E) on the periphery of the camp reveals his marginal position in Sauniqtuajuq society (Figure 37).

Maniapik's (2) *ilagit* stands in marked contrast to Attaguyuk's in being smaller and integrated neither on sibling nor father-son ties, but on multiple linkages created by a pair of opposite-sex siblings (11 and 12) marrying opposite-sex cousins (2 and 13). As 11 and 12 were siblings, the latter (Angnaqok), is both terminologically *sakiaq* and *ningauk* to Maniapik. It is interesting to note that a particularly close bond of affection existed between Maniapik and Angnaqok as they lived, travelled, and hunted together until Maniapik's death around 1940. Maniapik also has overlapping ties with 14 (Kopee); the latter is both his *uyuruk* (sister's son) and *nukaunnguk* (wife's younger sister's husband). While multiple kinship ties existed among Attaguyuk's *ilagit* (e.g., Attaguyuk and his brother married women who were cousins), residential solidarity was not founded on such ties. Alternatively, Maniapik's *ilagit* demonstrates one of the highest frequencies of multiple affinal ties among any kin group recorded in this study.

All three households comprising Maniapik's kin group (G, H, and I) are located in the centre of the site between *qammat* occupied by Attaguyuk's kinsmen. The core of Attaguyuk's *ilagit*, however, was located on lower ground in a small valley at the north end of the settlement. Here, Akatoogaq (C), Nukeeruaq (F), and Koseaq's (B) *qammat* looked out across the ravine upon those occupied by Attaguyuk (A) and Maniapik's *ilagit* (Figure 37). It is perhaps significant that, even though Maniapik and Attaguyuk maintained a competitive joking relationship, their dwellings were originally located on the same side of the camp facing in the same direction; it was their kinsmen who chose to construct their houses opposite each other. The inclusion of 7's household (C) within the physical core of Attaguyuk *ilagit* is explainable with reference to his kinship connection to 5, and to the fact that 7 appears to have been a man of substance, e.g., Akatoogaq often conducted church services in

Pangnirtung. In relative agreement with kinship directives, individuals 8 and 10 erected their houses (D and E, respectively) some distance from Attaguyuk, on the other side of Maniapik's *qammaq*.

Both Maniapik's and Attaguyuk's *ilagiit* functioned as independent socio-economic units so long as both possessed whaleboats. And, while *nekaishutu* sometimes occurred on a village-wide basis, especially when food was scarce, this particular sharing custom was normally performed by Attaguyuk and Maniapik among their respective kinsmen. Maniapik and Attaguyuk also took charge of transporting fox furs and blubber skins to Pangnirtung, and distributing trade goods on their return. Interestingly, Maniapik was generally considered the more generous as he gave away more white man's food. Unlike Idlungajung, or Nunaata for that matter, there was no central food cache at Sauniqtuajuq; most families managed their own food stores which they kept in the porches of their dwellings.

Every summer able-bodied men from each kin group would form crews on Maniapik's and Attaguyuk's whaleboats. However, as in other camps, winter hunting was a more individualistic activity carried out with other nuclear family members. For example, Attaguyuk and Nukeeruaq (4) hunted almost exclusively with their sons during the winter. If larger co-operative task parties were formed, it was usually with other Inuit from Idlungajung and sometimes Iqalulik, in order to hunt seal at *aglu*. Maniapik, on the other hand, continued to hunt with Angnaqok (12) throughout the year, though frequently he would be joined by his adolescent sons.

Of the six married couples within Attaguyuk's *ilagit*, five demonstrate patrilocal tendencies. Only Alaq (10), who was from Umanaqjuaq via Kivitoo, appears to exhibit a matrilocal arrangement. Matrilocality, however, prevails among Maniapik's kin group if the sibling core of 11, 12, and 15 is designated to be the structural foundation of this group. Of the two adoptions about which information exists, 7 and 9, both were arranged between close relatives of the same generation, i.e., siblings or cousins. Intergenerational adoptions appear to be non-existent at Sauniqtuajuq during the mid-1920s.

## *1937–1939*

By the late 1920s a number of substantive changes had taken place in Sauniqtuajuq's composition, the most significant of which was the relocation of Maniapik and his kinsmen to Iqalulik over the winter of 1925–26. At about the same time, families headed by 3, 7, and 8 moved to Pangnirtung, while 10's wife's brother and family took up residence at Sauniqtuajuq. Shortly thereafter, Nukeeruaq (4) acquired a whaleboat from the HBC and moved to a site in Nettilling Fiord. In a period of less than three years then, Sauniqtuajuq's population fell from a high of about 50 in 1923–24 to a low of 17 in 1927 (Table 8). Around 1929, Makitoq (5) died and Attaguyuk took 16 (Malukaitok), 10's resident sister, for a wife. Subsequent to these changes, however, Attaguyuk's kin group underwent a period of rapid growth. While Koseaq (3) returned to Sauniqtuajuq, by the late 1930s Attaguyuk's four sons (9, 17, 18 and 19) had married and were on the way to establishing large families of their own (Figure 38).

Clearly, Attaguyuk's relationship with his married sons formed the structural foundation of residential solidarity at Sauniqtuajuq during the 1930s (Figure 38). However, as Attaguyuk's *ilagit* grew in size and influence, a number of affinally-related families settled at Sauniqtuajuq. Although 10 no longer resided here, Attaguyuk has attracted another of Malukaitok's (16) older brothers, 20. Similarly, Attaguyuk's eldest son, Ekaliq (9), has attracted his wife's (21) older sister (22) and her husband (23), as well as the latter's mother (24). In addition to these affinal relations, Attaguyuk's sister's daughter (25) and her husband (26) moved to Sauniqtuajuq.

While Attaguyuk and his sons were the main unifying factors at Sauniqtuajuq during the 1930s, a second kin group, headed by 23 (Nowlalik) was beginning to emerge. This latter individual appears to have been a man of some influence as he owned his own boat, though it was somewhat older and smaller than Attaguyuk's HBC whaleboat. Both Nowlalik, who was 14's (Kopee) brother from Kekerten, and 9 married two sisters (21 and 22) from Umanaqjuaq. These latter two individuals were the adopted daughters of Pawla, the former *angajuqqaq* of whaling and trading activity at Umanaqjuaq. Further ties between Attaguyuk and other Umanaqjuarmiut are also evident. Koseaq's (3) current wife (27) apparently came from Umanaqjuaq, as did 16's mother (28) and brother (20). Individual 19's marriage to 29 and 9's adoption of 30 also created ties with Umanaqjuarmiut; both women were the daughters of 15, one of the original occupants of the camp. These marriages appear to support a consistent pattern, revealed in previous sections, whereby Kekerten men have taken Umanaqjuarmiut wives.

The strengthening of *irniriik* ties between Attaguyuk and his sons during the 1930s appears to have been accomplished at the expense of the erosion of *nukariik* ties with his older brother, 3 (Koseaq). While the latter originally settled within the geographic centre of Attaguyuk's kin group, he returned to occupy the west end of the site in the 1930s—the resultant spatial distance between Attaguyuk's and Koseaq's households (A and B) being greater than any other two dwellings at Sauniqtuajuq (Figure 39). Attaguyuk appears to have moved into Akatoogaq's *qammaq* (A) after the latter left, leaving his former abode (C) to his niece's (25) family. As Attaguyuk's sons married, and as Koseaq (3), Nukeeruaq (4), and Maniapik (2) abandoned their original house locations (i.e., D, E, and F, respectively), Ekaliq (9), Inosiq (19), and Qaqasiq (18) moved in. An additional dwelling (G) was constructed to accommodate Attaguyuk's second eldest son, 17 (Mosesie), who took Akatoogaq's widow for a wife. With the exception of 18's reoccupation of Maniapik's old house (F), which is located at a higher elevation, all dwellings inhabited by Attaguyuk's kinsmen in the first descending generation are located within the

180　Part Three: Cumberland Sound Inuit Social Structure

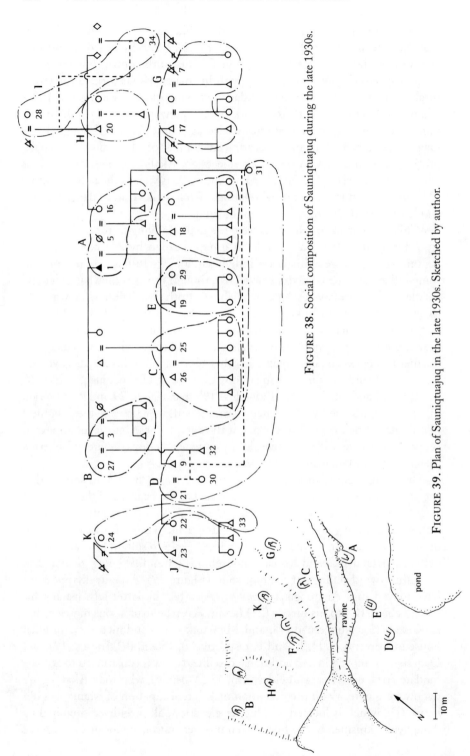

FIGURE 38. Social composition of Sauniqtuajuq during the late 1930s.

FIGURE 39. Plan of Sauniqtuajuq in the late 1930s. Sketched by author.

same small valley. Together these households comprised the economic, social, and political centre of the settlement. Predictably, affinally-related households are located geographically outside this core on higher ground, where, interestingly, dwellings inhabited by Attaguyuk's wife's brother (H) and mother (I) are further removed from the sociopolitical core of the settlement than those occupied by Ekaliq's affinal relations (J and K).

While Attaguyuk's marriage to Malukaitok (16) constitutes local group endogamy, six of the seven marriages within his kin group represent patrilocal arrangements, as does 20's marriage. Alternatively, Nowlalik's (23) and 26's marriages exhibit matrilocal tendencies. However, neither of these individuals apparently had the option of residing in their fathers' camps. While 26 was from Singnija (Cape Haven) with no immediate relations in the Sound, Nowlalik's birth father was a Scottish whaler who resided only briefly at Kekerten around the turn of the century.

Adoptions at Sauniqtuajuq during the late 1930s were more numerous and varied than that of a decade earlier. Not only has Ekaliq (9) adopted a daughter (30) from a couple (14 and 15) who had previously inhabited the site, but he has adopted his half-sister (31) and his paternal parallel cousin (32). Ekaliq's younger brother Mosesie (17) has given two daughters in adoption to relations in Pangnirtung and a son to his first wife's parents in Naujeakviq. While two individuals in the top generation (24 and 28) live with their grandchildren (33 and 34, respectively), only one of these arrangements constitutes an ideal grandparental adoption; the remaining one represents care of the elderly.

*Circa 1950*

During the 1940s the size of Nowlalik's (23) *ilagit* gradually expanded as his children began to raise families of their own. Conversely, households belonging to Attaguyuk's *ilagit* gradually shrank in number. Although Attaguyuk's sister (35) settled at Sauniqtuajuq in the early 1940s, Attaguyuk's brother (3) moved away permanently. By 1950, only two of Attaguyuk's sons (18 and 19) remained; Malukaitok (16) has died, and Attaguyuk has taken a third wife (36). By the end of the decade, Nowlalik's kin group rivalled that of Attaguyuk's in both size and influence (Figure 40).

The dissolution of Maniapik's *ilagit* at Iqalulik after his death around 1940 directly encouraged two families to settle at Sauniqtuajuq. One of these was headed by 12 and 13, two of the original inhabitants of the site. The other was headed by Maniapik's son, 37 (Towkie). Towkie's wife (38) soon died, however, and he married his wife's younger sister (39). Although the sororate was not practised extensively among the Oqomiut, it was not frowned upon either. Nowlalik's eldest son (40) has married Attaguyuk's younger brother's daughter (42), while Nowlalik's daughters (43 and 44) have married, respectively, Attaguyuk's sister's son (45) and grandson (46). These marriages created extensive ties between Nowlalik's and Attaguyuk's kinsmen. However, only 44 and

182 Part Three: Cumberland Sound Inuit Social Structure

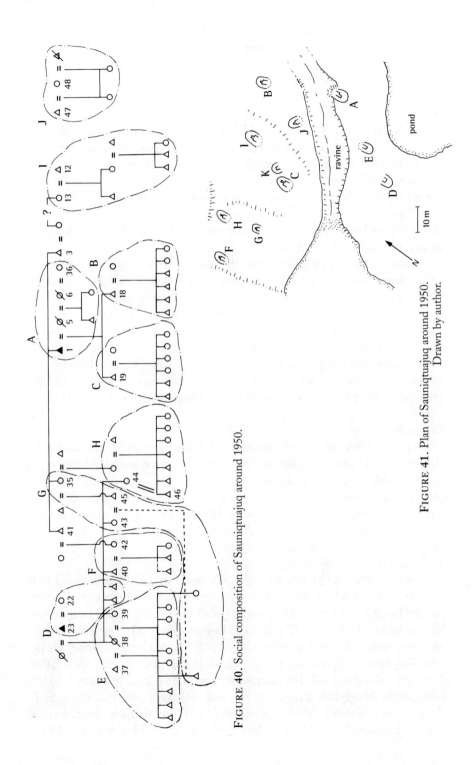

FIGURE 40. Social composition of Sauniqtuajuq around 1950.

FIGURE 41. Plan of Sauniqtuajuq around 1950. Drawn by author.

46's union represents a group endogamous marriage; 45 moved to Sauniqtuajuq in the mid-1940s to perform bride-service for Nowlalik (23). The remaining couple (47 and 48) apparently had no direct relations at Sauniqtuajuq.

The erosion of Attaguyuk's *ilagit* and subsequent expansion of Nowlalik's kin group, and the frequency of intermarriages between the two, has resulted in a very different arrangement of households than that recorded for the late 1930s. Attaguyuk and two resident sons (18 and 19) still live, respectively, in households A, B, and C (Figure 41). However, the area occupied by Attaguyuk's immediate relations has shrunk dramatically as Nowlalik's *ilagit* has expanded across the ravine, where Nowlalik now resides with his daughter (39) and her family in households D and E. The families of Nowlalik's remaining children, 40, 43, and 44, live on the opposite side of the ravine in *qammat* F, G, and H—dwellings formerly occupied by Attaguyuk's kinsmen. The last two families, who had only distant or no relations at Sauniqtuajuq, occupy houses I and J. Household K was not inhabited during the late 1940s as it was abandoned by 37 after his first wife died.

Attendant with the increase in the size and influence of Nowlalik's kin group, Attaguyuk's role as *angajuqqaq* diminished. While Attaguyuk was still recognized to be the leader of Sauniqtuajuq, his declining health and support base undermined his authority and decision-making ability.

Patrilocality was still the dominant form of post-marital living arrangement among Attaguyuk's kinsmen during the 1940s. Alternatively, matrilocality appears to as common as patrilocality among Nowlalik's kinsmen; 45 was performing bride-service for Nowlalik, while 37 was forced to move to Sauniqtuajuq after the death of his father, Maniapik.

Adoptions between close relatives of the same generation continued to be carried out with some frequency at Sauniqtuajuq during the 1940s. For example, as 45 and 43 were unable to have children, they adopted the latter's sister's (39) youngest daughter and youngest step-son, intensifying affectional ties among this sibling group.

*1960–1962*

During the 1950s, Attaguyuk and Nowlalik as well as their spouses died, changing the structural basis of residential solidarity. *Irniriik* ties between Attaguyuk and his sons, and between Nowlalik and his children, gave stability to Sauniqtuajuq's population throughout the 1930s and 40s. Yet, by the mid-1950s behavioural directives implicit in these bonds were supplanted by *nukariik* relationships among each *ilagit*. Apart from this natural evolution, there appears to have been few other significant changes in Sauniqtuajuq's social composition between the late 1940s and early 1960s.

Even though a few of Attaguyuk and Nowlalik's descendants intermarried, each kin group retained some degree of autonomy well into the 1960s. Whereas leadership of Nowlalik's kin group was left in the hands of his eldest son, 40 (Mosesie), Attaguyuk's eldest resident son, Inosiq (19), assumed Attaguyuk's position as *angajuqqaq*. In spite of the fact that Inosiq was considered to be the more substantive of the two, leadership does not appear to have been as well-developed as it was when Attaguyuk and Nowlalik were alive; my informants had difficulty recalling the leader of each group during the 1960s. The erosion of group leadership at Sauniqtuajuq appears to be part of a broader trend that was occuring throughout Cumberland Sound, and one that may have been related to the fact that, as many of the old leaders died off, various contact agents and institutions came to assume many of the former roles and responsibilities of leadership (Chapter 4).

During the late 1950s *nukariik* ties between Inosiq (19) and two of his brothers constituted the basis of this kin group's solidarity. However, the growth of Inosiq's own family and the marriage of his three daughters weakened this foundation. Similarly, while Mosesie's (40) relationships with his brother and two sisters represent the core of residential stability within this kin group, at least one of his children was beginning to head a large family of his own.

Whereas the marriage of 38's son to 19's daughter represents a village endogamous union, only the marriage of one of Attaguyuk's sons to Mosesie's wife's (42) sister constitutes a possible kin endogamous marriage (Figure 40). These sisters were considered, by one informant, to be the daughters of the brother of Attaguyuk's sister (35). Whether this individual and Attaguyuk were brothers is something my informants could not agree upon, although they doubted whether the marriage would have been permitted had their fathers been so related.

For the most part, the social configuration of the camp as reflected in the spatial relationship of dwellings changed little from the late 1940s. Nowlalik's and Attaguyuk's descendants still occupied roughly the southern and northern areas of the site, respectively. While Inosiq (19) continued to occupy Maniapik's old *qammaq*, his brothers lived further to the north in feature B and a newly constructed adjacent dwelling.

Once again, patrilocality is the predominant living arrangement among Attaguyuk's descendants; matrilocality exists only in the form of bride-service. Although matrilocality remains more common among Nowlalik's kinsmen, owing to the special circumstances noted above, patrilocality is now the dominant form of living arrangement among this kin group.

Adoption assumed a more varied format during the early 1960s. The adoptions arranged between Nowlalik's daughters (38, 39 and 43) have been noted previously. However, two other types of adoptions are also present. One of these entailed Mosesie's adoptive son giving his *ataatasaq* his youngest daughter, recalling the common practice noted by Guemple (1979) of an adoptee giving his/her parents a child out of a sense of respect and duty. Two other couples have adopted children from distant relatives and/or non-kin living in other camps.

Two kin groups occupied Sauniqtuajuq throughout much of its history. During the mid-1920s Maniapik and Attaguyuk's rough joking relationship, the absence of marriages between their kinsmen, and the occupation of different areas of the site helped to maintain the distinctiveness of each group. However, concomitant with the marriage of Attaguyuk's sons and Maniapik's move to Iqalulik, another kin group headed by Nowlalik (23) settled at Sauniqtuajuq. This latter *ilagit* began to expand and marry into Attaguyuk's kin group, while the latter gradually shrank in size. Although Attaguyuk's kinsmen maintained their dominance throughout much of the contact-traditional period, by the 1950s Nowlalik's *ilagit* rivalled that of Attaguyuk's in size and influence. The integrity of each kin group, though not as evident as earlier times, was maintained until the site was abandoned in 1966.

## Naujeakviq

The settlement of Naujeakviq is located on a small bay opposite the western end of the Kaigosuit Islands, which form the southern shore of Nettilling Fiord (Figure 18). Whereas Naujeakviq provides year-round access to caribou in the vicinity of Nettilling Lake and ringed seal in the Kaigosuit and Kaigosuiyat Islands, fox was a major determinant in the selection of this site. We know that the prominent Qikirtarmiut hunter, Veevee, lived at Naujeakviq during the mid-1930s,[21] and that his son (Padluq) was commuting between this camp and Pangnirtung as early as 1932.[22] However, it is not known whether they were the first to occupy Naujeakviq; Veevee travelled extensively beyond Cumberland Sound, living in many different places, and on-site investigations suggest that two kin groups originally may have lived there.

The Pangnirtung Post Journals indicate that Oshutapik and Aksayuk (indiv. 4 at Idlungajung) headed small camps in Nettilling Fiord during the late 1920s. While Aksayuk lived at Koangoon, near Sarbukjuarloo, these records do not mention the location of Oshutapik's camp, though they indicate some degree of autonomy between the two camps. Whereas men from Maniapik's and Oshutapik's camps often appear to have travelled to and from Nettilling Fiord together, men from Idlungajung, particularly Angmarlik, seem to be associated almost exclusively with Aksayuk's camp.

### 1935–1937

Oshutapik may have been responsible for one of the two older groupings of *qammat* at Naujeakviq. However, he could not have occupied this site very long before Veevee moved in. Veevee may, in fact, have joined his son (Nukeeruaq, indiv. 6 at Sauniqtuajuq) as the latter apparently left Sauniqtuajuq for Nettilling Fiord soon after his mother's death around 1929. Whatever the case, Veevee (1) and his married sons, 2 (Nukeeruaq) and 3 (Padluq) constituted the basis of residential solidarity at Naujeakviq during the mid-1930s (Figure 42). Also living at Naujeakviq during this time was Kopalee (4),

186 Part Three: Cumberland Sound Inuit Social Structure

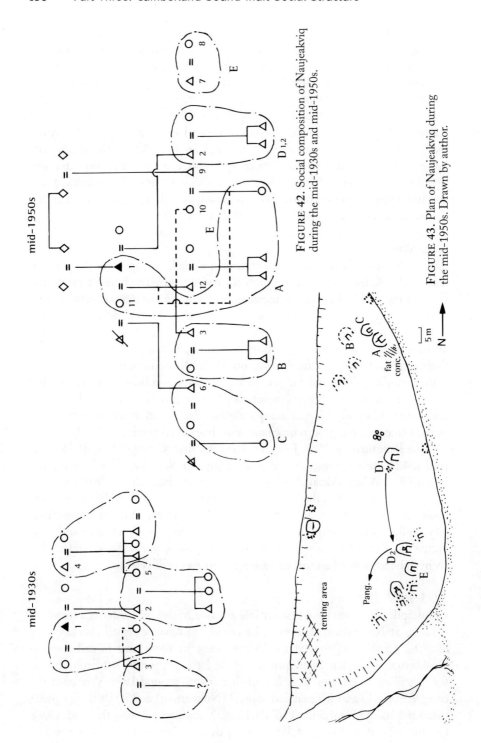

FIGURE 42. Social composition of Naujeakviq during the mid-1930s and mid-1950s.

FIGURE 43. Plan of Naujeakviq during the mid-1950s. Drawn by author.

Nukeeruaq's father-in-law, who also was the head of his own small extended family. Nukeeruaq (2) apparently lived at the opposite end of the site from his father (see below), which raises the possibility that the two major clusters of features at Naujeakviq may have originally been occupied by small groups headed by Veevee (1) and Kopalee (4).

Unfortunately, I was unable to confirm or reject this notion as the informant familiar with this earlier occupation was not able to visit or draw a map of the site.[23] It is not known what motivated Nukeeruaq to build his *qammaq* so far from his father's house. Perhaps he was performing bride-service for 4. Regardless, Nukeeruaq's relationship with Veevee was probably not very close as he was raised in his mother's household at Sauniqtuajuq where he maintained an *irniriik*-like relationship with his step-father, Attaguyuk. Moreover, Nukeeruaq also owned his own whaleboat, which may have allowed him a certain degree of economic independence not enjoyed by Veevee's other married son, Padluq (3). Finally, a falling out between Veevee and Kopalee, which may have ultimately led to Kopalee's suicide in 1936, might have also created a certain amount of tension between Nukeeruaq and Veevee.[24]

Veevee and his sons continued to live more or less permanently at Naujeakviq for the next two decades. Although Veevee's personal history remains sketchy for this period, he does not appear to have lived at any settlement other than Naujeakviq.

*1954–1956*

Veevee, despite his declining productivity, remained the leader of the camp into the mid-1950s (Figure 42). *Irniriik* relationships between Veevee and 3 (Padluq) and now resident step-son (6) continued to provide the basis for residential solidarity among this kin group. Nukeeruaq (2) has also remained at Naujeakviq, though he would soon depart for Pangnirtung to become a special constable for the RCMP. The only other family to live at Naujeakviq during the mid-1950s was a young couple (7 and 8) who were the grandchildren, respectively, of Eevic at Idlungajung and Angnaqok from Sauniqtuajuq. Although Oshutapik (9) does not appear to have lived at Naujeakviq during the mid-1950s, his marriage to 10 deserves further scrutiny as it has the appearance of a kin endogamous union. Oshutapik (9) was apparently Veevee's first cousin, while 10 (Mary) was Veevee's adopted daughter. While first and second cousin marriages were generally frowned upon in Cumberland Sound, this union could be explained by the possibility that 10 was beyond infancy at adoption, thus preventing any significant degree of affectional bonding to develop between her and her adoptive parents (1 and 11), subsequently sanctioning the marriage. Whether this union was regarded as a kin endogamous marriage, however, and whether Oshutapik and Veevee's kinship tie was explicitly recognized or not, remains uncertain.

There are no matrilocal living arrangements among this aggregation. With the exception of 7 and 8, who demonstrate a neolocal situation, all marriages represent patrilocal arrangements. Excluding 7, who is descended from some Umanaqjuarmiut, all married adults at Naujeakviq are either from Kekerten or are descended from Qikirtarmiut. The marriage of 7 and 8 represents a rare instance whereby an Umanaqjuarmiut man has married a Qikirtarmiut woman, although 9 and 10's marriage might also be similarly interpreted.

The distance between dwellings at Naujeakviq is only a moderate reflection of the social distance among individuals as prescribed by the kinship system (Figure 43). Veevee, his eldest son from Emakee (11) (i.e., 3), and his step-son (6) form a small cluster of *qammat* (A, B, and C, respectively) at the north end of the camp. Nukeeruaq (2), on the other hand, continues to live at the other end of the camp where he built amongst a second cluster of *qammat* (D). While a number of possible explanations for Nukeeruaq's living arrangement have been provided, a particularly close bond between 6 and his mother, 11 (Emakee)—the latter referred affectionately to 6 as *anik* as he was named after her younger brother—allowed him to settle within Veevee's core group. In accordance with kinship directives, 7 and 8 built their house (E) at the opposite end of the site from Veevee.

In the 1950s Veevee (1) and Emakee (11) adopted two grandsons from their non-resident son (12) and a granddaughter from their adopted daughter (10). These adoptions constitute one of the highest rates of such adoptions in Cumberland Sound, second only to Angmarlik and Ashivak's adoption of four *inngutat* at Idlungajung.

*1960–1962*
Sometime during the mid-1950s, Padluq (3) died. Soon after, Veevee passed away. About the same time, Nukeeruaq (2) relocated to Pangnirtung, while Oshutapik (9) moved to Naujeakviq, where he assumed leadership of the camp. Interestingly, Oshutapik, not one of Veevee's sons, inherited the latter's motorized whaleboat. While 3 and 6 apparently predeceased their father, Oshutapik was married to 10, Veevee's adopted daughter, and the mother of one of Veevee's adopted grandchildren. The only other whaleboat at Naujeakviq during the 1960s was owned by the son of Shorapik, a prominent *aggutik* at Kekerten during the early 20th century. However, it was this individual's wife's brother (no. 37 from Idlungajung) who operated this motorized boat, with the boat-owner serving as the operator's assistant. Apparently, the operator owned the engine. The latter, who was Eevic's son from Idlungajung and 8's father, was superior to the boat-owner in both age and the kinship system.

Uncharacteristic of most occupations described to this point, social and spatial distance as represented, accordingly, by kinship relations and the placement of dwellings were not positively correlated at Naujeakviq during the early 1960s. Although a cross-sibling relationship may have provided an element of stability to this aggregation, the brother lived at opposite ends of the site from his sister, indicating a weak sibling tie. Similarly, 8 and her husband (7) lived at the other end of the site from her parents' household. Conversely, Oshutapik (9) settled in the same area of the camp as the boat operator, to whom he was unrelated, concentrating most of the wealth and political power in one location.

The cumulative effect of these residential arrangements left the camp divided along a number of lines including those of age, wealth, and status. Families headed by in-marrying males (*ningaut*) lived at the south or opposite end of the camp to the boat operator, to whom they may have been providing bride-service. Alternatively, the latter individual joined Oshutapik (9) at the north end of the camp. Despite the fact that one of the in-marrying males was the spouse of the younger sister of the boat operator, opposite-sex kinship ties could not have been important in determining the placement of dwellings at Naujeakviq during the early 1960s. Although small, this occupation represents a possible case whereby factors other than kinship appear to have been more significant in determining the spatial arrangement of households.

Around 1962, or thereabouts, Oshutapik (9) died and the three remaining families moved to Keemee. Naujeakviq continued to be used as a summer camp for a few years by families from Iqalulik (Haller et al. 1966: 72–3).

## Other Qikirtarmiut Camps
### Keemee
Keemee is located several kilometres SSW of Kudjak Island near the northern entrance to Brown Inlet (Figure 18). Although this site may have been occupied briefly (seasonally?) prior to the early 1960s, it did not become a permanent settlement until three families settled here around 1962 in order to be closer to the *sina* and the productive sealing grounds around Brown Inlet.[25] This group was comprised of the same brother-sister core that lived at Naujeakviq. The economic partnership between the two brother-in-laws—one owned the boat, the other its motor—was another unifying element to this aggregation, with the latter being regarded generally as the camp leader. This group was soon joined by another family from Naujeakviq, headed by Oshutapik's daughter and her husband.

Families apparently did not often share at Keemee as seal was plentiful. As in other settlements on the southwest shore of the Sound, meat caches were placed in numerous locations away from camp, and food was most often shared when game was scarce or after a bearded seal or small whale had been caught. Keemee was abandoned in 1966.

At least three other camps were occupied by Qikirtarmiut between 1920 and 1970. These include Tesseralik at the mouth of Kingnait Fiord, Koangoon up Nettilling Fiord, and Kingnait in Kingnait Fiord. While we know that Aksayuk and Ooneasagaq (4 and 9 from Idlungajung) and their adult sons comprised the structural foundation of Tesseralik and Koangoon, respectively, from the mid-1920s to the mid-1930s, we also possess some knowledge of the social history of Kingnait.

## Kingnait

The contact-traditional settlement of Kingnait was first occupied in 1929 when, under the encouragement of the RCMP and HBC, several families moved there from Pangnirtung. Access to the relatively untapped trapping grounds of Kingnait Fiord appears to have been a major determinant in the selection of this settlement, though its location also afforded access to the excellent sealing grounds at the mouth of the Fiord.

Two brothers, the sons of a Scottish whaler and an Inuit woman, moved to Kingnait in 1929. However, their relationship was not the predominant foundation of residential stability; one brother, Nowlalik, appears to have had an *irniriik*-like relationship with his maternal grandfather (Pudjun). While Nowlalik often travelled and hunted with Pudjun,[26] he appears never to have formed a comparable economic relationship with his brother. Other than Ussualung, where they may have been performing bride-service (see below), they rarely lived together in the same settlement at the same time. Two other families, also headed by two brothers, soon attached themselves to this group.

As Nowlalik owned the only boat at this camp, it is likely that he assumed the role of *angajuqqaq*, though Pudjun may have also served in this capacity. Kingnait's population reached 25 in 1930. However, it appears to have decreased steadily throughout the 1930s (Table 8). While Nowlalik left for Sauniqtuajuq sometime around the mid-1930s, a sudden increase in the population of Kingnait around 1944 (Table 8) forced the relocation of this camp to Tesseralik and its better sealing grounds the following year.

UMANAQJUARMIUT SETTLEMENTS

### Ussualung

The settlement of Ussualung is located on the northeast shore of Cumberland Sound in American Harbour between Kekertelung and the Sunigut Islands (Figure 18). As this harbour was used occasionally as an over-wintering site by whalers during the 1860s (Goldring 1984: 512), several 19th century *qammat* foundations are found near the old whalers' shack on the island. Inuit did not settle permanently here, however, until the German-American whaler/trader, William Duval, established a post on the mainland in 1918 on behalf of H.T. Munn's Arctic Gold Exploration Syndicate. Access to ringed seal and beluga whale for both domestic use and commercial markets appear to have been the primary reasons for the selection of this location; seal denning sites are abundant off the west shore of the Sunigut Islands in the spring, while concentrations of beluga are found each July at the head of a nearby inlet and at Milurialik. Duval, or Sivutiksaq ('the harpooner's apprentice'), had lived more or less permanently among the Oqomiut since the mid-1870s. As Duval was thoroughly integrated into Inuit society, his obligations lay primarily with his wife's kin rather than with any of the whalers and traders he represented over the years.

Ussualung served as the major Inuit and trading centre in the upper half of the Sound until the HBC established a post in Pangnirtung Fiord in 1921. For the next two years Munn and the HBC trader were locked in an intense competition for furs, oil, and Inuit labour, a fact that the Inuit were quick to turn to their advantage. Although Munn's holdings at Ussualung were purchased by the HBC in the fall of 1923, Sivutiksaq continued to work on and off for the Company until his death in 1931. During the winter of 1923–24 the entire population of Ussualung, as well as that of Nunaata, moved to the head of Kingua Fiord, where, under the direction of Sivutiksaq, they trapped for the HBC. Although poor sealing forced Kingua to be abandoned the following spring, only Sivutiksaq and his immediate kinsmen seem to have returned to Ussualung.[27]

## *1921–1923*

Ussualung was occupied initially by Inuit from settlements both inside and outside Cumberland Sound. However, most occupants appear to have either originated from or had strong connections to Umanaqjuaq. Sivutiksaq (1) and his wife, Aulaqeak (2), and the latter's siblings (3, 4, and 5) formed the basis of residential solidarity at Ussualung during the early 1920s (Figure 44).

It is apparent that both 4 (Veevee; indiv. 1 from Naujeakviq) and 5 (Maniapik; indiv. 2 from Sauniqtuajuq) benefited socially as well as economically from their relationship with Duval. Both were men of substantial means, although Veevee was the only Inuk at Ussualung to own a whaleboat. While Duval was this camp's chief trader, there was no overall Inuit leader at Ussualung. Veevee and Maniapik, and even 6 (Agaliq; indiv. 1 from Avatuktoo) and 7 (Tautuajuk) possessed some influence, but Sivutiksaq, who spoke fluent Inuktitut, was recognized to be the camp boss as he represented the interests of his kinsmen in all dealings with outsiders. Yet, he did not oversee game distribution in the community. Rather, *nekaishutu* would be performed by Veevee or Maniapik when beluga whales were scarce. When game was plentiful individual families exchanged food among themselves (*piutuq*). Unlike some settlements at the head of Cumberland Sound, there was no central food cache at Ussualung. Although Angmarlik, the *angajuqqaq* of Idlungajung, stored beluga whale products in the old whalers' shack on the island across from the settlement, most hunters maintained their own caches, which were located some distance from the village.

The personal histories of Sivutiksaq's affinal relations are difficult to trace as he lived at a number of whaling/trading stations on Baffin Island during the late nineteenth and early twentieth centuries. For example, towards the end of

192　Part Three: Cumberland Sound Inuit Social Structure

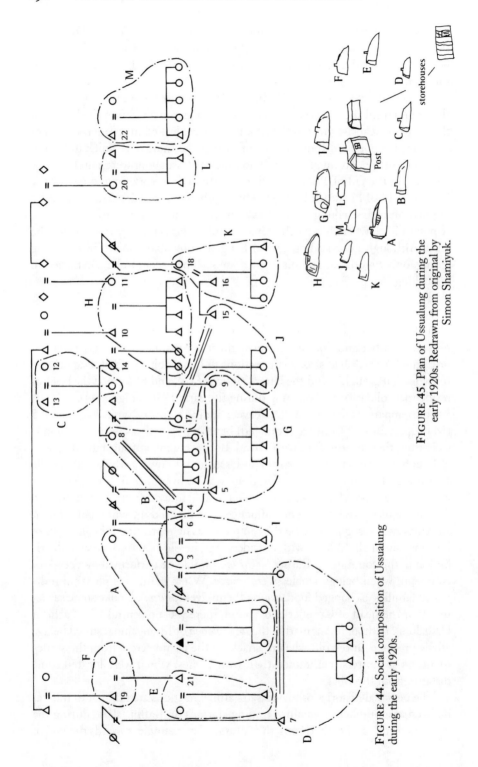

FIGURE 44. Social composition of Ussualung during the early 1920s.

FIGURE 45. Plan of Ussualung during the early 1920s. Redrawn from original by Simon Shamiyuk.

the first decade of this century, Duval appears to have maintained residences at both Kekerten and Durban Island on the east coast of Baffin Island. While historical sources indicate that his wife (2) came from Frobisher Bay,[28] my informants believe that she and her brother, Veevee (4), originated from Pond Inlet. Even so, Aulaqeak's half-brother, Maniapik (5), was regarded as an Umanaqjuarmiut. We know that Veevee moved to Ussualung from Padloping Island. Yet, both he and his wife (8) were considered by my informants to be Qikirtarmiut. Alternatively, Greenshield implies that a man named Veevee, whom he met near Padloping Island in 1910, might have been from Umanaqjuaq.[29] Although Veevee may have been performing bride-service for his wife's parents at Umanaqjuaq, it is important to point out that there is a connection between Inuit from Pond Inlet and Umanaqjuaq. In 1903, James Mutch took two whaleboat crews and their families from Umanaqjuaq to Pond Inlet in order to set up a shore-based whaling station (Mutch 1906).

The confusion over these personal histories is a function, in part, of the importance that the Cumberland Sound Inuit attach to locality. Although kinship appears to play the central role in determining local group composition in most camps, individuals are usually regarded as members of a specific local group if they reside permanently with that *nunatakatigit*, regardless of their birth place or primary kinship connections. In other words, people become members of a particular social aggregation if they have been accepted by and have made socioeconomic commitments as well as emotional attachments to that group—an occurrence that normally occurs with birth or marriage. Thus, when an individual is labelled a Nunaatarmiut, it denotes the fact that he or she resides permanently at Nunaata, irrespective of whether or not he or she is related consanguineally to others in the settlement, although there is an expectation of some degree of relatedness. Therefore, when my informants state that Veevee (4) is a Qikitarmiut, while his half-brother Maniapik (5) is an Umanaqjuarmiut, it means that these individuals developed into, and came to be regarded as, contributing members of society at these respective settlements, even though they may have been born or raised elsewhere.

The sibling core of 2 through 5 formed the basis of residential unity at Ussualung during the early 1920s. These individuals were either from Umanaqjuaq or had strong kinship ties to people that were. Veevee (4) and Maniapik (5) took wives (8 and 9, respectively) who, in turn, have attracted their parents. While 10, 11, and 12 were Umanaqjuarmiut, 13 is said to have been from Kekerten, although the latter moved around a lot. Both Maniapik's and Veevee's wives and parents-in-law were related; 10's first wife (14) was 12 and 13's daughter. Thus, 8 and 9 are maternal aunt-niece, or *aiyak-nubak*, to each other. Here, we have a situation where half-brothers have married a couple's daughter and granddaughter. Another instance of brothers marrying into the same *ilagit* occurs in the case of 15 and 16, whereby two brothers from Kekerten have married terminological cousins (17 and 18) from Umanaqjuaq.

Families headed by individuals related through marriage to Maniapik and Veevee (i.e., 10 through 18) form a second subgroup at Ussualung strengthened by a number of kinship ties. The couple that headed the largest family in this group (10 and 11) has, in turn, attracted same-sex first cousins, 19 and 20. The latter individual's husband came from Padloping, had few other relations in Cumberland Sound, and appears to have resided at Ussualung only briefly before moving to Idlungajung. Individual 19 was 10's paternal parallel cousin from Umanaqjuaq. While both his sons (7 and 21) have started families of their own, his eldest son (7) has married into the dominant sibling core. Local group endogamy, however, is not indicated as this marriage likely took place prior to 1918. Neither is kin endogamy present in this aggregation, in spite of the occurrence of multiple kinship ties such as that between 15 (Nowlalik) and 16 (Kopee). Whereas 15 is 16's *angayuk*, he is also his brother's wife's cousin's husband. Only one unattached family appears to have lived at Ussualung during the early 1920s. The individual that headed this family (22) was from Frobisher Bay and had few kin relations in the Sound. During the 1920s, 22 lived at three different camps including Nunaata and Avatuktoo.

The spatial arrangement of dwellings at Ussualung during the early 1920s agrees, for the most part, with the degree of social distance between individuals as prescribed by the kinship system (Figure 45). In other words, the arrangement of houses appears to be more a function of kinship directives than a combination of other factors. Thus, the *qammat* of closely related individuals tend to be grouped together. Sivutiksaq and Aulaqeak (2) lived together as well as separately in either the trading post or *qammaq* A. The latter feature and that of Veevee's (B) and his *sakik* (C) form a series of adjacent *qammat* on the west side of the settlement. Similarly, the houses of Tautuajuk (D), his half-brother (E), and his father (F) form a group of three dwellings integrated on the basis of *irniriik* relationships at the back of the camp. Maniapik's house (G) and those of his wife's parent's (H) and half-sister's (I) are located on the east side of the site. The *qammat* of 15 and 16 (J and K, respectively) comprise a small group of houses founded on *nukariik* ties at the entrance to the site. The fact that these latter dwellings are located between houses occupied by 15 and 16's mother-in-laws (i.e., households B and H, respectively) suggests that *panniriik* (mother-daughter) relations played a role in their placement.

While kin relatedness and the location of dwellings appear to be positively correlated at Ussualung, two *qammat* (L and M) directly contradict this pattern. That individuals most distant to Sivutiksaq and Aulaqeak in the kinship system would reside closest to them is reminiscent of Idlungajung during the early 1930s. It is possible that Sivutiksaq suppressed *ungayuq* behaviours in his relationships with his wife's male kin, to whom he was terminologically subordinate, while exerting his authority. Alternatively, and more plausibly, the area in which L and M were built was open space, implicitly recognized to be

Sivutiksaq's by his affinal relations, but available for use by site visitors and temporary occupants. Both 20's husband and 22 were the most nomadic of Ussualung's residents. In light of this observation, it is interesting to observe that the smallest houses in the 1921–23 plan of Ussualung drawn by Simon Shamiyuk (Figure 45) belonged to the most transient residents. Conversely, the inhabitants of the largest illustrated *qammat* were more proximate to Aulaqeak in terms of the kinship system.

Of the 13 married couples in this aggregation, five exhibit matrilocal living arrangements, two of which (the heads of households J and K) may represent cases of bride-service. Only two unequivocal examples of patrilocal residence were recorded. While three couples demonstrate neolocal living arrangements, another three couples live in the same camp as both sets of parents, although death or divorce appears to account for the absence of at least one parent in each case. Again, this may be a reflection of not so much local group endogamy, but rather the composite nature of the aggregation.

*1937–1939*

With the purchase of Munn's holdings in the fall of 1923, the HBC became the sole supplier of goods and purchaser of Inuit produce and labour in the Sound. This precipitated the break-up of Ussualung and, after the closure of the Kingua post in 1924, Veevee (4), Maniapik (5), and Agalik (6) moved to other camps, where they represented the interests of the HBC trader as well as their kinsmen. In a period of less than two years, then, Ussualung's population fell from 55 or so to less than 15. The death of Duval in 1931 notwithstanding, the composition of Ussualung remained remarkably stable between the mid-1920s and early 1950s. Throughout this period *ungayuq* behaviours implicit in the *nukariik* relationship between Aulaqeak's daughters, Towkee and Alukie, provided the unifying basis for this aggregation. Affectional bonds between these sisters were further intensified by Towkee's adoption of three sons and one daughter from Alukie. While the Inuit practised a number of types of adoption (Guemple 1979, Hickey and Stevenson 1990), this case represents the highest rate of sibling adoption recorded in Cumberland Sound.

The only other family to reside permanently at Ussualung during the 1920s and 30s was headed by Mike (indiv. 23 from Nunaata), who came to Cumberland Sound from Coral Harbour. Two generations of this family have married into Ussualung's original sibling core. Wheres Mike married Aulaqeak's niece, Mike's son, Akpalialuk, married Aulaqeak's daughter, Alukie. As Akpalialuk was not Aulaqeak's niece's natural son, but the son of Mike's former wife in Coral Harbour, this union was not regarded an endogamous marriage. Nonetheless, the fact that a father and son married into the same *ilagit* undoubtedly increased affectional solidarity among this group, while facilitating their acceptance into Oqomiut society. Mike was only gradually accepted by most

Cumberland Sound Inuit, a fact reflected in his constant shifting of residences. By 1940 Mike had moved to Nunaata.

The spatial arrangement of dwellings at Ussualung during the late 1930s reflects the dominance of Alukie and Towkee's sibling tie over Akpalialuk and Mike's *irniriik* relationship. Alukie resided in Aulaqeak's old *qammaq*, while Towkee and her adopted children lived beside her in the old trading post, where they were supported by Akpalialuk. Alternatively, Akpalialuk's and Mike's dwellings were located at opposite ends of the site—a distance somewhat greater than that predicted by kinship directives and logistical considerations. In this regard, Akpalialuk, not Mike, was regarded as the *angajuqqaq* of Ussualung after the death of Sivutiksaq. Akpalialuk apparently inherited Duval's whaleboat, whereas Mike was regarded by the authorities as one of the most shiftless men in the district.[30] Finally, Aulaqeak, perhaps because of the size of her daughter's families, lived alone in what once was the blubber storehouse (Figure 45).

*1957–1959*

During the mid-1950s Ussualung was temporarily abandoned. In response to Aulaqeak's death and the break-up of the Kingmiksoo camp on the southwest side of the Sound, Akpalialuk, Alukie, and Towkee moved to Kingmiksoo site in 1954. In the late 1950s, Ussualung was reoccupied intermittently by three of Akpalialuk's sons, as well as Towkee's adopted daughter and their respective families. These families seem to have shifted residences a number of times between Ussualung and Kingmiksoo. Also living at Ussualung during this time, albeit briefly, was the brother of Keenainak from Nunaata. Akpalialuk's second eldest son was the son-in-law (*ningauk*) of this individual. The Akpalialuk and Keenainak families were related in other ways, e.g., Aulaqeak adopted Keenainak's third eldest son's son. Ussualung appears to have been abandoned permanently around 1960 as Akpalialuk's sons never resided there after this date.

**Iqalulik**

Iqalulik is located on a small island on the north shore of Nettilling Fiord near its entrance (Figure 18). Ringed seals are found near this site throughout the winter at breathing holes, the *sina*, and at *sarbut*, which occur in a group of small islands at the mouth of Nettilling Fiord. Fox also played a role in the selection of this site. During the mid-1920s, the HBC provisioned a number of prominent Inuit to move to the Nettilling Fiord area to trap fox, including Maniapik (indiv. 2 from Sauniqtuajuq) who settled at Iqalulik in 1925–26. Although access to both fox and ringed seal encouraged people to settle at the mouth of Nettilling Fiord, a species of land-locked char, which inhabit a small inland lake on an adjacent island (Auneavikuluk?), was the primary reason why Iqalulik was selected as a campsite.[31]

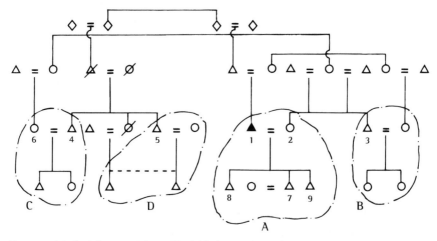

FIGURE 46. Social composition of Iqalulik during the early 1930s.

*1931–1933*
Maniapik (1) and the same opposite-sex sibling core that lived at Sauniqtuajuq in 1924–25, 2 and 3, were the first Inuit to live at Iqalulik during the 20th century (Figure 46). Although Maniapik and 3's *nukariik*-like relationship constituted the basis of residential solidarity at Iqalulik throughout the late 1920s, they were soon joined by two brothers, 4 and 5, from Nuvujen. These individuals were the sons of the prominent hunter and whaler from Umanaqjuaq, Kaka. After the mysterious murder-suicide of their parents in 1927, 4 (Oshutapik) and 5 (Petaosie) left Nuvujen with their father's whaleboat for a small camp in Nettilling Fiord (Naujeakviq?), where they lived briefly before settling at Illutalik (see below) and then Iqalulik sometime in the early 1930s.

The above sibling cores were related in a number of ways. While Oshutapik's wife (6) and Maniapik's wife (2) were parallel cousins, Maniapik and the sibling core of 4 and 5 were second cousins. Oshutapik (4) and 2 (Kakatunaq) were also apparently blood-related, although my informants could not specify the exact nature of their relationship. If 4 was a cousin of 2's, as was suggested, it is possible, although far from certain, that 4's marriage to 6 represents a kin endogamous union.

Despite the fact that two whaleboat owners lived at Iqalulik, the relatively small size of the population precluded the formation of two whaleboat crews. Thus, only one boat, frequently captained by Maniapik, would be used at a time. It was often the case, however, that Oshutapik (4) and Petaosie (5) would go caribou hunting rather than joining Maniapik's crew for the beluga whale hunt. When Maniapik and Angnaqok (3) found vacancies on their crew, they normally filled them with those men willing to join them from settlements such as Sauniqtuajuq and Idlungajung. Even though Oshutapik sometimes assembled his own crew for the whale hunt, and was considered to be the head of his own *ilagit*, Maniapik was generally regarded to be the *angajuqqaq* of the camp.

There appears to be little difference in the spatial separation of dwellings at Iqalulik. Even so, Maniapik and Oshutapik constructed their houses at different elevations, where they were joined, respectively, by Angnaqok and Petaosie. With the exception of Maniapik and his resident married son (7), individual families maintained their own caches, which were located some distance from camp.

## 1938–1940

Sometime during the mid-1930s this aggregation moved to a nearby island (Auneavikuluk?) 1.5 km north of Iqalulik as its protected waters offered better seal hunting. By the end of the decade, Maniapik's two eldest sons (7 and 8) had married, respectively, Nowlalik's (indiv. 23 from Sauniqtuajuq) and Keenainak's (indiv. 1 from Nunaata) daughters and begun to raise families of their own. Not only do these marriages represent hypergamous unions, but they constitute rare instances of Umanaqjuarmiut men marrying Qikirtarmiut women. While 7 (Towkie) continued to live with Maniapik, 8 (Karpik) moved into his own house. Around the same period, Oshutapik (4) and his brother (5) left Iqalulik. A short time later, Maniapik died and the camp disbanded. While 7 moved to Sauniqtuajuq, the camp of his wife's parents (no.'s 22 and 23 at Sauniqtuajuq), 8 moved to Nunaata with his mother (2), where his wife's father (Keenainak) was the camp leader. In accordance with tradition, 8 inherited his father's whaleboat.

## 1957–1959

Only those families headed by 3 (Angnaqok) and his nephew, Maniapik's youngest son (9), appear to have remained at Iqalulik after Maniapik's death. However, they were soon joined by Aksayuk, 10 (indiv. 4 at Idlungajung). Aksayuk, who quickly assumed the role of *angajuqqaq*, appears to have moved from Idlungajung to Koangoon in Nettilling Fiord sometime prior to 1927, where he lived more or less continuously for the next decade. It is not known what motivated Aksayuk to move to Iqalulik, though his advancing age may have played a role in his decision. Whatever the case, Aksayuk was soon joined by his son (11) and daughter-in-law (12) from Kipisa and Kingmiksoo (Figure 47). This couple moved often between Kipisa and Iqalulik, as did 12's parents (13 and 14) and sister (15). With the death of 13 (Koodlooaktok), however, 14 moved to Iqalulik permanently where she lived with her eldest daughter (15). A strong sibling tie between 12 and 15 was an important unifying factor in this aggregation as it provided Aksayuk's only kinship connection to 3 and 9, the original occupants of the camp. Aksayuk also appears to have attracted his sister (16) and her *sakkiik* (17 and 18). Similarly, 12 and 15's paternal aunt (19) settled at Iqalulik in the same *qammaq* that Angnaqok's younger brother (20) occupied before he moved to Pangnirtung in the late 1950s. Like Sauniqtuajuq, Inuit from both Kekerten and Umanaqjuaq lived at Iqalulik. However, the only marriage that appears to have taken place between these two groups was that between Aksayuk's son (11) and 14's daughter (12).

My informants were uncomfortable with drawing maps of the settlement from memory as people apparently moved around a lot at Iqalulik. However, the map produced for the period just before 1960 shows some degree of spatial separation between the two kin groups (Figure 48). Aksayuk (10), his son (11), and sister (16) occupied the same area, if not *qammat* foundations, previously inhabited by 4 and 5. These include dwellings A, B, and C, respectively. Located behind household C and some distance away from Aksayuk's *qammaq* (A) is the house (D) of his sister's *sakkiik* (17 and 18). While Aksayuk's son (11) originally lived with his wife's kin group, where he may have been performing bride-service, he soon moved his household (B) to a location beside his father's *qammaq* (A). This move illustrates the greater strength of *irniriik* relationships over *nukariik* ties among the Qikirtarmiut, even when sisters constitute the foundation of the latter bond. Members of Maniapik's *ilagit* appear to have remained in the same area with 9 occupying his late father's *qammaq* (E), where he was sometimes joined by his wife's parents (13 and 14). Angnaqok (3) inhabited more or less the same dwelling (F) he occupied in the 1930s. Another house (G) was built in this area for Angnaqok's younger brother (20). Although he lived in this dwelling throughout much of the 1950s, by the end of the decade it was occupied by 12 and 15's paternal aunt (19).

Aksayuk owned the only whaleboat at Iqalulik during the 1950s. However, while he may have occasionally directed activity during the annual whale and caribou hunts, his son (11) appears to have assumed charge of the boat for everyday hunting. Aksayuk's *uyuruk* or sister's sons, and other able-bodied men, usually formed the core of Iqalulik's whaleboat crew. Although Aksayuk's productivity declined throughout the 1950s, he remained in charge of *nekaishutu*, the most common form of food sharing at Iqalulik. Yet, there was no central food cache at Iqalulik; most households maintained individual stores outside the immediate area of the camp.

The fact that patrilocality was the dominant form of post-marital residence at Iqalulik during the 1950s is owing in a large measure to Aksayuk's presence, e.g., 11's residence shifted to a patrilocal arrangement once his father (10) moved to Iqalulik. Even so, Aksayuk has attracted his sister (16) and her husband's (21) parents (17 and 18), who benefited by their son's matrilocal living arrangement. Ekalujuaq's (21) residence can be explained by the fact that, although he was the son of Shorapik (17), a prominent *aggutik* at Kekerten during the whaling period, he was regarded as Aksayuk's helper. In

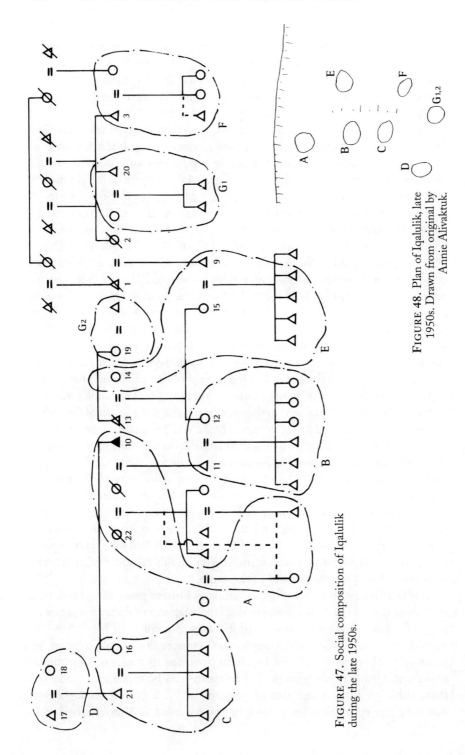

FIGURE 47. Social composition of Iqalulik during the late 1950s.

FIGURE 48. Plan of Iqalulik, late 1950s. Drawn from original by Annie Alivaktuk.

a similar vein, 14 also appears to have benefited from her daughters' (12 and 15) patrilocal arrangements.

Of the four cases of adoption at Iqalulik during the 1950s, two represent grandparental adoptions by Aksayuk and his current wife (22). The remaining adoptions appear to have been arranged with families living in other camps.

Like most settlements in Cumberland Sound, Iqalulik was evacuated temporarily during the dog epidemic of 1961–62. While most people moved back to their camps over the winter of 1962–63, only Aksayuk's kin group appears to have returned to Iqalulik. Shortly before Iqalulik was abandoned in 1966, its population fell from a high of 35 in 1961 to fewer than 18 (Haller et al. 1966: 150).

## Kingmiksoo

The contact-traditional site of Kingmiksoo is located on the southwest shore of Cumberland Sound on the southern end of Nimigen Island (Figure 18). A larger and considerably earlier settlement is situated 3 km north on a narrow isthmus in the middle of the island. This is the well-known late prehistoric/early contact site of Kingmiksoo (Figure 11), which in 1840 was reported to be the largest and principal aboriginal whaling settlement in Cumberland Sound (M'Donald 1841). In the fall of 1846 Sutherland (1856: 213) counted 111 Inuit living at Kingmiksoo in 16 sod houses. Although large by most standards, the size of this settlement may be a fairly accurate reflection of the pre-1840 population of the site. As M'Donald (1841) found over 60 individuals at Kingmiksoo in the late summer of 1840—a season when many of the site's inhabitants were still inland caribou hunting—the size of this camp later in the fall was undoubtedly larger. Some 24 or so sod houses are visible at this location today (Figure 11). Using this figure and Sutherland's (1856) statistics, up to 180 Inuit may have lived here during the early 1850s when the crew of the *McLellan* over-wintered on the island.

Bowhead whales appear to have been a major reason for Kingmiksoo's original selection as a site. This area of Cumberland Sound was known to be a popular calving ground for bowheads (Wareham 1843: 24). As soon as the ice broke in June female bowheads and their calves entered the Sound where they congregated throughout the early summer among the islands between the mouths of Nettilling Fiord and Chidlak Bay. A number of other large pre-contact villages in the immediate area of Kingmiksoo attest to the former abundance of this animal in these waters. Ringed seal, however, appears to have been the primary reason that Kingmiksoo was chosen as a campsite during the twentieth century. The large expanses of associated fast ice and the complexity of the coastline between Brown Inlet and and Chidlak Bay provide ideal winter habitat for ringed seals. In addition, during the late spring silver jars gather in large numbers near Nuvujen and in Brown Inlet (Haller et al. 1966).

*1926–1928*
From the mid-1850s to the early 1920s Kingmiksoo remained virtually uninhabited. Only after a small group of Umanaqjuarmiut settled here in 1923 did the site once again become the principal settlement on the southwest shore.[32] Kingmiksoo was led by the well-known hunter and catechist, Tooloogakjuaq. No better description of this individual exists than that of Hantzsch's 1910 character sketch (1977: 29–30):

> He was a remarkable man ... and a conspicuous character among his own folk, quick, brisk, accurate in all he did, swift as a youth, in spite of his 50–55 years, excellent as a shot and a fearless hunter of those two dangerous Arctic animals, the walrus and the bear. How many adventures in hunting could he have related had he been prone to self-advertisement, but that was not his nature! In those days when he was travelling to Singnija with Mr Greenshield, he encountered a great she-bear with her cub. He dashed right at the savage creature as she was rearing up, and stabbed her to the heart with a pocket knife. But afterwards he made little of it and refused to give himself airs over such adventures. How many others would so act in such a case! Small in appearance, almost beardless, of the true Eskimo type, with small crafty eyes and broad cheekbones, he was one of the most intelligent persons of the region, genuinely devoted to Christianity, and himself a zealous and fluent preacher at Blacklead Island and wherever his journeys took him. Listening to him, I often thought, 'What a great orator as a parliamentarian or judge he would have been in my own country.' He spoke too fast and too passionately for the pulpit. Everyone respected the worthy Tullugakdjuak, for he was always the most generous, always ready to give, or to help when opportunity offered.

Tooloogakjuaq, who was also a prominent *aggutik* in the bowhead fishery at Umanaqjuaq, was responsible for converting most Cumberland Sound Inuit to Christianity. Despite the efforts of Peck and others, few men accepted Christianity until Tooloogakjuaq converted in 1902 (see Chapter 4), after which more and more hunters began to adopt the new belief system.

Families headed by Tooloogakjuaq (1) and his son-in-law, 2 (Koodlooaktok, indiv. 13 from Iqalulik), were the first to settle at Kingmiksoo during the contact-traditional period (Figure 49). These families soon attracted others and by 1924, 12 families, totalling 60 people in all, lived at Kingmiksoo.[33]

During the late 1920s, Kingmiksoo was composed of two interwoven kin groups. While Tooloogakjuaq (1) and his married daughters (3, 4, and 5) constituted the basis of residential solidarity of one group, Koodlooaktok (2), and the latter's younger brother (6) and sister (7) formed the structural core of the other. This latter kin group, which was held together by *nukariik* directives, had strong ties to Tooloogakjuaq as both Koodlooaktok (2) and his brother (6) have married Tooloogakjuaq's daughters, 4 and 5. Although Koodlooaktok was terminologically and politically subordinate to Tooloogakjuaq and was considered the lesser of the two leaders, he too was a man of considerable substance. Like Tooloogakjuaq, Koodlooaktok owned a whaleboat and often led prayer services and performed *nekaishutu* for the community. Although these individuals sometimes hunted together, their relationship was not as close as that of Angmarlik and Eevic at Idlungajung.

Both Tooloogakjuaq and Koodlooaktok attracted a number of more distant relatives. For example, Koodlooaktok's sister's husband (8) and the latter's half-brother (9) and step-mother (10) lived at Kingmiksoo. This third subgroup was founded principally on sibling ties between 8 and 9. Tooloogakjuaq has also attracted his *uyuruk* (11), the well-travelled indiv. 20 from Idlungajung. Two marriages have taken place among members of these more distantly related families and Tooloogakjuaq's grandchildren. Specifically, while 12 has married his father's cousin's (11) wife's sister (13), 14 has married her affinal uncle's brother's son (15). Although neither marriage were regarded as a kin endogamous union—the individuals involved were outside each other's respective terminological frameworks—both served to intensify affectional bonds within the group, especially the union of 14 and 15, which was a village endogamous marriage. Matrilocality appears to have been particularly well-developed at Kingmiksoo during the 1920s as, with the exception of Tooloogakjuaq's marriage to 16 (Angnalik), all principal core group marriages exhibit matrilocal living arrangements. However, two of these may represent extended periods of uxorilocal residence. Alternatively, the marriages of more socially peripheral members (i.e., 9, 11, and 12) demonstrate patrilocal arrangements.

Three adoptions are evident in this aggregation. While 10's relationship to 17 appears to represent care of the elderly more so than it does a grandparental adoption, 7 adopted her brother's youngest daughter (18). Individual 19 appears to have lived first in Koodlooaktok's and then Tooloogakjuaq's households. While the fact that 19 retained his father's name (Akulujuk) suggests he was adopted after infancy, his membership in both households may have intensified relations between Tooloogakjuaq and Koodlooaktok.

Perhaps the most puzzling feature about this aggregation is that leadership was not as well-developed as Tooloogakjuaq's hunting prowess, generosity, and other personal attributes would predict. For example, Tooloogakjuaq could not retain all his children as at least one son and daughter lived in other camps. Kingmiksoo's population also appears to have been substantially more unstable than most camps discussed to this point; less than half the family heads enumerated in the 1927 RCMP census of Kingmiksoo were thought by my informant to have lived there during the same period.[34] While Tooloogakjuaq was still economically and politically superior to Koodlooaktok, the gulf between them with respect to these spheres was not substantial. Recall that Koodlooaktok often divided and distributed game for the community, and that

204 Part Three: Cumberland Sound Inuit Social Structure

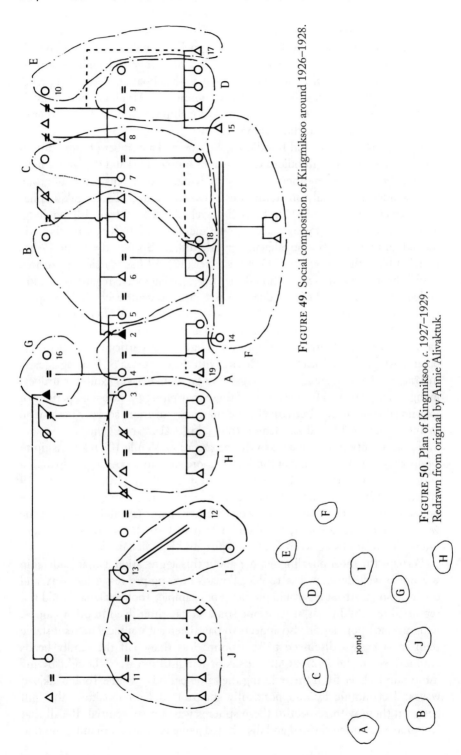

FIGURE 49. Social composition of Kingmiksoo around 1926–1928.

FIGURE 50. Plan of Kingmiksoo, c. 1927–1929. Redrawn from original by Annie Alivaktuk.

his kin group was similar to Tooloogakjuaq's in size. Although Tooloogakjuaq appears to have attracted a fair number of people, his advanced age (*c.* 65 years) and declining productivity may have undermined his economic capability and political leadership while fostering Koodlooaktok's rise to prominence. Whatever the case, Tooloogakjuaq appears never to have attained the power or influence of Angmarlik, Kanaaka, and a handful of other Oqomiut. While the reasons for this are varied, a number of explanations merit consideration.

Both Tooloogakjuaq's modest personality and Christian values may have hindered the development of authoritarian tendencies as 'everyone was equal in the eyes of God'. Alternatively, Tooloogakjuaq's early adoption of Christianity may have undermined his ability to lead in the eyes of other hunters; only women and the disenfranchised initially accepted the new religious system. Finally, unlike Kekerten, individual leadership at Umanaqjuaq during the early twentieth century appears to have been suppressed in favour of group decision-making. In contrast to Kekerten, where Angmarlik usually had the last say in community matters, major decisions at Umanaqjuaq were normally made by a group of the settlement's most prominent hunters.[35] If there was an *angajuqqaq* at Umanaqjuaq, it was Pawla, the educated son of Paul Roche, Umanaqjuaq's American whaling station manager during the late 19th century. After the death of his father, Pawla returned to his birth place, Umanaqjuaq, where he bargained with the whites for furs on behalf of the community and piloted the fall whale hunts (Hantzsch 1977: 93–4). Ittirq was another prominent Umanaqjuarmiut who undertook similar responsibilities during the early twentieth century. Tooloogakjuaq's main role in commercial whaling, on the other hand, was to instruct the inexperienced. While people sought Tooloogakjuaq's advice on all sorts of matters, he was perhaps less an *angajuqqaq* than an *isumataq*.

The spatial arrangement of dwellings at Kingmiksoo reflects the existence of distinct, albeit overlapping, kin groups (Figure 50). Koodlooaktok (2) and his brother (6) and sister (7) lived at the southwest end of the settlement in houses A, B, and C, respectively. The fact that Tooloogakjuaq's daughters (4 and 5) built their *qammat* some distance from their father's house would seem to suggest a strengthening of sibling bonds between 2 and 6 at the expense of parent-child ties. House C also appears to be part of a second spatially defined cluster of *qammat* founded principally on *nukariik* relationships. While this dwelling was occupied by 8, houses D, E, and F were inhabited by 8's half-brother (9), the latter's mother (10), and 9's son (15). Located nearest to Tooloogakjuaq's dwelling (G) are houses inhabited by his eldest daughter (H), his deceased son's eldest son (I), with whom he maintained an *irniriik*-like relationship, and his nephew (J). It would be a mistake to assign any insularity to the latter cluster of houses as Tooloogakjuaq had close kinship relations with individuals in all households except C, D, and E. Accordingly, this latter subgroup was more geographically separated from Tooloogakjuaq than any other at Kingmiksoo; their *qammat* were built on an elevation across a small pond from Tooloogakjuaq's house.

## 1936–1938

Although Tooloogakjuaq appears to have lived briefly at Umanaqjuaq sometime during the early 1930s, where he represented the HBC, he soon returned to Kingmiksoo (Figure 51). Over the winter of 1935–36, Koodlooaktok (2) and his wife (4) moved to Kipisa along with Koodlooaktok's sister (7) and her family. It is not known whether Koodlooaktok's brother (6) died before or after Koodlooaktok's departure. Whatever the case, the fact that this male sibling core lasted for close to a decade attests to the strength of the bond between 2 and 6. Whereas 5 remained at Kingmiksoo with her mother (16) after the death of 6, Tooloogakjuaq's eldest daughter's (3) family moved to another camp. The death of 6 may have influenced Koodlooaktok's decision to change residences, but Tooloogakjuaq's declining productivity may have been at least partially responsible for both moves.

The composition of Kingmiksoo during its first decade seems to have fluctuated considerably. Even so, Kingmiksoo's population appears to have remained between 39 and 53 people (Table 8) owing partly to the addition of three families (Figure 51). Individuals 20 and 21, though originally from Umanaqjuaq, had no close relations at Kingmiksoo. Similarly, 22 and 23 had no immediate relations here, although 22 and 24 may have been distant relatives (second cousins?). The latter, who hunted frequently together, apparently came from Iqaluit (Frobisher Bay) on 24's whaleboat to join 24's mother-in-law (5). Historical sources indicate that several families left Singnija in the spring of 1936 for Kingmiksoo, and that other families had also done this in the past.[36] Intermarriage between Inuit from Umanaqjuaq and Frobisher Bay appears to have been a fairly common occurrence; Koodlooaktok (2) was originally from the head of Frobisher Bay, as was Tooloogakjuaq's wife, Angnalik (16). The wives of other prominent Umanaqjuarmiut, e.g., Ittusarjuaq, also came from Iqaluit (Hantzsch 1977). How far back this pattern extends back in time is uncertain, though it likely intensified when 120 Nugumiut from Singnija moved temporarily to Umanaqjuaq in 1896–97 (see Chapter 3).

During the 1930s a number of individuals in the bottom (youngest) generation married and began to start families of their own. While 17 took a wife from outside the community, after the death of 6, 26 married the latter's daughter (27). Apparently, 26 performed bride-service for 5 for one year before moving to another camp. Although the marriage of 28 to 29 was likely not regarded as kin endogamy, this marriage, as well as that of 30 and 31, took place within the local group

Compared to the earlier occupation of Kingmiksoo, residential solidarity during the mid-1930s appears to be somewhat weakened and disjointed. Only

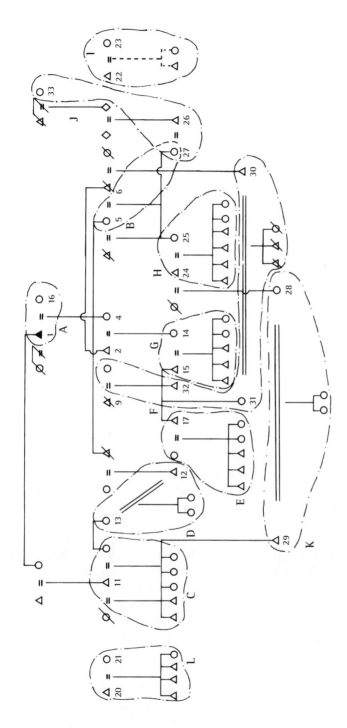

FIGURE 51. Social composition of Kingmiksoo during the mid- to late 1930s.

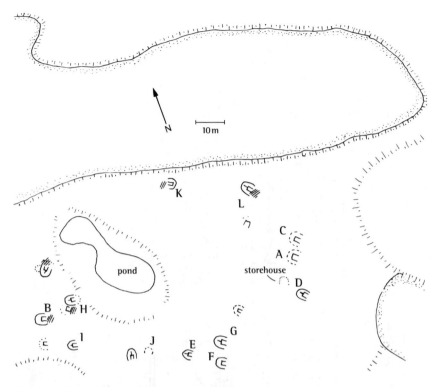

FIGURE 52. Plan of Kingmiksoo, c. 1936–1938. Drawn by author.

one of Tooloogakjuaq's natural children (5) remained at Kingmiksoo. Yet she resided across the pond from her father's household (A) in dwelling B (Figure 52). If the arrangement of houses is any indication, Tooloogakjuaq appears to have maintained closer relations with his grandson (12), whom he regarded as his adoptive son, and his nephew (11). Accordingly, 11's and 12's houses (C and D, respectively) were located on either side of Tooloogakjuaq's *qammaq* (A). The sibling core of 15, 17, 31, and 32 form perhaps the most cohesive group of kinsmen at Kingmiksoo during the mid-1930s. The spatial proximity of the former three individuals' houses (E, F, and G), in turn, reflects the closeness of this group. Individual 25 along with her mother (5) and sister (27) form the basis of a less integrated spatial subgrouping (H, B, and J, respectively). Also living in the latter dwelling is 26's grandmother (33). Associated with this *panniriik/nukariik* structured core is the household of 22 (I), who may have been a distant relative of 24. The possibility that 24 and 22's relationship was as close as that of 25 and 5's *panniriik* bond is hinted at by the closer spatial proximity of their houses, H and I. The *qammaq* of 28 and 29 (K) is one of the most isolated on site. While the latter are related to Tooloogakjuaq, albeit distantly, the inhabitants of L (20 and 21), which is located closer to Tooloogakjuaq's household, trace no kinship connections to other Kingmiksurmiut.

Material wealth appears to have been fairly evenly distributed throughout the settlement during the 1930s. While every hunter possessed his own kayak, at least four individuals owned whaleboats. Tooloogakjuaq and Adla (6) owned boats that more closely resembled sailing pinnaces than whaleboats, while 20 (Inosiq) and 25 (Nowdluk) owned whaleboats that they may have acquired from the HBC.

Individual households maintained a number of food caches on nearby islands that they relied upon during the winter. While some families at Kingmiksoo stored food in the porches of their *qammat*, only Tooloogakjuaq owned a separate facility specifically for this purpose. The sharing of game followed the same conventions as elsewhere in Cumberland Sound. However Tooloogakjuaq, possibly because of his declining productivity, did not always distribute food to the community. Rather, in times of scarcity consensus often selected a hunter who was skilful at butchering and generous in distributing meat. Individuals that performed *nekaishutu* were perhaps more democratically, and thus variably, appointed at Kingmiksoo than elsewhere in the Sound.

Of the 12 married couples at Kingmiksoo in the late 1930s occupation, four exhibit matrilocal living arrangements, while another four demonstrate patrilocal tendencies. Three marriages represent group endogamous unions, and one appears to be neolocal in character. Adoption seems to be relatively uncommon during this period of occupation, particularly among those families with extensive kinship ties. Of the three adoptions recorded, two were adopted into the group, while one was adopted out.

## *1954–1956*

Sometime during the mid-1940s Tooloogakjuaq died. Starvation, which ravaged the settlement in 1945, might have been a cause or consequence of Tooloogakjuaq's death. Whatever the case, with no resident sons or sons-in-law to assume leadership, the camp disbanded. Although one to three families may have lived intermittently at Kingmiksoo throughout the late 1940s and early 1950s, the settlement remained sparsely occupied until 1954 when a group of Inuit headed by Akpalialuk, 34 (indiv. 27 from Ussualung) settled here. Whereas sibling ties between Akpalialuk's wife and her sister (indiv.'s 24 and 23 from Ussualung), and their adopted brother contributed to group stability, the basis of residential solidarity lay principally in Akpalialuk's relationships with his five resident sons and one daughter. Individuals 26 and 27 were the only inhabitants remaining from the previous recorded occupation of Kingmiksoo (Figure 51).

Akpalialuk apparently controlled two central meat caches at Kingmiksoo: one for the dogs and the other for the community, which he located in a metal/wood shed beside his *qammaq*. Every hunter in the settlement regularly contributed to and withdrew from both these stores. Akpalialuk also owned the only boats at Kingmiksoo during the 1950s. These included a motorized whaleboat and two small skiffs or dories, which were used primarily as floe edge boats. Again, as with food, every hunter had access to these craft.

With *irniriik* ties so well-developed in this occupation, patrilocality appears to be particularly pronounced; all five of Akpalialuk's sons have taken wives from outside the residential and kin group. Adoption has intensified bonds of affection within this group among both generations. In the top generation, Akpalialuk and Alukie have given their youngest son to the latter's sister, Towkie. Similarly, Akpalialuk's eldest son gave his youngest son to his younger brother.

Akpalialuk's *qammaq* was flanked by the houses of three of his sons and his wife's cousin. The latter dwelling was located closer to Akpalialuk's house than either the *qammat* of his eldest son and only resident daughter. However, it important to bear in mind that the residents of this dwelling, indiv.'s 26 and 27, were already living at Kingmiksoo when Akpalialuk and his kinsmen arrived. Further away from Akpalialuk were the dwellings of his youngest son and wife's sister and his wife's adopted brother.

### 1963–1965

Although some of Akpalialuk's sons may have temporarily lived at Ussualung, Kingmiksoo's social composition remained virtually unchanged from the mid-1950s to the mid-1960s. However, during the late 1950s Akpalialuk died. In accordance with behavioural directives, Akpalialuk's eldest son, Charlie, assumed the role of camp leader. At the same time, 26 and 27 moved back to their place of origin, Iqaluit (Frobisher Bay). Another notable change to occur during this period was Tooloogakjuaq's adopted grandson's (19) decision to move back to Kingmiksoo. Akpalialuk's death heralded a structural shift in the basis of residential solidarity from one maintained by *irniriik* ties to one founded on *nukariik* relationships. Sibling ties also continued to give stability to the top (eldest) generation, while allowing 19 to reside permanently at Kingmiksoo—19's wife was Akpalialuk's daughter's husband's sister.

As with Kingmiksoo's social composition, the arrangement of houses did not change significantly between the mid-1950s and mid-1960s. While Akpalialuk's two youngest sons continued to live their father's *qammaq* after his death, 19 built a new *qammaq* beside that of his wife's brother.

Like many other camps in Cumberland Sound, Kingmiksoo was abandoned permanently in 1966.

### Opinivik

Opinivik is located on the southwest shore of a small island mid-way between Ikpit Bay and Robert Peel Inlet, 25 km northwest of Kingmiksoo (Figure 18).

This settlement is situated in good year-round ringed seal hunting habitat, near the mouth of an excellent char-fishing river known as Iqaluit. Although earlier occupations may have occurred here, the contact-traditional site of Opinivik was settled originally in 1923 by a small group of Umanaqjuarmiut.[37]

## 1926–1928

While every man at Opinivik possessed a kayak, only Nakashuk (1) owned a whaleboat (Figure 53). Nakashuk's adoptive father, Apiluk (2), was also regarded as a camp leader. Three other families, including those headed by 9 and 3 (Tooloogakjuaq's daughter) originally settled at Opinivik. However, by 1927 these families had moved to Kingmiksoo, and residential solidarity at Opinivik shifted from an *irniriik* core to one based on Nakashuk's relationship with his older brother, Tooleemaijuk (3), the natural son of Tooloogakjuaq. Although Tooloogakjuaq was not considered to be Nakashuk's natural father, Kokopaq (4) was the natural mother of Nakashuk and his brother (3).

## 1937–1939

Throughout the 1930s, Nakashuk's (1) relationship with his older brother (3) constituted the principal basis of camp solidarity, though his *irniriik* tie with his increasingly infirm *ataatasaq* (2) also contributed to residential stability. Horizontal ties within this group were strengthened, however, when Nakashuk's wife's (5) younger sister (6) and her husband (7) moved to Opinivik from Illutalik (Figure 53). The addition of this family and the death of 2, after which Nakashuk became the sole leader of the camp, were the principal changes that occurred at Opinivik in the 1930s.

During Apiluk's (2) last years, Nakashuk's eldest son moved into his grandmother's (8) household to take care of her. About the same time, Nakashuk's natural widowed mother (4) took up residence in her eldest son's (3) household. Ties among this group were fortified through Nakashuk's adoption of his wife's sister's *inngutaq* (10), who was also his cousin's (9) granddaughter (Figure 53).

The arrangement of houses at Opinivik conforms well with expectations derived from kinship ties and camp history (Figure 54). Nakashuk (1) and his *ataatasaq* (2) originally placed their houses (A and B, respectively) perpendicular to each other, where they shared a common work space. Individual 3 later built his house (C) on the other side of Nakashuk's *qammaq*, while Nakashuk's sister-in-law (6) and husband (7) constructed their house (D) at the opposite end of the camp.

Sharing customs at Opinivik were similar to those at Kingmiksoo, while every household maintained a meat cache in the porches of their dwellings, except Nakashuk, who built a storage facility beside his house. As with other Umanaqjuarmiut, caches from fall kills were placed on numerous islands along the southwest shore of the Sound and retrieved over the winter.

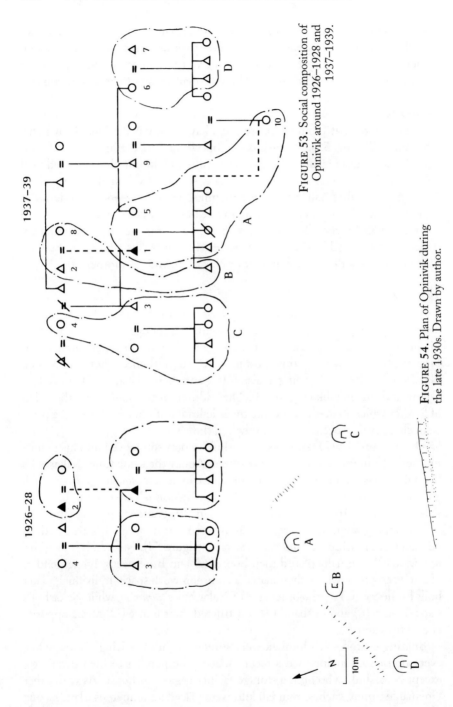

FIGURE 53. Social composition of Opinivik around 1926–1928 and 1937–1939.

FIGURE 54. Plan of Opinivik during the late 1930s. Drawn by author.

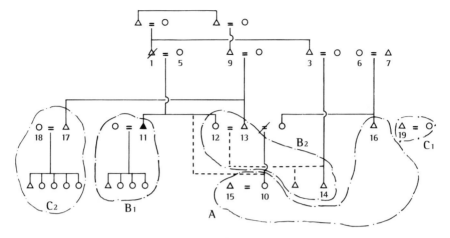

FIGURE 55. Social composition of Opinivik during the mid-1950s.

*1954–1956*
With the deaths of Nakashuk, his *anaana* (4), *anaanasaq* (8), and *nukaunnguk* (7), Nakashuk's widow (5), brother (3), and sister-in-law (6) moved elsewhere. By the mid-1950s, Opinivik's social composition had changed markedly (Figure 55). Nakashuk's eldest son (11) married and started a large family of his own, while Nakashuk's daughter (12) married her father's cousin's (9) son (13). This latter union was apparently a kin endogamous marriage as 13 called Nakashuk, *akka*. This latter couple has adopted two sons, including 12's cousin (14) from her uncle (3). Nakashuk's adopted daughter (10), the natural daughter of 13, has also remained at Opinivik, where she resided with 15, who was from Kingmiksoo. Living with this couple as their helper was 16, who was 10's *angak*. Individual 13 has attracted his older brother (17) and his wife (18), i.e., indiv.'s 26 and 27 from Kingmiksoo. Also living briefly at Opinivik during the mid-1950s was the well-travelled and now elderly 19 (indiv. 11 from Kingmiksoo) and his wife. Although Opinivik's population appears to have been in a state of flux from the mid-1940s to mid-50s (e.g., 10 and 15 also lived at Kingmiksoo, Opinivik, and Kipisa during this period), the sibling core of 11 and 12 constituted the basis of residential solidarity during this time. However, after 11 moved to Seegatok, group stability was founded largely on 13 and 17's *nukariik* relationship. Strong leadership, either under 11's or 13's tenure, was not characteristic of Opinivik during later occupations.

During the mid-1950s Nakashuk's eldest son (11) occupied his grandparents' *qammaq* (B), while his adopted daughter (10) remained in Nakashuk's house (A) (Figure 54). After 11 moved to Seegatok, however, Nakashuk's other daughter (12) and her family moved into dwelling B. Similarly, after 19 briefly occupied Nakashuk's brother's (3) old house (C), it was reoccupied by 17 and 18. Dwelling D appears to have been largely unoccupied after it was vacated by 6 and 7. Opinivik remained sparsely populated until it was abandoned around 1961.

### Kipisa

The contact-traditional settlement of Kipisa is situated on the northwest shore of Robert Peel Inlet near its entrance c. 10 km southeast of Opinivik (Figure 18). As with Opinivik and Kingmiksoo, ringed seal was a major determinant in the selection of this campsite. Also like the last two settlements, Kipisa was originally settled by a small group of Umanaqjuarmiut.[38]

#### 1936–1938

Kipisa became a permanent camp when two families headed by Koodlooaktok (1) moved here from Kingmiksoo around 1936 (Figure 56). While the opposite-sex sibling core of 1 and 2 constituted the foundation of group solidarity among this aggregation, these individuals also maintained close ties with their cousin, 3 (indiv. 9 from Opinivik). Relations between members of this founding generation were reinforced through adoption. Whereas 3's son, 4 (indiv. 13 from Opinivik), was adopted into 2's household, Koodlooaktok's youngest daughter (5) was adopted by his niece (6). The remaining family during this occupation was headed by Koodlooaktok's daughter (7) and Aksayuk's son (8) (indiv.'s 12 and 11 from Iqalulik). While this couple may have been performing bride-service for Koodlooaktok, they apparently moved often between Iqalulik and Kipisa. Individual 10 also moved to Kipisa from Iqalulik.

Movement of personnel between Iqalulik and Kipisa appears to have been commonplace. For example, around 1940, in response to Maniapik's death, Maniapik's son and Koodlooaktok's daughter (indiv.'s 9 and 15 at Iqalulik) moved to Kipisa, strengthening sibling ties among the descending generation. By the end of the decade this couple as well as Koodlooaktok (1), his sister (2) and brother-in-law (9), and other married daughter (7) and son-in-law (8) had moved, albeit not *en masse*, back to Iqalulik. Koodlooaktok (1) was the recognized leader at Kipisa during the late 1930s and 1940s. While he possessed the only boat, a pinnace, all men owned kayaks. Food was apparently shared freely among the inhabitants of Kipisa, while capital equipment was not. Matrilocal residence was the dominant arrangement at Kipisa during the late 1930s as individuals 8, 9, and 10 had no blood relatives there.

The spatial arrangement of dwellings at Kipisa accurately reflects social relationships as prescribed by the kinship system (Figure 57). The closest house to Koodlooaktok's (A) belongs to his eldest daughter and son-in-law (D), while the household of his niece (E) is further removed than any other. Intermediate in spatial distance are the households of his sister (B) and cousin (C).

#### 1947–1949

As Koodlooaktok became increasingly infirm during the late 1940s, leadership passed from his hands into those of his son-in-law, 11 (indiv. 15 at Kingmiksoo),

Cumberland Sound Inuit Kin and Local Groups, 1920–1970 215

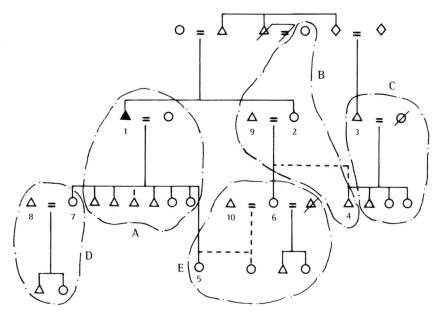

FIGURE 56. Social composition of Kipisa during the late 1930s.

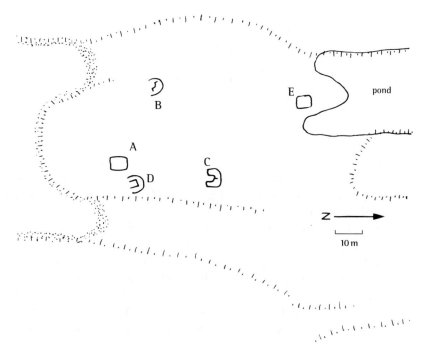

FIGURE 57. Plan of Kipisa during the late 1930s. Sketched by author.

who moved to Kipisa sometime in the 1940s (Figure 58). The fact that leadership was not transferred to one of Koodlooaktok's sons may have been related to the possibility that none were resident during the late 1940s. Whatever the case, this change in leadership altered the foundation of residential solidarity from one based on an opposite-sex sibling core to one founded primarily on ties between 11 (Nowyook Nicketimoosie) and his married sons (12 and 13) and daughter (14). Also contributing to group stability was Koodlooaktok's *panniriik* ties to his two resident daughters, 7 and 15. Although Koodlooaktok's sister (2) no longer resided at Kipisa during the late 1940s, his niece (6) and her husband (10) did. Adding stability to the dominant generation was the marriage of 7 and 15's adopted brother, 16 (indiv. 19 from Kingmiksoo), to their second cousin, 17. As 16 was adopted into Koodlooaktok's household from Tooloogakjuaq as an infant, this union likely represents second cousin marriage, even if it was not explicitly recognized as such—recall that 16 took his natural father's name, Akulujuk, as his own. Also continuing to reside at Kipisa during the late 1940s was Koodlooaktok's cousin's son (4), along with his daughter and son-in-law (indiv.'s 10 and 15 from Opinivik).

The positioning of dwellings at Kipisa during the late 1940s clearly reflects the emergence of one *ilagit* and subsequent decline of another. While Koodlooaktok (1) and his eldest married daughter (7) continued to reside at the south end of the site in the same houses they had occupied a decade earlier (*qammat* A and B, respectively), the central area of the camp was occupied primarily by Nowyook (11) and his married children (12, 13, and 14), where they lived, accordingly, in dwellings C, D, E, and F (Figure 59). Whereas households G and B may also be considered part of the core of the settlement, the houses of Nowyook's wife's second cousin (4) and adopted brother (16), dwellings H and I, respectively, are located on the periphery of the settlement some distance from feature C.

*1957–1959*
By 1950 Koodlooaktok, his sister (2) and brother-in-law (9), and daughter (7) and son-in-law (8) resided at Iqalulik. Though 16 and 17 remained at Kipisa, Nowyook (11) and his married sons (12, 13, and 19) and daughter (14) formed the basis of residential solidarity (Figure 60). This aggregation became even more vertically structured after 16's sister (15) died. However, 16's ties with Nowyook were maintained when his son (20) married Nowyook's *inngutaq* (21). While this union may verge on kin endogamy, the possibility that 16's status may have been somewhat negotiable might have sanctioned it. Another case of local group endogamy occurred when 19 died and his brother (12) married 19's widow (22) after 12's wife's (23) death. Multiple ties of affinity were also created when Nowyook married his son's wife's mother (24), the widow of Akpalialuk.

Cumberland Sound Inuit Kin and Local Groups, 1920–1970 217

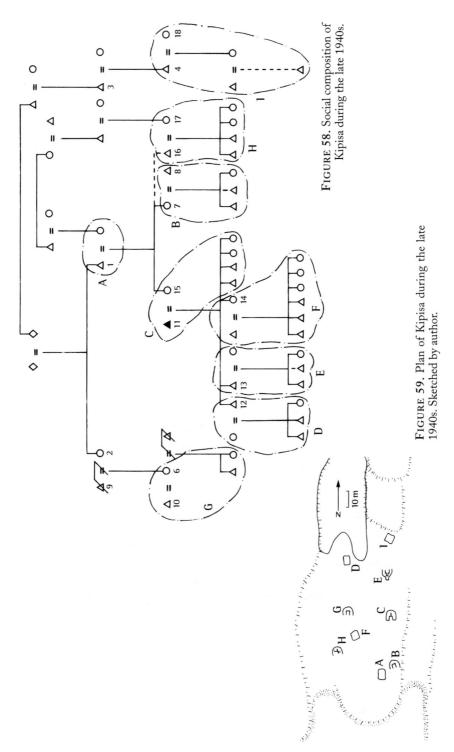

FIGURE 58. Social composition of Kipisa during the late 1940s.

FIGURE 59. Plan of Kipisa during the late 1940s. Sketched by author.

Adoption appears to have been somewhat limited during this occupation; only 14 and 25 have adopted children, who were their *inngutat*, into their household. However, these adoptions may have been initially more a form of 'caretakership' than an actual transferral of rights and obligations as the natural parents of these children lived and worked in Pangnirtung.

Under Nowyook's (11) leadership, Kipisa became known as a fairly prosperous camp. He owned and controlled the use of two whaleboats, one of which was motorized. The other was apparently operated by his eldest son (12). While other households maintained a food cache in their porches, Nowyook built a shack especially for this purpose. The practice of freely sharing food, while restricting access to capital equipment, such as boats, continued throughout this occupation.

Nowyook's *qammaq* (A) is flanked by those of his three married sons (B, C, and D), while the household of his daughter (E) is located some distance away (Figure 61). In accordance with kinship directives, the *irniriik* core of 16 and 20 occupy dwellings (F and G), which are further removed from feature A than those of Nowyook's sons. Individuals 20 and 21 have set up residence (F) mid-way between those of their parents (B and G).

## *1960–1984*

Throughout the 1960s and 1970s, Nowyook remained the *angajuqqaq* of Kipisa. Although other families came and went during this period, Nowyook's ties to his married daughters and sole resident son (12), who also lived at Seegatok for part of this period, remained the basis of residential solidarity. We are fortunate that the anthropologist, Jean Briggs, spent 16 months among the Kipisamiut in the 1970s, at which time she made a number of observations about Kipisamiut social organization:

> (The Kipisamiut) are composed of bilaterally related kin—a core of close relatives together with a few other families that are related to the core in various ways.... The core comprises an old man, widowed, ... his married daughters and their families, and his unmarried adult and adolescent children of both sexes. Numbers fluctuate from year to year as the less centrally related families join or separate from the main group. They also fluctuate seasonally, as families tend to disperse in spring and summer and rejoin one another in the autumn at a central winter camp. (1982: 111)

In regard to group leadership, Briggs observed that:

> sons and sons-in-law may continue to defer in certain matters to the wishes of their fathers and fathers-in-law even after they marry and set up their own households, especially if they live in the same camp. In the case of the Qipisamiut, the elder—the father of them all—is recognized as an authority

Cumberland Sound Inuit Kin and Local Groups, 1920–1970 219

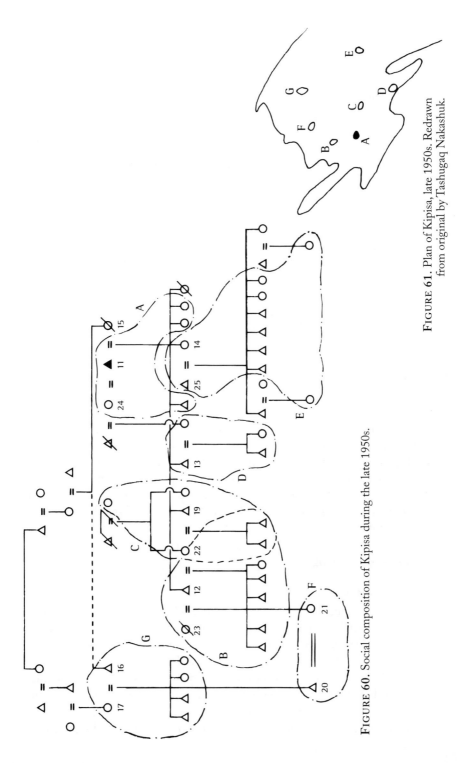

FIGURE 60. Social composition of Kipisa during the late 1950s.

FIGURE 61. Plan of Kipisa, late 1950s. Redrawn from original by Tashugaq Nakashuk.

of this kind by everyone in the camp. In everyday matters of whether or not to hunt and where and what, he exercises authority only over his own household members; but in long-range decisions, such as whether to move to Pangnirtung or not, people tend to defer to his wishes. No household head is sanctioned if he makes his own decisions in such matters; deference is voluntary, but phrased as loyalty, it is nevertheless often there. (1982: 112)

Sometime during the 1960s, Nowyook's *ilagit* was joined by another headed by individual 37 from Kingmiksoo. This man, who was originally adopted by Aulaqeak at Ussualung from Ishulutaq (indiv. 8 from Idlungajung), appears to have joined his two sons, who have married Nowyook's youngest daughter and eldest granddaughter. Others, most notably Nowyook's third wife's son and granddaughter have married Nowyook's grandchildren, again intensifying and complicating affinal ties within the group.

Matrilocality appears to have been far more pronounced than patrilocality during the last decade and a half at Kipisa. Although other factors may have contributed to this situation (see next chapter), the promise of a better life in Pangnirtung may have attracted Nowyook's younger son(s).

Kipisa remained the only permanently occupied contact-traditional camp in Cumberland Sound throughout the 1970s and early 1980s. It was not the demise of the seal skin market nor the attraction of Pangnirtung, but the ill health of Nowyook, that eventually forced Kipisa to be abandoned in 1984.

### Illutalik

The contract-traditional settlement of Illutalik is located *c.* 17 km NNE of Opinivik on the northwest end of a small island (Figure 18). Illutalik was first settled around 1930 when a small group of Umanaqjuarmiut moved here. As with other settlements on the southwest shore of the Sound, ringed seal was a major determinant in the selection of this camp. While the sheltered waters and calmer seas around Illutalik reduced the dangers of seal hunting from kayaks, Illutalik appears to have been in an especially advantageous position to procure immature seals (*netsiavinik*) when they congregated at the mouth of Brown Inlet each spring.[39]

#### *1931–1933 and 1937–1939*

Individuals 1 (Pitsualuk) and 2 (Kudlu) left Etelageetok for Illutalik around 1931, where they joined a family headed by 1's sister (3) and 2's brother (4) (indiv.'s 6 and 7 at Opinivik). Two other families headed by the male sibling core of Oshutapik and Petaosie (indiv.'s 4 and 5 from Iqalulik) also apparently lived briefly at Illutalik around 1930. However, they soon departed and residential solidarity at Iqalulik during the early 1930s was cemented in the bonds between the above intermarrying opposite-sex sibling cores (Figure 62). Pitsualuk and his sister (Unaq) were the adopted son and natural daughter of Pawla (5), the

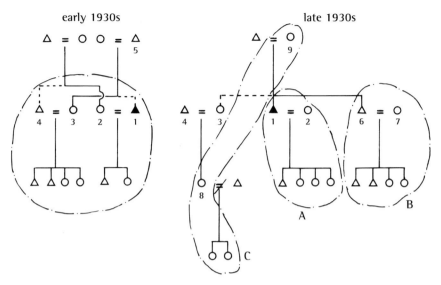

FIGURE 62. Social composition of Illutalik during the early and late 1930s.

whaling and trading foreman at Umanaqjuaq. Although 4 (Akulujuk) was older, Pitsualuk was regarded generally as the *angajuqqaq* of this camp, as he owned the only whaleboat, which he inherited from Pawla. While both families initially lived in the same house, 3 and 4 moved to Opinivik after a few years, where they joined Unaq's older sister. About the same time, Pitsualuk's younger brother (6) and wife (7) moved to Illutalik (Figure 62). Also living at this camp during the late 1930s was 3 and 4's daughter (8), the former wife of 13 from Opinivik, along with her two daughters and grandmother (9), who was 1 and 6's natural mother. Whereas 8 stayed at Illutalik only for a few years, 1 and 6's *nukariik* relationship formed the basis of residential solidarity at this camp for the next two and a half decades. Very strong bonds of affection also existed between 2 and 7—they had lived in the same house together at Etelageetok.

## 1952–1954

During the 1940s, Pitsualuk and Kudlu (2) took in Pitsualuk's step-sister's sons (10 and 11), who soon married their daughters, 12 and 13 (Figure 63). Despite the strong resemblance to first cousin marriage, these unions apparently were not considered as such because of Pitsualuk's adopted status. These couples remained at Illutalik and started large families of their own. Pitsualuk also attracted his widowed half-sister (14) and her son (15) during the early 1950s. The latter has, in turn, married 6 and 7's eldest daughter (16). This union was also likely not regarded as first cousin marriage by the principals involved, since the relationship between 6 and 14 was not considered to be an especially close one, a fact reflected in 14 and 15's brief stay here during the early 1950s.

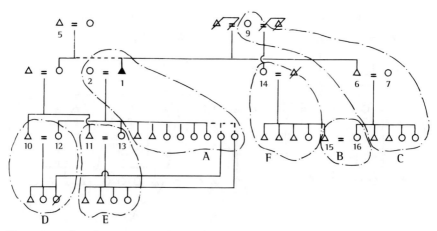

FIGURE 63. Social composition of Illutalik during the early 1950s.

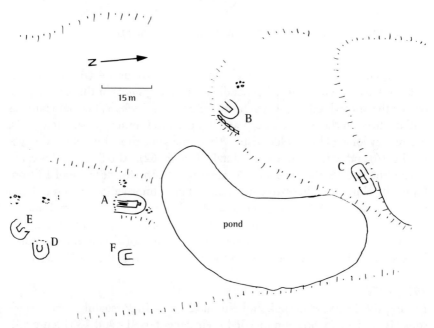

FIGURE 64. Plan of Illutalik during the early 1950s. Sketched by author.

Pitsualuk and Kudlu remained in the same house (A) that they had occupied during the early 1930s (Figure 64). However, after 15 and 16 married and moved into 6's old house (B), the latter built a new *qammaq* (C) at the north end of the site. As might have been predicted, the intermarrying brother and sister sibling cores of 10 and 11, and 12 and 13, built their houses (D and E) perpendicular to each other to form a common work space near Pitsualuk's *qammaq* (A). Pitsualuk's half-sister (14) moved into a dwelling (F), albeit briefly, adjacent to Pitsualuk's house.

*1964–1966*
Individuals 10 through 13's multiple consanguineal and affinal linkages constituted an important element of stability at Illutalik during the 1950s and early 1960s. Even so, Pitsualuk's *nukariik* bond with his younger brother (6), as well as 2 and 7's sibling-like relationship provided the foundation of residential solidarity at Illutalik throughout its history—these two families would often be the sole residents of Illutalik.[40] The only major development to occur at Illutalik during this period was Kudlu and Pitsualuk's adoption of two grandsons from their daughters, 12 and 13. Illutalik was permanently abandoned during the winter of 1966–67.

## Other Umanaqjuarmiut Camps
Not every camp occupied by the Umanaqjuarmiut between 1920 and 1970 has been described in the foregoing presentation. For example, there appear to be several small settlements about which very little information could be extracted, either from archival sources or my informants. Such camps include Etalik (1936), Kasigejut (*c.* 1930), Iglulik (1930), Neakunggoon (1927–1936), and Mamukto (1930). Fortunately, we have better information on the occupational history and social composition of three other Umanaqjuarmiut camps: Seegatok, Nuvujen, and Etelageetok.

*Seegatok*
The settlement of Seegatok was located on a small island near the mouth of Brown Inlet. As with other nearby camps, Seegatok provided access to good seal hunting grounds. Unlike most camps, however, Seegatok appears to have been inhabited only intermittently, particularly in the mid-1940s and late 1960s.[41]

Seegatok might have been inhabited in the mid-1940s, or perhaps earlier. This occupation, however, must have been brief as no dwellings or any other signs of human activity apparently remain from this time period. The foundation of group solidarity among this aggregation lay principally in the relationship between two brothers. The older brother and his wife (indiv.'s 24 and 25 from Kingmiksoo) lived with his eldest daughter and her husband (indiv.'s 28 and 29 at Kingmiksoo) in the same household, where the latter may have been preforming bride-service. Though the latter was the nephew of the older brother's wife, his marriage was not recognized generally as an endogamous union. It is not certain, however likely, whether 1 was the leader of this group or whether he owned a whaleboat.

Sometime during the late 1960s Nowyook's eldest son from Kipisa (indiv. 12 at Kipisa) appears to have moved to Seegatok along with his second cousin, who was the daughter of the founding inhabitants of the site. Although

Nowyook's son owned a boat, there was no overall leader among this small group, as *angajuqqat* were apparently not a common feature of small camps on this side of the Sound (Pauloosie Nowyook, personal communication, 1989). While this individual and his family appear to have soon returned to Kipisa sometime in the early 1970s, his second cousin and his family, along with their married children, moved permanently to Pangnirtung.

## *Nuvujen*

The mid-nineteenth century spring floe whaling base of Nuvujen was occupied in the 1920s by the well-known Umanaqjuarmiut hunter, Kaka, and his two sons, Oshutapik and Petaosie (indiv.'s 4 and 5 from Iqalulik). Akulujuk from Illutalik and Opinivik also lived at Nuvujen in 1923–24, but his exact kin relationship to Kaka is unknown. Although Kaka's two sons appear to have resided elsewhere during the 1920s, at Nuvujen they occupied the same snowhouse in the winter and same *qammaq* in the fall as their father. A falling-out between Kaka and Petaosie apparently precipitated the murder-suicide of Kaka and his wife, and subsequent abandonment of the camp in 1927.[42]

## *Etelageetok*

Located on a small island between Umanaqjuaq and Kingmiksoo, Etelageetok was occupied briefly for a year or so around 1930 by a small group of Umanaqjuarmiut.[43] The sibling core of 1 (Pitsualuk) from Illutalik, 6 from Illutalik, and 21 from Kingmiksoo appear to have provided the basis of solidarity among this group. However, as the latter's husband (indiv. 20 from Kingmiksoo) was the eldest related hunter, and owned the only boat in camp, he may have been regarded as the leader of this aggregation. The only other family to live here during this time was headed by individual 3 from Kipisa. The close *nukariik* tie between Pitsualuk and his younger brother is revealed in the fact that they lived in the same house. As discussed above, their sister was actually the step-sister of Pitsualuk, whose sons later married Pitsualuk and Kudlu's daughters at Illutalik. Although Pitsualuk's step-sister and her husband left for Kingmiksoo, Pitsualuk and his brother settled at Illutalik as seals and fox were more abundant and Pitsualuk wished to live with his other sister.[44]

# Chapter 6

## The Structure of Cumberland Sound Inuit Social Organization

Not all contact-traditional camps in Cumberland Sound were organized in the same way. In fact, given the size of the area and population under consideration, the variability exhibited in local group organization is surprising. Nevertheless, neither were most settlements eclectic clusters of pragmatically connected individuals. Rather, strong kinship ties and consistent arrangements of social features within most local groups suggest that Inuit in Cumberland Sound employed some underlying logic in the formation of their residential groups and the propagation of their productive relationships. Specifically, incongruities between Qikitarmiut and Umanaqjuarmiut aggregations in leadership, marriage, adoption, and a host of other variables indicate that two different strategies of affiliation or principles of group formation existed in Cumberland Sound.

Some Umanaqjuarmiut aggregations adopted, at certain times and places, characteristics normally associated with Qikitarmiut camps, and vice versa. Even so, dissimilarities between these two subregional populations are congruous enough as to entertain the possibility that two fundamentally different social formations operated in Cumberland Sound during the contact-traditional period. While the ecological basis for these different structural tendencies will be explored, much of the variability observed in social features during the early contact period, and presumably late prehistoric times, appears simply the result of an emphasis on one or the other strategy of affiliation. These two structural orientations provide the bases for closer examination of Cumberland Sound Inuit prehistory, variability in Central Inuit social organization, Canadian Arctic prehistory, and the anthropological and political implications of Central Inuit social structure.

Yet, no matter how comprehensive local group profiles presented in the previous chapter might appear, they document only the composition of residential aggregations at certain times. In this regard, most profiles fail to chronicle completely the dynamics and complexity of local group organization at specific locations over the full duration of occupation. Rather, what most camp histories record are gross patterns of continuity and change in social structure, particularly among core group members. All local groups in Cumberland Sound during the contact-traditional period possessed a core group of relatives around which more peripherally related individuals and families attached themselves. However, the primary kinship ties binding these central cores together differed depending on whether they were Qikirtarmiut or Umanaqjuarmiut. How local groups associated with each division maintained and reproduced their productive relations and activity is the subject of what follows.

Each local group in Cumberland Sound was confronted with the same basic problem: maintaining its own forces and relations of production from one year to the next. No more forcefully did this come to the fore as after the loss of a predominant group member, such as the head of a central core family. What were the principal social features and institutions that facilitated the perpetuation of these respective structures under such circumstances, and how did they operate in support of each other to maintain the underlying structure of each group?

Employing such anthropological constructs as leadership, marriage, and territoriality in our examination of Cumberland Sound Inuit social organization is attractive insofar as they offer the advantages of simplicity and clarity of presentation. However, the use of such constructs also brings together so many disparate relationships and categories that precision in meaning and distinctiveness is sacrificed (Trott 1982). For example, leadership influences and is itself influenced by group size, post-nuptial residence patterns, residential stability, etc. Thus one cannot undertake an analysis of these constructs without considering their interrelationships and effects upon each other. Even so, their use may be the most parsimonious way to examine Cumberland Sound Inuit social organization so long as we acknowledge that our goal in this chapter is to document the operation of two structural tendencies.

Perhaps the best way to describe Qikirtarmiut and Umanaqjuarmiut social organizational differences from the outset is to consider the former as vertically structured and the latter as horizontally structured. Simply stated, the former favoured adjacent-generation kin relationships, while the latter valued same-generation arrangements. Sibling cores provided the structural foundation of most Umanaqjuarmiut aggregations, while parent-child cores formed the basis of most Qikirtarmiut camps. Examining various social features and their systemic associations will help determine how they define and support the operation of each structure.

## STRUCTURAL TENDENCIES

As Inuit abandoned Kekerten and Umanaqjuaq for their traditional settlements and hunting grounds at the beginning of the contact-traditional period they faced the problem of forming groups that could maintain and perpetuate their productive forces and relationships. While the size of aggregations varied considerably, the minimum local group has been estimated by Etuangat Aksayuk to consist of four to five nuclear families:

> I remember a family of four qammaq (sic) would often form a camp . . . when they moved to another area outside Kekerten. They would be able to form a community with about four or five qammaqs (sic). Better to have more people around when they were hunting seals at their breathing holes, although they would also hunt together at the floe edge.[1]

Even though this anecdote refers specifically to Qikirtarmiut group formation, Hantsch (1977), RCMP censuses taken in the 1920s and data presented in the previous chapter suggest that four was also the minimum number of productive couples required to form viable local groups among the Umanaqjuarmiut. How each regional subdivision achieved this goal, however, differed. Specifically, primary kinship ties among Qikirtarmiut groups were based largely on parent-child relationships, whereas Umanaqjuarmiut groups were established mainly on sibling cores (Table 13). Even though social solidarity in each local group was accomplished largely through primary kinship ties between central core group members, secondary kin relationships also contributed to the stability of the local group. For example, despite the fact that a weak, opposite-sex sibling core characterized the top generation among Angmarlik's kinsmen at Idlungajung during the early 1920s, *uyurugiik* ties with his ZS and ZD and *ningaugiik* relationships with his DH and ZH gave this aggregation an underlying vertical structure. Similarly, while strong parent-child bonds dominated kin ties among Nowyook's kin group at Kipisa (late 1940s), well-developed sibling ties among the top generation contributed horizontal stability to this aggregation.

What seems apparent is that a good mixture, or balance, of both vertical and horizontal kin relationships was required in the formation, maintenance, and reproduction of Qikitarmiut and Umanaqjuarmiut local groups. In this regard, a negative correlation was observed between primary and secondary kinship ties among central cores in both subregional populations. If primary kin relationships among central core group members were vertically structured, then there was a 86.4 per cent probability among the Qikirtarmiut and a 68.4 per cent probability among the Umanaqjuarmiut that secondary kin ties would be horizontally structured. In other words, for those local groups exhibiting both primary and secondary kinship relationships, only three out of twenty-two cases among the Qikirtarmiut and six out of nineteen cases among the Umanaqjuarmiut exhibited the same structure in both categories (Table 13). A simple statistical measure, the sign test, where $Z=3.44$, indicates that this tendency is significant beyond the .0003 level of confidence. When the strength of these kinship relationships is considered (see below), only one kin group demonstrated strong, similarly structured primary and secondary ties.

TABLE 13. Structure and strength of kinship ties among Qikirtarmiut and Umanaqjuarmiut central group cores. (M=mother, F=father, P= F+M, S=son, D=daughter, B=brother, Z=sister, GP=grandparents, GS=grandson, GD=granddaughter, C=cousin)

| Camps/ Occupations | Occupation Date and Kin Group Head | Primary Kinship Ties and Structure of Core Adults/Families | Strength and Generation of Primary Ties | Strength and Generation of Secondary Ties |
|---|---|---|---|---|
| **Qikirtarmiut** | | | | |
| 1) Nunaata | mid-20s (Ken) | B-B,B (male sib. c.) | weak same-gen. | mod. adj.-gen. |
| 2) " | early 40s ( " ) | F-S,S,D,D (par.-child c.) | strg. adj.-gen. | weak adj.-gen. |
| 3) " | early 60s ( " ) | B-B,B (male sib. c.) | mod. same-gen. | weak adj.-gen. |
| 4) Idlungajung | early 20s (Ang) | B-B-Z (opp.-sex sib. c.) | weak same-gen. | strg. adj.-gen. |
| 5) " | mid-30s (Ang) | P-S,S,D,D (par.-child c.) | strg. adj.-gen. | strg. adj.-gen. |
| 6) " | mid-40s (Ang) | P-S,S,D,D,D (par.-child) | strg. adj.-gen. | weak same-gen. |
| 7) " | mid-40s (Evi) | P-S,D (par.-child c.) | strg. adj.-gen. | weak same-gen. |
| 8) Avatuktoo | early 30s (Aga) | Z-Z (female sib. core) | mod. same-gen. | mod. adj.-gen. |
| 9) " | early 40s (Sok) | P-D,D,D,S (par.-child c.) | strg. adj.-gen. | weak same-gen. |
| 10) " | late 50s (Sou) | M-D,D,D (par.-child c.) | strg. adj.-gen. | weak same-gen. |
| 11) Tuapait | mid-50s (Kis) | B-B (male sib. c.) | mod. same-gen. | — |
| 12) " | mid-50s (Kon) | P-D (par.-child c.) | mod. adj.-gen. | — |
| 13) " | early 60s (Kis) | B-B (male sib. c.) | mod. same-gen. | — |
| 14) " | early 60s (Kon) | P-S,D,D (par.-child c.) | strg. adj.-gen. | weak same-gen. |
| 15) Sauniqtuajuq | mid-20s (Atg) | P-S,S (par.-child c.) | mod. adj.-gen. | mod. same-gen. |
| 16) " | late 30s (Atg) | F-S,S,S (par.-child c.) | strg. adj.-gen. | weak same-gen. |
| 17) " | late 40s (Atg) | F-S,S (par.-child c.) | strg. adj.-gen. | mod. adj.-gen. |
| 18) " | late 40s (Now) | F-S,D,D,D (par.-child c.) | strg. adj.-gen. | — |
| 19) " | early 60s (Atg) | B-B,B (male sib. c.) | mod. same-gen. | mod. adj.-gen. |
| 20) " | early 60s (Now) | B-B,Z,Z (opp. sex sib. c.) | mod. same-gen. | mod. adj.-gen. |
| 21) Iqalulik | mid-50s (Aks) | F-S (par.-child c.) | strg. adj.-gen. | mod. same-gen. |
| 22) Naujeakviq | mid-30s (Vee) | P-S,S (par.-child c.) | strg. adj.-gen. | weak same-gen. |
| 23) " | mid-50s (Vee) | P-S,S,S (par.-child c.) | strg. adj.-gen. | mod. same-gen. |
| 24) " | early 60s (Vee) | B-Z (opp.-sex sib. c.) | weak same-gen. | mod. adj.-gen. |
| 25) Keemee | early 60s (Evi) | B-Z (opp.sex sib. c.) | weak same-gen. | — |
| 26) Kingnait | early 30s (Now) | B-B (male sib. c.) | weak same-gen. | mod. adj.-gen. |
| **Umanaqjuarmiut** | | | | |
| 1) Ussualung | early 20s (Aul) | Z-Z,B,B (opp.-sex sib. c.) | mod. same-gen. | mod. adj.-gen. |
| 2) " | late 30s (Akp) | Z-Z (female sib. c.) | strg. same-gen. | mod. adj.-gen. |
| 3) Sauniqtuajuq | mid-20s (Man) | B-Z,Z (opp.-sex sib. c.) | strg. same-gen. | mod. adj.-gen. |
| 4) Iqalulik | early 30s (Man) | B-Z (opp.-sex sib. c.) | strg. same-gen. | — |
| 5) " | early 30s (Osh) | B-B (male sib. c.) | strg. same-gen. | — |
| 6) Kingmiksoo | mid-20s (Too) | P-D,D,D (par.-child c.) | mod. adj.-gen. | mod. adj.-gen. |
| 7) " | mid-20s (Koo) | B-B, Z (opp.-sex sib. c.) | strg. same-gen. | mod. adj.-gen. |
| 8) " | late 30s (Too) | GP-GS,GD (gr.-ch. c.) | mod. altrn.-gen. | mod. adj.-gen. |
| 9) " | mid-50s (Akp) | P-S,S,S,S,D (par.-child ) | strg. adj.-gen. | mod. same-gen. |
| 10) Opinivik | late 20s (Ape) | P-S (par.-child c.) | weak adj.-gen. | mod. same-gen. |
| 11) " | late 30s (Nak) | B-B (male sib. c.) | mod. same-gen. | mod. same-gen. |
| 12) " | mid-50s (Sek) | B-Z (opp.-sex sib. c.) | mod. same -gen. | mod. same-gen. |
| 13) Kipisa | late 30s (Koo) | B-Z (opp.-sex sib. c.) | strg. same-gen. | mod. adj.-gen. |
| 14) " | late 40s (Noy) | P-S,S,S,D (par.-child c.) | strg. adj.-gen. | mod. same-gen. |
| 15) " | late 50s (Noy) | F-S,S,S,D (par.-child c.) | strg. adj.-gen. | mod. same-gen. |
| 16) " | late 70s (Noy) | F-D,D,D,S (par.-child c.) | strg. adj.-gen. | mod. adj.-gen. |
| 17) Illutalik | early 30s (Aku) | B-Z (opp.-sex sib. c.) | mod. same-gen. | mod. same-gen. |
| 18) " | late 30s (Pit) | B-B (male sib. c. ) | strg. same-gen. | mod. adj.-gen. |
| 19) " | late 50s (Pit) | B-B (male sib. c.) | strg. same-gen. | mod. adj.-gen. |
| 20) Seegatok | mid-40s (?) | B-B (male sib. c.) | mod. same-gen. | mod. adj.-gen. |
| 21) " | late 50s | C-C (male cousin c.) | mod. same-gen. | — |
| 22) Nuvujen | mid-20s (Kak) | P-S,S (par.-child c.) | mod. adj.-gen. | strg. adj.-gen. |
| 23) Etelageetok | early 30s (Ino) | B-B,Z (opp.-sex sib.c.) | mod. same-gen. | — |

TABLE 14. Umanaqjuarmiut and Qikirtarmiut differences in structure of primary kinship ties among central core adults/families.

|  | VERTICALLY STRUCTURED OR ADJACENT-GENERATION CORES | HORIZONTALLY STRUCTURED OR SAME-GENERATION CORES | TOTALS |
|---|---|---|---|
| Umanaqjuarmiut | 8 | 15 | 23 |
| Qikirtarmiut | 15 | 11 | 26 |
| Totals | 23 | 26 | 49 |

Perhaps not surprisingly, this was Angmarlik's kin group at Idlungajung during the mid-1930s, where both primary and secondary kin ties were governed by strong adjacent-generation kinship relations.

Differences between Qikirtarmiut and Umanaqjuarmiut, as reflected in primary kin relationships among central core group members, can be tested statistically by referring to Tables 14 and 15. While Umanaqjuarmiut aggregations were based largely on sibling or sibling-like cores (15 same-generation cores as opposed to eight adjacent-generation cores), there was a more even but opposite distribution of horizontally and vertically structured cores among Qikirtarmiut kin groups (11 as opposed to 15), with parent-child cores predominating (Table 14). Using another simple measure, the chi-squared test, this difference is judged to be significant at the .15 level of confidence, but not at the .10 level of confidence ($X^2$=2.57, df=1).

A more accurate test of the structural differences between these two subregional populations is obtained, however, when the strength of the relationship among central core group members is considered (Table 15). This characteristic was determined by a variety of contextual data including the duration of cohabitation, location of households, structure of productive relationships (e.g., hunting partnerships and the composition of trading parties), number and continuity of personnel in the central core group over time, as well as other variables not usually accessible to ethnographic observation (see Chapter 5). In fact, had an ethnographer undertaken a study of Cumberland Sound Inuit social organization during the contact-traditional period, limiting research to a few field seasons, s/he might well have attributed any differences in social organization observed between kin groups associated with each subregional division to be the result of the inevitable cyclical tendency in social structure that all local groups must undergo to remain viable over the long term (Goody 1966). In other words, all kin groups must cope with the loss of central core members in the top generation, and leaders in particular. The original structure of a local group, and whether it adopted a more horizontal form (such as the Nunaatarmiut did after Keenainak died), or split apart after the death of its leader (such as the Kingmiksormiut did upon the death of Tooloogakjuaq), normally can only be determined through consideration of diachronic information. For example, while most local groups associated with both populations

TABLE 15. Umanaqjuarmiut and Qikirtarmiut differences in strength and structure of primary kinship ties among central core adults/families.

|  | STRONG SAME GEN. | MODERATE SAME GEN. | WEAK SAME GEN. | WEAK ADJACENT GEN. | MODERATE ADJACENT GEN. | STRONG ADJACENT GEN. |
|---|---|---|---|---|---|---|
| Qikirtarmiut | 0 | 6 | 5 | 0 | 2 | 13 |
| Umanaqjuarmiut | 8 | 7 | 0 | 1 | 3 | 4 |
| Totals | 8 | 13 | 5 | 1 | 5 | 17 |

were founded originally on sibling cores, rarely did such arrangements last for more than a few years among the Qikirtarmiut, as most camps reverted at the earliest opportunity to a more vertical format with parent-child ties providing the basis of residential solidarity. In turn, the spatial arrangements of dwellings at most Qikirtarmiut settlements tend to reflect the predominance of parent-child relationships over others.

When the strengths of various kin ties among central core group members are considered we find that differences observed between Umanaqjuarmiut and Qikirtarmiut kin groups to be significant beyond the .005 level of confidence ($X^2=19.13$, df=5) (Table 15). These differences in structure are even more pronounced if we examine what may be considered anomalies for these subregional groups. For example, Akpalialuk and his kin group's occupation of Kingmiksoo has been included with other Umanaqjuarmiut camps, even though he originated from Coral Harbour. While Akpalialuk was undoubtedly influenced by the values, customs, and traditions of his wife's kinsmen, the Umanaqjuarmiut, he was nonetheless raised on Southampton Island and carried the cultural baggage of another regional group (the Aivilingmiut?).

If we exclude this case for the time being, only three Umanaqjuarmiut occupations are dominated by strong adjacent-generation relationships, and all three occupations are associated with only one kin group, that headed by Nowyook at Kipisa. Yet the last of these occupations was based on *panniriik*, as opposed to *irniriik*, ties. While some Qikirtarmiut kin groups were founded on parent-daughter relationships, it is interesting to observe that the most prominent of all Umanaqjuarmiut males, Tooloogakjuaq and Koodlooaktok, were unable to form groups based on father-son ties. Although accidental or premature death might have contributed to this situation, it should be recalled that *irniriik* cores were among the most *naalaqtuq*-directed of any in Central Inuit society. From this perspective, then, the possibility should be considered that the lack of strongly developed *irniriik* cores among Umanaqjuarmiut kin groups may represent a systemic rejection of respect-obedience directives as a basis for forming groups. Clearly, the lack of such cores has important ramifications for Umanaqjuarmiut group structure, insofar as camp leaders had to form economic relationships with individuals other than their sons—a niche that their same-generation kinsmen and sons-in-law appear to have assumed.

TABLE 16. Umanaqjuarmiut and Qikirtarmiut differences in strength of primary kinship ties among parent-child cores with resident adult sons and/or children of both sexes present.

|  | STRONG/WELL-DEVELOPED TIES | NOT STRONG/POORLY DEVELOPED TIES | TOTALS |
| --- | --- | --- | --- |
| Qikirtarmiut | 9 | 1 | 10 |
| Umanaqjuarmiut* | 2** | 2 | 4 |
| Totals | 11 | 3 | 14 |

\* Does not include Akpalialuk and his kin group's occupation of Kingmiksoo.
\*\* Only one kin group, i.e., that of Nowyook's at Kipisa, is represented in this category.

By way of contrast, the sons of Qikirtarmiut camp leaders, including Attaguyuk and Veevee, tended to remain permanently with their fathers. Even stronger leaders, such as Angmarlik and Keenainak, however, retained both sons and daughters within the local group, while at the same time attracting sons-in-law and other individuals. This phenomenon has implications for the relationship between leadership and residence in Oqomiut society. If we take into account the strength of parent-child cores where sons and/or children of both sexes remained with their parents after marriage (Table 16), we find that 90 per cent of these cores among the Qikirtarmiut to be strongly developed, while only 50 per cent of such cores among the Umanaqjuarmiut are similarly structured. This difference is significant at the .10 level of confidence, where $X^2=2.71$, df=1.

The observed tendency for Umanaqjuarmiut kin group heads to retain their daughters, but not their sons, may also reflect an emphasis on female relationships, particularly between sisters and between mothers and daughters, whereby *ungayuq* directives are especially well-developed. In this light, statements made by Valentine (1952) regarding the matrilocal tendency he saw in Boas' descriptions of Oqomiut society might possibly attain greater credibility than previously accorded (see Chapter 2). The strength of female relationships in Umanaqjuarmiut society may also account for Cardno's (Ross 1985c: 235) observation that marriages were arranged principally by mothers and grandmothers. Similarities between Qikirtarmiut and Umanaqjuarmiut social structures notwithstanding, their differences indicate that we can no longer paint historic Cumberland Sound Inuit, let alone aboriginal Oqomiut, social organization with the same brush.

The scarcity of well-developed *irniriik* cores among the Umanaqjuarmiut, as well as an apparent emphasis on sibling, female, and other *ungayuq*-dominated relationships, may also inform the tragic murder-suicide of Kaka and his wife at Nuvujen in 1927. Petaosie, Kaka's second eldest son, left his father at the end of March of that year.[2] Less than two weeks later, Kaka shot his wife and himself. Although we cannot be certain that the loss of his son drove Kaka to this extreme, we know that Kaka's camp was an anomaly among Umanaqjuarmiut local groups insofar as he attempted to maintain a vertical structure in the midst of other, more horizontally structured settlements. In this regard, it is noteworthy that Kipisa under Nowyook's leadership also appears to have been isolated socially from other Umanaqjuarmiut groups.[3] These examples lead us to consider the possibility that too much *naalaqtuq*, especially in father-son relationships, may have introduced systemic contradictions into Umanaqjuarmiut society.

The fact that the Qikirtarmiut attached far greater importance to father-son relationships than the Umanaqjuarmiut finds additional support in the Pangnirtung Post Journals. No fewer than 31 Inuit parties that traded at Pangnirtung between 1922 and 1935 were composed of *irniriik* cores.[4] Of these, 26 parties were composed of prominent Qikirtarmiut and their adult and/or adolescent sons, nearly half of which were headed by Angmarlik and Keenainak. Alternatively, the remaining five father-son trading parties were split between Umanaqjuarmiut and Saumingmiut. While distance may have factored into this finding, over 80 per cent of all Umanaqjuarmiut parties that traded at Pangnirtung during this period were composed of more anomalously or horizontally structured parties. For example, typical Umanaqjuarmiut trading parties recorded for this period, and their dominant kin relationship, include the following:

'Pudjut, Kelabuk, Kupee and Angnakuk . . . from Maniapik's camp', (indiv.'s 3 and 4 = husbands of two sisters);
'Kukkik, Merkusah, Kanajuk and Pitchulak . . . from Blacklead', (indiv.'s 2 and 4 = husbands of aunt and niece);
'Peterlosey and Akkulujuk . . . for Iglootalik', (same generation, but exact tie unknown);
'Nakasook, Kohnayoke, Mosesie from Blacklead', (indiv.'s 2 and 3 = husbands of two sisters); and
'Pitsualak, Kukkik, Petusie and Novakik from Blacklead', (indiv.'s 1 and 4 = two brothers).[5]

Clearly, father-son trading parties were uncommon among the Umanaqjuarmiut.

Another line of evidence that suggests different structural tendencies existed in Cumberland Sound historically involves the adoption of surnames. While ministers bestowed a Christian name upon most converts at baptism, Government authorities later requested that Inuit provide them with a surname so as to avoid confusion and to better administer to their perceived needs. While Umanaqjuarmiut men usually gave the name they received at birth as their surname (e.g., Jim Kilabuk), Qikirtarmiut normally adopted their fathers' names as their surname (e.g., Etuangat Aksayuk). A cursory examination of the Pangnirtung telephone directory reveals that Qikirtarmiut surnames dominate the listings, while Umanaqjuarmiut surnames are less noticeable—Maniapik being a notable exception. From this brief survey we may conclude that, while the sons of more horizontally-structured groups did not take their fathers' names, the sons of most vertically-structured kin groups did, explicitly recognizing the strength of the *irniriik* tie as well as a latent tendency towards patrilineal descent. In accordance with the customs of Euroamerican society, women of both subregional groups took their husband's surname as their own. The issues of naming and descent in the context of acculturation are complex ones that warrant more attention than can be devoted to them here. Nevertheless, differences between Qikirtarmiut and Umanaqjuarmiut in these customs are of the order that might be predicted from previous discussion.

The RCMP and other government agents were also well aware of differences between the Qikirtarmiut and Umanaqjuarmiut. While the former had 'more dealings with the white population', were 'the ones that (were) employed as servants', and received 'more of their share of what money (was) spent on relief', the latter apparently 'devoted all their time to hunting' and did not have 'the feeling that they (were) living close to the whiteman'.[6] Yet, Qikirtarmiut aggregations were normally regarded as 'healthy, contented and prosperous', while Umanaqjuarmiut groups 'seemed the reverse'.[7] And, although Umanaqjuarmiut camps were 'on the whole well enough off', they were felt to be composed of 'the more miserable type of native'.[8] While some officials mistakenly blamed these differences on 'laziness' or 'indolence',[9] others attributed these differences, perhaps more correctly, to leadership or lack thereof:

> It is noticed that the camps that have the leadership of a good headman seem to get by very well indeed. These camps (Idlungajung and Sauniqtuajuq) are in good shape. Other native camps in the Gulf do not come up to the standards of the above mentioned.[10]

Most Umanaqjuarmiut, of course, were not lazy, indolent, or 'inferior to others in the Sound',[11] though they must have seemed that way in comparison with the more entrepreneurial, vertically-structured Qikirtarmiut. Rather, they employed different principles of group formation or strategies of affiliation which were reflected in a host of features.

## Leadership

It is axiomatic, if not redundant, at this juncture to state that leadership was considerably more developed among the Qikirtarmiut than the Umanaqjuarmiut. Nonetheless, marked differences in authority and decision-making were not products exclusively of the contact-traditional period; they can be traced back to the turn of the century, if not before. At Umanaqjuaq, decisions affecting the community would be taken jointly by a group of the settlement's most prominent hunters: 'the men would get together as to where they were to go hunting and discussed other things to be decided as a community'.[12] Conversely, at Kekerten, Angmarlik was responsible for leading the community in productive activity and taking care of its interests. No other hunter attained Angmarlik's social position or level of decision-making, and his leadership was never an issue:

> Although the main leader wasn't able to led (sic) all of the tasks. . . , Angmarlik would lead them all and they would follow him faithfully. They (the sub-leaders) would follow what the leader wants them to do as to what has to be done. . . . They didn't have any problems as they followed him the way they were supposed to. . . . As he was able to look after everything when he was in the community . . . the sub-leaders didn't even seem to exist when he was around. No one questioned his ability or leadership at all. And when the helpers lead (sic) others they would not question them either, I've never known anyone questioning their responsibilities.[13]

As the preceding quote implies, there appears to have been a well-defined hierarchy of productive relationships at Kekerten during the early twentieth century: an overall leader (Angmarlik), several lesser or subgroup leaders chosen by Angmarlik, and those hunters who formed the core of most work parties (e.g., whaleboat crews).[14] The possibility that Angmarlik's position as *angajuqqaq* was not solely the result of his own initiative or abilities, but was part of an existing structure is suggested by the fact that he apparently did not so much actively campaign for the position of leader as he was asked by several of the community's more prominent hunters to lead them in productive enterprise.[15] While a similar system existed at Umanaqjuaq, it was nowhere near as developed or hierarchical as that at Kekerten. For example, although Pawla was generally regarded as the leader of productive activity at Umanaqjuaq during the first quarter of this century, his leadership 'only affect(ed) those working under him and the others were free to make their own decisions. . .'.[16] Moreover, camp leaders among the Umanaqjuarmiut did not select their assistants as such, 'it didn't work that way, as it was open to anyone who wished to help the leader'.[17] Finally, during the late nineteenth and early twentieth centuries the prominent Saumingmiut, Kanaaka, appears to have overseen most secular and sacred activities at Umanaqjuaq.

Whereas leaders, their responsibilities, and the criteria upon which they were selected (or more appropriately, followed), were remembered unequivocally by Qikirtarmiut informants, such was not the case among numerous elderly Umanaqjuarmiut interviewed by the author or Jaypeetee Akpalialuk on behalf of Parks Canada. Some Umanaqjuarmiut could not recall the leaders of some of the camps in which they had lived, nor their responsibilities, but guessed that so-and-so was the leader because he 'had a boat', was not 'too bossy', was 'older', or because of 'his ability to hunt and his ownership of needed things'.[18] Conversely, among the Qikirtarmiut, leaders were 'the ones the people felt comfortable with and who were able to manage others to do different tasks', while their responsibilities included looking after 'trading supplies, such as ammunitions (*sic*), tea, biscuits, and sugar' as well as the 'catch of foxes, seals, and also after the dogs'.[19]

The fact that leadership differed between the Umanaqjuarmiut and Qikirtarmiut is indisputable. While each division may have been subject to different acculturative forces, this explanation does not account for the order or magnitude of differences observed in leadership since commercial whaling activity was concentrated originally on the southwest shore of the Sound. In other words, even though leadership was more strongly developed among the Qikirtarmiut, the Umanaqjuarmiut had a longer and arguably more intensive history of association with Qallunaat. The Christian doctrine of 'everyone is equal in the eyes of God' may have played a role in promoting egalitarian relations among the Umanaqjuarmiut, but, as will be recalled, the adoption of Christianity did little to alter the nature of productive relationships, at least initially (see Chapter 4). Not until the old leaders died off did the new religious ideology began to supplant the secular and sacred functions of the old religious order.

### Group Size, Residential Stability, and Individual Mobility

Leadership is intimately interwoven with many other features of Oqomiut society, especially group size, residential stability, and individual mobility. It would be wrong, however, to assign any causality or directionality to these influences, since leadership simultaneously shapes and is shaped by these and other variables. Nonetheless, differences between Umanaqjuarmiut and Qikirtarmiut aggregations in terms of these features are correlated with leadership, and suggest that different strategies of affiliation were employed by each.

Another indication that different principles of group formation once operated within Cumberland Sound comes from federal government censuses of Idlungajung and Kingmiksoo for the period 1923 to 1944 (Figure 65). While Kingmiksoo was originally settled by Tooloogakjuaq and Koodlooaktok over the winter of 1923–24, by early spring its population had ballooned to 60. The population of Kingmiksoo remained large though variable, as it fluctuated between 39 and 70, until 1944. Shortly thereafter, subsequent to the death of Tooloogakjuaq, the camp was abandoned. Conversely, Idlungajung's population started and remained small at around 20 until the early 1930s when Angmarlik's children began to marry and raise families of their own.

Population size and the level of socioeconomic organization most societies worldwide attain are thought by many anthropologists to be positively correlated. The larger the population, the greater the need to organize and control people, resources, information, etc., or so the reasoning goes (e.g., Johnson 1982). However, that facts that Idlungajung, at the height of Angmarlik's leadership and productive capabilities, was among the smallest in the Sound, while

Kingmiksoo with its poorly developed leadership would be the largest, challenges this assumption when applied to Central Inuit society.

Idlungajung was not the only Oqomiut camp where leadership, not to mention prosperity, was associated negatively with group size. Under the direction of Kingudlik, Padli was considered to be 'far richer than most camps' owing to the fact that he made 'the camp get out and hunt at all times, even when there (was) a fair reserve on hand'.[20] Yet, precisely because of Kingudlik's strong direction and authoritarian hand 'the average native (did) not remain long at the camp'.[21]

Among a strongly independent people such as the Oqomiut, the reasons why strong leadership would predict small group size are obvious: people have a tendency to 'vote with their feet' when they are subjugated to the extent that the disadvantages of such arrangements (e.g., loss of, or failure to gain, prestige or influence) outweigh the benefits (e.g., increased economic security). Thus, settlements dominated by social relationships that are too asymmetrical tend to be small. Not surprisingly, where leadership is well-developed parent-child relationships form the predominant bond among core group members. However, it follows that small, productive social units may also experience local labour shortages which could sometimes require the incorporation of unrelated individuals or distant kinsmen into the local group (e.g., recall Idlungajung's composition during the mid-1930s).

Figure 65 reveals that Idlungajung and Kingmiksoo differed not only in group size, but also in stability of group size and membership over time. While Idlungajung's population grew more or less at a constant rate from the mid-1920s up to the early 1940s, when it surpassed Kingmiksoo's population for the first time, the latter appears to have ebbed and surged on a number of occasions during this period. In contrast to Kingmiksoo's demographic history, Idlungajung's evolution appears to be one of consistent growth and stability, wherein new individuals were added to the local group primarily through adoption, marriage, and birth. The possibility that group membership was more volatile among the Umanaqjuarmiut became apparent when my informants thought that fewer than 50 per cent of the people enumerated in the 1927 RCMP census of Kingmiksoo actually lived there at that time. Conversely, group membership among Qikirtarmiut central cores tended to be more stable. Unrelated, distantly attached, and/or destitute families moved frequently between camps occupied by both Umanaqjuarmiut and Qikirtarmiut. However, a close reading of Chapter 5 reveals that core group members shifted residences more often among the Umanaqjuarmiut. In particular, individuals appear to have moved with some regularity between Kingmiksoo, Kipisa, Opinivik, and Illutalik. Most of my elderly Qikirtarmiut informants, after leaving Kekerten and before moving to Pangnirtung, had never lived anywhere for more than a year at a time than at their permanent winter quarters. On the

# The Structure of Cumberland Sound Inuit Social Organization 237

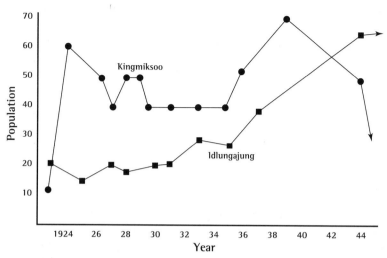

FIGURE 65. Populations of Idlungajung and Kingmiksoo between 1923 and 1944.

SOURCES: PAC RG85/1044, file 540–3 [3B], 19 Oct. 1935, 10 March 1933, 9 May 1934, 30 May 1930, 25 Feb. 1930, 31 Oct. 1928, 27 April 1930, 9 April 1927, 14 Feb. 1925, 20 April 1924, 30 June 1930; PAC RG85/64 file 164–1, 3 March 1925, 10 Sept. 1927, 31 March 1933; PAC RG85/6954 [1], 21 Jan. 1935, 14 Sept. 1936; PAC RG85/6954 [3], 1 Sept. 1937.

other hand, most of my elderly Umanaqjuarmiut informants had lived in at least three different camps during their adult life—one individual changing residences 18 times before settling permanently in Pangnirtung.[22]

The facts that group membership was less stable and individuals were more transient among the Umanaqjuarmiut appear to be a direct function of the way local groups belonging to each subregional group were organized. Parent-child cores are generally more enduring and stable than sibling cores because both *naalaqtuq* and *ungayuq* directives are stronger. Conversely, for *nukariik* cores to remain viable over the long term, *ungayuq* behaviours must be emphasized at the expense of *naalaqtuq* behaviours. Simply put, for groups founded on sibling cores to perpetuate their forces and relations of production, egalitariansim must be promoted, while hierarchy must be suppressed. While this accounts for the lack of strong leadership among most Umanaqjuarmiut groups, it also explains why individual mobility was greater and group membership less stable.

We know from the previous chapter that among the Qikirtarmiut male sibling cores rarely formed the basis of co-operative activity or endured for any length of time. Conversely, Umanaqjuarmiut sibling cores tended to last considerably longer, sometimes a generation or more. Yet most families involved in these arrangements eventually aligned themselves with other families in other camps. The reasons for this, while varied, are related. First and foremost, truly egalitarian and democratic relations are virtually impossible to maintain in a society where one's social standing and success are so dependent upon economic productivity. Thus, systemic contradictions arise in local groups where *naalaqtuq* relationships are suppressed in favour of *ungayuq* behaviours. These contradictions, in turn, may generate social tensions which require a shift in residences or a realignment of personnel to resolve. At the same time, movement of individuals between camps is substantially facilitated and group acceptance is made easier when egalitarian behaviours are well-developed and override super-subordinate relations. Yet absence of leadership and direction may lead to organizational dysfunction, a general lack of prosperity, and a tendency to look for 'greener pastures' elsewhere.

**Territoriality**

Different group formation principles or strategies of affiliation appear to have contributed directly to differences in Qikirtarmiut and Umanaqjuarmiut leadership, group size, residential stability, and individual mobility. Based on these findings we might also predict that the degree of territoriality exhibited by local groups associated with each division differs as well. Specifically, Qikirtarmiut groups would be anticipated to demonstrate a greater degree of territoriality than the Umanaqjuarmiut. Even though information regarding territoriality was not directly accessed by the writer from informants, at least two lines of evidence support this proposition.

Continual references to Angmarlik's, and to a lesser extent Keenainak's, 'country' in the Pangnirtung Post Journals from the 1920s denote an element of territoriality not associated with other camp leaders. While Angmarlik's exploitive zone is also variously referred to as his 'hunting ground', 'fiord', or 'land', no other individual's hunting territory is alluded to, let alone described in these terms, though reference is made to Kanaaka's, Kaka's, and Aksayuk's 'place'. Thus, differences in territoriality are indicated, both between and within regional subdivisions.

A clearer picture of Qikirtarmiut and Umanaqjuarmiut territoriality emerges when we consider Haller's study of late contact-traditional human geography in Cumberland Sound (Haller 1967, Haller et al. 1966). Haller was able to record the summer and winter hunting ranges of those camps occupied in 1965–66. Figure 66 unequivocally establishes the fact that the winter hunting territories of men from Tuapait, Idlungajung, Sauniqtuajuq, and Iqalulik were mutually exclusive, while those from Keemee, Illutalik, Kipisa, and Kingmiksoo overlap. Not unexpectedly, Kingmiksoo under Akpalialuk's leadership demonstrated the greatest territorial tendencies of any Umanaqjuarmiut camp. In regard to summer hunting ranges, the same basic pattern appears to have existed (e.g., see Haller et al. 1966: 79–80, Map 17).

One camp, Keemee, represents somewhat of an anomaly in that it exhibited an overlapping hunting area characteristic of other Umanaqjuarmiut camps,

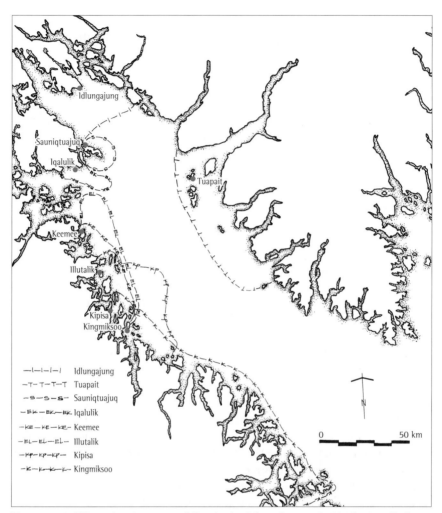

FIGURE 66. Winter hunting areas of Cumberland Sound camps in 1965–1966. Redrawn from Haller et al. (1966: 75–6, Map 16).

but was occupied predominantly by Qikirtarmiut from Idlungajung. A lack of strong leadership, which was symptomatic of most camps in Cumberland Sound during the late contact-traditional period, including those occupied by Qikirtarmiut, and the fact that this aggregation was founded principally on an opposite-sex sibling core, may have contributed to the lack of strong leadership and absence of an exclusive hunting territory in this case. However, the nature of the ecological base, which clearly differed from that around Idlungajung, may have also accounted, in part, for this difference. We will return to this example when we consider the ecological underpinnings of Umanaqjuarmiut

240  Part Three: Cumberland Sound Inuit Social Structure

and Qikirtarmiut social structure. It will suffice for now to conclude that territoriality, or lack thereof, appears to be correlated positively with leadership, group size, residential stability, and individual mobility.

## Marriage Patterns

The above variables serve to delineate the operation of two different structural tendencies in Cumberland Sound. However, they do not exhaust the number of social features that distinquished the Umanaqjuarmiut and Qikirtarmiut. As briefly noted in Chapter 5, marriage patterns appear to have differed between each division.

### *Marrying Out*

It was observed previously that, although men and women of each subregional group intermarried, Qikirtarmiut men tended to marry Umanaqjuarmiut women far more than the converse. Of the 19 inter-divisional marriages documented, 13 represent cases of Qikirtarmiut males marrying Umanaqjuarmiut females. These differences are even more pronounced when one differentiates between pre- and post-1940 intermarriages. The Fisher's Exact Test, another simple statistical test, where $p=.00287$, indicates that the probability of this result being random is very remote indeed (Table 17). While this evidence suggests that the Qikirtarmiut practised regional group exogamy to a greater extent than the Umanaqjuarmiut, it would also appear to negate the existence of dual exogamous organization. The absence of reciprocity in terms of the wife-giver/wife-taker relationship might be expected to generate a certain amount of social tension between these two divisions, and in this regard it is noteworthy that young, marriageable Umanaqjuarmiut females apparently feared Qikirtarmiut males—the latter were considered 'different' from Umanaqjuarmiut men.[23] However, the basis of this fear may have been simply a function of the fact that marriage to a Qikirtarmiut male normally meant leaving the comfort and safety of one's family and camp for those of her husband's, as the Qikirtarmiut tended towards patrilocal residence.

It is interesting to note that after 1940 more Umanaqjuarmiut men appear to have married Qikirtarmiut women than the converse (Table 17). These data could be misconstrued as delineating the existence of a dual exogamous system wherein imbalances created by those whom affinity pulled asunder in one generation to form productive socioeconomic arrangements were mended in the next, when in fact, with an overall decline in leadership during the late contact-traditional period, Qikirtarmiut marriage and residence preferences may not have been enforced to the extent they were previously. The facts that Qikirtarmiut men also took spouses from the east coast of Baffin Island, while Qikirtarmiut women generally did not, and Umanaqjuarmiut men and women intermarried equally with Inuit from Frobisher Bay and Hudson Strait, would seem to obviate the existence of dual exogamy. At the same time, these differences

TABLE 17. Marriage arrangements between Umanaqjuarmiut and Qikirtarmiut.

|  | QIKIRTARMIUT MEN AND UMANAQJUARMIUT WOMEN | UMANAQJUARMIUT MEN AND QIKIRTARMIUT WOMEN | TOTALS |
|---|---|---|---|
| pre-1940 | 12 | 1 | 13 |
| post-1940 | 1 | 5 | 6 |
| Totals | 13 | 6 | 19 |

would also appear to illuminate a stronger patrilineal bias among the Qikirtarmiut. The possibility that Qikirtarmiut and Umanaqjuarmiut marriage arrangements were founded on different conventions is evident from an examination of local and kin group endogamy.

Of the 19 local group endogamous marriages recorded, 12 (63 per cent) occurred among the Umanaqjuarmiut (Table 18). The tendency for Umanaqjuarmiut to marry more often within the residential unit is even more apparent when one considers that their population throughout most of the contact-traditional period numbered less than half that of the Qikirtarmiut. No marriage within either subregional group could be classified unequivocally as a kin endogamous union, at least to the first degree of collaterality. However, a number of cases of second cousin marriage and several unions between more distantly-related kinsmen of the same or adjacent generations were documented among the Umanaqjuarmiut. *Quasi*-kin endogamous unions, whereby the adoptive status of one or both partners or the death of a key linking individual appears to have sanctioned some marriages, were also more prevalent among the Umanaqjuarmiut. Alternatively, with rare exception (e.g., Nunaata during the early 1960s), Qikirtarmiut marriages were less subject to negotiation and/or individual interpretation; only four *quasi*-kin and distant kin endogamous unions were recorded for the Qikirtarmiut as opposed to 11 for the Umanaqjuarmiut. The number of marriages creating multiple ties of affinity between consanguines within local groups were also more common among Umanaqjuarmiut. It was not unusual, for example, for two Umanaqjuarmiut siblings to marry two siblings, cousins, or other consanguines in the same or adjacent generations. Of the 39 instances where multiple affinal ties were established among consanguines within the same local group, 29 (74 per cent) occurred within Umanaqjuarmiut camps.

Local group endogamy, *quasi*-kin and distant kin group endogamy (including second cousin marriage), as well as the establishment of multiple affinal relations among co-resident consanguines were more characteristic of the Umanaqjuarmiut than the Qikirtarmiut. Even though the small size of their population may have limited the pool of eligible marriage partners among the Umanaqjuarmiut, these tendencies undoubtedly served to multiply and intensify relations within local aggregations. Considered together, these differences suggest that the Umanaqjuarmiut employed somewhat different marriage rules and, thus, strategies of affiliation than the Qikirtarmiut.

242  Part Three: Cumberland Sound Inuit Social Structure

TABLE 18. Local group endogamy, marriages resulting in multiple affinal ties among co-resident consanguines, and hypergamy within Qikirtarmiut and Umanaqjuarmiut camps.

| Camps/ Occupations | Approximate Date of Occupation | Cases of Local Group Endogamy | Number and Description of Marriages Resulting in Multiple Affinal Ties Among Group Members | Number of Marriages between Inuit of High Status |
|---|---|---|---|---|
| **Qikirtarmiut** | | | | |
| 1) Nunaata | mid-1920s | — | — | — |
| 2) " | early 1940s | 1 | — | 2 |
| 3) " | early 1960s | 2 | 2 (St.S+St.D=ZS+ZD)* | 3 |
| 4) Idlungajung | early 1920s | — | — | 3 |
| 5) " | mid-1930s | — | — | — |
| 6) " | mid-1940s | — | 2 (S+D=B+Z) | 1 |
| 8) Avatuktoo | early 1930s | — | — | — |
| 8) " | early 1940s | — | — | — |
| 9) " | late 1950s | — | — | — |
| 10) Tuapait | mid-1950s | — | — | — |
| 11) " | early 1960s | 1 | — | 1 |
| 12) Sauniqtuajuq | mid-1920s | — | 2 (B+B=cousins) | — |
| 13) " | late 1930s | 1 | — | 3 |
| 14) " | late 1940s | 1 | 3 (M. Ego=bw; B+Z=cousins) | 1 |
| 15) " | early 1960s | 1 | — | 1 |
| 16) Iqalulik | mid-1950s | — | — | 2 |
| 17) Naujeakviq | mid-1930s | — | — | 1 |
| 18) " | mid-1950s | — | 1 (Apt.D=1st cousin's S)** | — |
| 19) " | early 1960s | — | — | — |
| 20) Keemee | early 1960s | — | — | — |
| 21) Kingnait | early 1930s | — | — | — |
| **Umanaqjuarmiut** | | | | |
| 1) Ussualung | early 1920s | — | 2 (F.Ego + ZD=St.B's) | — |
| 2) " | late 1930s | — | 2 (F+S=cousins) | — |
| 3) Sauniqtuajuq | mid-1920s | — | 4 (B+Z=cousins; M.Ego+ZS=Z's) | — |
| 4) Iqalulik | early 1930s | — | 2 (2nd cousins=1st cousins)? | — |
| 5) Kingmiksoo | mid-1920s | 1 | 4 (B+B=Z+Z; GS+ZS=Z's) | 1 |
| 6) " | late 1930s | 3 | — | — |
| 7) " | mid-1950s | — | — | — |
| 8) Opinivik | late 1920s | — | — | — |
| 9) " | late 1930s | — | 1 (WZD=FBSS) | — |
| 10) " | mid-1950s | — | 1 (2nd cousins marry) | — |
| 11) Kipisa | late 1930s | — | — | 1 |
| 12) " | late 1940s | — | — | 1 |
| 13) " | late 1950s | 2 | 4 (F+S=F.Ego+D; M.Ego=BW; GD=WBS) | — |
| 14) " | late 1970s | 3 | 4 (D+GD=B+B; GD=WS; GS=WGD) | 1 |
| 15) Illutalik | early 1930s | — | 2 (B+Z=B+Z) | — |
| 16) " | late 1930s | — | — | — |
| 17) " | late 1950s | 3 | 3 (B's=Z's; D=St.ZS) | — |
| 18) Seegatok | mid-1940s | — | — | — |
| 19) " | late 1950s | — | — | — |
| 20) Nuvujen | mid-1920s | — | — | — |
| 21) Etelageetok | early 1930s | — | — | — |

\* read: stepson and stepdaughter marry (=) sister's son and daughter.
\*\* read: adopted daughter marries 1st cousin's son.

## Marrying Up

The propensity to marry up is to be expected in a society where prestige is so dependent upon productivity. Therefore, we might expect families of high socioeconomic standing, unrelated through blood, to marry amongst each other. While this does not constitute hypergamy *per se*, the propensity to marry high status individuals nonetheless represents a hypergamous tendency. In this regard, it is instructive to recall Angmarlik's intention to select spouses for his grandchildren that met his standards (Chapter 4). Table 18 reveals, not unexpectedly, that hypergamous tendencies were particularly well-developed among the Qikirtarmiut. Of the 22 recorded marriages exhibiting hypergamous tendencies, 82 per cent (n=18) occurred among Qikirtarmiut. To this may be added several other examples not represented in this table, such as Niaqutsiaq's (Kivitoo's headman) and Ishulutaq's (Keenainak's brother) marriages to Angmarlik's sisters (Kowna and Mequt, respectively). Etuangat Aksayuk's marriages to Kingudlik's daughters would be another. Conversely, only four (18 per cent) of all hypergamous oriented marriages documented involved Umanaqjuarmiut. While some component of this difference may be owing to the general absence of prosperous families with well-developed leadership among the Umanaqjuarmiut, the tendency for families of high socioeconomic standing to marry amongst themselves is irrefutable. As a matter of note, this proclivity seems to have been especially prevalent among the first-born children of Qikirtarmiut camp leaders.

## Marital Residence

A detailed analysis of post-nuptial residence patterns would seem perfunctory, if not suspect, given the multiplicity of factors involved in this decision. Nonetheless, a close reading of Chapter 5 reveals that the Qikirtarmiut tended towards patrilocality, whereas the Umanaqjuarmiut were more accepting of matrilocal arrangements. Bride-service was common among both subregional groups. However, as we have seen, where leadership is well-developed sons tended to remain in the camps of their fathers. At the same time, even stronger leaders retained both their sons and daughters, while attracting sons-in-law. Consequently, while Qikirtarmiut camps such as Sauniqtuajuq and Naujeakviq exhibited pronounced patrilocal tendencies, Idlungajung and Nunaata possessed a more balanced ratio of both types of residence. Although larger aggregations among both groups, such as Idlungajung (mid-1940s), Ussualung (early 1920s), and Kingmiksoo (late 1930s), demonstrated neolocal and bilocal forms of residence, matrilocality was more prevalent among the Umanaqjuarmiut, especially the more established camps and continuously occupied camps (e.g., Kingmiksoo and Kipisa). Opposite examples to these tendencies exist, e.g., Avatuktoo after the 1930s and Kingmiksoo under Akpalialuk's tenure. However, special circumstances appear to account for their anomalous character

(Chapter 5). In conclusion, then, while the Umanaqjuarmiut sanctioned matrilocality, the Qikirtarmiut preferred patrilocal residence.

**Adoption**

Adoption, like marriage, established socioeconomic alliances between families by redistributing individuals among local groups so as to facilitate, maintain, and perpetuate productive forces and relationships. As adoption assumed both a social and an economic role, it was practised widely in the central and eastern Arctic. In Cumberland Sound, 22 of the 108 (nuclear) families in 1938 had adopted children.[24] As noted previously, there appears to have been a tendency in Cumberland Sound for childless couples to adopt children of both sexes. This proclivity seems to have been particularly well-developed among the Qikirtarmiut. Alternatively, the available evidence suggests that there may have been a slight preference for male adoption among the Umanaqjuarmiut (Chapter 5). The possibility that the high rate of adult male death recorded among the Umanaqjuarmiut during the early contact period contributed to this preference seems likely (see Chapter 2 and below).

Even though most adoptions took place between closely related kinsmen in either the same or adjacent generations, some adoptions occurred between more distantly related families. This penchant was noted especially among the Qikirtarmiut, wherein ties between families of high socioeconomic standing were sometimes established through adoption. For example, during the mid- to late 1920s Angmarlik adopted boys from Kookootok and Kopalee, two prominent Qikirtarmiut who played leading roles in commercial whaling activity at Kekerten. Whereas the adoption of an infant son or daughter was the most popular form recorded (n=31), infrequently couples would adopt individuals (n=7) of the same generation (e.g., siblings or cousins) or ascending generation (e.g., uncles).

By far the two most common types of adoptions, for which the birth parents of the adoptee were known, involved the transfer of a newly born infant from one sibling to another (sibling adoption) or from an adult child to his/her parents (grandparental adoption). While both served to intensify bonds of co-operation and affection among the exchanging parties, grandparental adoptions possessed an added *naalaqtuq* element. Recall that an adoptee often gave his/her adoptive parents a child out of a sense of duty and respect. *Naalaqtuq* directives also appear to have motivated such adoptions between other consanguines, though convenience and the obligation to provide comfort for one's parents in old age undoubtedly were also important (Guemple 1979). Thus, it is perhaps not too surprising that grandparental adoptions were more commonplace among the Qikirtarmiut, while sibling adoptions were more prevalent among the Umanaqjuarmiut (Table 19). This difference is judged to be significant beyond the .01 level of confidence, where $X^2=7.3$, df=1. Differences between the Umanaqjuarmiut and Qikirtarmiut in terms of adoption,

TABLE 19. Adoptions among Umanaqjuarmiut and Qikirtarmiut.

|  | GRANDPARENTAL ADOPTIONS | SIBLING ADOPTIONS | TOTALS |
|---|---|---|---|
| Qikirtarmiut | 13 | 3 | 16 |
| Umanaqjuarmiut | 5 | 10 | 15 |
| Totals | 18 | 13 | 31 |

then, would seem to support the proposition that different social formations operated in Cumberland Sound historically.

**Caching and Sharing**
Other social features that distinguished the Umanaqjuarmiut from the Qikirtarmiut include food storage and, to a lesser extent, food sharing. Headmen of local groups among both subregional groups possessed special facilities for the storage of meat and blubber. However, large, strategically located community food caches maintained and controlled by camp leaders appear to have been characteristic only of the Qikirtarmiut. This was especially so at Idlungajung, Nunaata, and Avatuktoo under Angmarlik's, Keenainak's, and Nukinga's leadership, respectively. Game would normally be deposited in community stores, which were located adjacent to the camp leader's house, irrespective of the hunter responsible for taking the animal or the amount of reserves on hand. In times of abundance, people freely helped themselves to these provisions, while in times of scarcity game would be distributed to the community by its headman. Although this regime describes the ideal pattern of community food storage among the Qikirtarmiut, in actual fact it appears to have persisted throughout the year only at Idlungajung. At Nunaata, for example, hunters deposited game in the community cache only during the open-water hunting season when they served aboard Keenainak's whaleboat. During the rest of the year individual hunters deposited food in their own caches located in the porches of their houses.

In contrast, the Umanaqjuarmiut usually maintained, in addition to individual household stores, caches on islands away from their camps, retrieving meat and blubber when needed. Although free-roaming dogs were cited to be the reason for the latter practice, two other explanations—one logistic, the other social—warrant consideration. First, during the open-water season Umanaqjuarmiut men normally hunted in small groups from kayaks, while Qikirtarmiut hunted from whaleboats. Whereas virtually every Umanaqjuarmiut hunter possessed a kayak, comparatively fewer Qikirtarmiut owned, or hunted during the open-water season from, such craft. Rather, until they were replaced by row boats, kayaks were used predominantly at the floe edge. The caching of game away from camps may have been characteristic of the Umanaqjuarmiut simply because the use of kayaks precludes the immediate transportation of large amounts of game back to camp. Thus, Umanaqjuarmiut seal and caribou hunters were often forced to cache their kills for later retrieval, usually by dog team in early winter. Second, the placement of caches away from camp may have served to retard the development of hierarchical ideology, while promoting egalitarian relations among Umanaqjuarmiut camps. Stated simply, such caching strategies do not acknowledge the prowess of the hunter to the same extent as the immediate transportation of the animal back to camp whereby the productivity of the hunter is overtly recognized in public display and celebration.

Umanaqjuarmiut headmen, like Qikirtarmiut leaders, normally butchered and distributed game in hard times, as they were the most skilled and proficient at such tasks. However, in good times Umanaqjuarmiut women usually assumed responsibility for game sharing: 'The women would be the ones who cut up the animal and distribute it (as) the man was mainly concerned with hunting and the women felt responsible for others as far as sharing was concerned. . . .'[25] Though Qikirtarmiut women, like their Umanaqjuarmiut counterparts, removed blubber from hides, the greater exposure of the latter to the Christian value of providing for the needy could account, in part, for this difference. Yet, even after Qikirtarmiut women assumed responsibility for the welfare of the destitute, they rarely oversaw the butchering and distribution of game in the community. Rather, Qikirtarmiut women gave food to the needy within the context of *piutuq*, and rarely performed *nekaishutu* for the community. Thus, caching and sharing of game, like other social features, appear to have differed between the Umanaqjuarmiut and Qikirtarmiut.

**Kinship**
Considering the many different, yet systemically associated, social features that distinguish the Qikirtarmiut and Umanaqjuarmiut, it would be surprising if these differences were not somehow reflected in kinship terminology. As noted in Chapter 4, the terminology in use today differs first and foremost from the one recorded by Morgan in that no longer is MBW merged with MZ (*aiyak*). Rather, both MBW and FBW are now regarded as *ukuaq*, together with in-marrying females in Ego's and the descending generations. It was reasoned previously that the MBW=MZ equation was representative only of the Kingmiksormiut/ Talirpingmiut, and not the Cumberland Sound Inuit as a whole. It was further suggested that this feature disappeared in the context of increased Inuit/Qallunaat interaction and Inuit regional group mixing after contact. The numerous differences between the Qikirtarmiut from Umanaqjuarmiut illuminated throughout this chapter serve to strengthen these arguments. Thus, we may speculate that the feature whereby MBW=MZ, but FBW≠FZ, in the terminology of the early contact Kingmiksormiut was commensurate with many characteristics described above for the Umanaqjuarmiut, especially a recognition of sibling (both cross-sex and female) relationships.

TABLE 20. Summary of Qikirtarmiut and Umanaqjuarmiut structural tendencies.

| SOCIAL FEATURE/CHARACTERISTIC | QIKIRTARMIUT TENDENCIES | UMANAQJUARMIUT TENDENCIES |
|---|---|---|
| Structure of primary kinship ties | parent-child core | sibling core |
| Strength of primary kinship ties | strong | moderate |
| Structure of group relationships | vertical | horizontal |
| Strength of F-S relationship | stronger | weaker |
| Leadership | well-developed | poorly developed |
| Decision making | more authoritarian | more egalitarian |
| Group size | smaller | larger, more variable |
| Residential stability | higher | lower |
| Individual mobility | lower | higher |
| Local group hunting territories | mutually exclusive | overlapping |
| Distant/*quasi*-kin endogamy | rare | more prevalent |
| Local group endogamy | rare | more prevalent |
| Multiple affinal ties among consanguines | rare | more prevalent |
| Hypergamy | frequent | rare |
| Marital residence | patrilocal | matrilocal |
| Adoption pattern | grandparental | sibling |
| Food distribution in good times | family heads, leaders | women, family heads |
| Caching | community and individual caches | local and scattered individual caches |

## *NAALAQTUQ* AND *UNGAYUQ* SOCIAL STRUCTURE

'Every camp was different' is a truism well-recognized by my informants. Yet, they were also aware that the Umanaqjuarmiut differed from the Qikirtarmiut and described several differences in social organization between the two. However, it also seems clear that the Umanaqjuarmiut and Qikirtarmiut shared many social features and customs. The occurrence whereby aggregations associated with each subregional division assumed characteristics normally associated with the other (e.g., Nunaata and Kipisa during the late contact-traditional period) seems to suggest that Qikirtarmiut and Umanaqjuarmiut social formations were not so invariant to the extent that local groups could not adopt alternative social configurations. Nonetheless, the facts that significant dissimilarities existed in core group kin relationships, leadership, territoriality, marriage, adoption, etc., and that these differences appeared to be systemically related, indicate the operation of two distinguishable social structures within Cumberland Sound historically.

Table 20 summarizes the major structural tendencies distinguishing these two divisions. Some variables, e.g., leadership, may have played a larger role than others in shaping and supporting these structural tendencies. Even so, an exclusive reliance on any one feature fails to account adequately for not only most differences observed between these regional subdivisions, but for their systemic associations as well. Rather, dissimilarities between Qikirtarmiut and Umanaqjuarmiut local groups may be more parsimoniously accommodated within an explanatory framework that considers the former to be *naalaqtuq*-structured and the latter to be *ungayuq*-structured. Most Central Inuit are governed by *naalaqtuq* and *ungayuq* directives implicit within their age, gender, and kinship relationships to others. Respect-obedience and affection-closeness, however, are not just flip sides of the same coin, but work in complementary fashion to structure and maintain productive relationships and activity. A balance between these axes of interpersonal relations exists only in theory. In reality, the nature of any given relationship between two people for any given situation must be governed by one or the other directive, even between parent and child, which exhibited the strongest, most developed, and invariant *naalaqtuq* and *ungayuq* behaviours of any in Central Inuit society. It thus follows that core members of local groups, and ultimately entire camps, must also be governed predominantly by one or the other directive, and thus structural tendency.

We have seen how the structure of a local group might alternate from one generation to the next as prominent men become infirm or pass away, leaving leadership and decision-making in the hands of resident sons. However, we have also observed that such configurations rarely endure among the Qikirtarmiut—Nunaata being the notable exception—as super-subordinate relationships overshadow any bonds of affection between male siblings. Conversely, both same- and opposite-sex sibling cores among the Umanaqjuarmiut, and the families which attached themselves to such groups, not only tended to reside considerably longer together, but they appear to have moved more freely from one local group to another. Stated somewhat differently, while the subregional group appears to have attained greater importance among the Umanaqjuarmiut, the Qikirtarmiut seem to have placed more emphasis on the local group. The absence of leadership among the Umanaqjuarmiut was not the result of happenstance, but was an integral ingredient in a strategy designed to maintain egalitarian/democratic relationships and socioeconomic interdependence among people spread through a number of local groups. Clearly, emphases on *ungayuq* behaviours among co-resident Umanaqjuarmiut, and *naalaqtuq* directives among co-resident Qikirtarmiut, account for foregoing data and interpretation in ways other explanations cannot. While the specific reasons why each subregional group chose to emphasize one set of behavioural directives over the other may never be known, the environmental factors contributing to these different structural orientations are considered below.

Contained within each structural tendency are systemic contradictions that undermine the maintenance and reproduction of productive forces and relationships from one generation to the next. Here, we come to what may be a central problem in Central Inuit society. Local groups dominated by super-subordinate relationships tend to split apart after the loss of the leader, as *ungayuq* behaviours are not developed sufficiently enough to hold the camp together. Under such circumstances, ways must be found to recruit new members, either to maintain the existing structure of the local group or to reproduce the same structure anew through the formation of another productive unit. Local and kin group exogamy serve these purposes well, as they bring individuals into the group (e.g., *ningaut* and *ukuat*) who are subordinate to resident extended family members. Conversely, local groups characterized by egalitarian relationships generally weather losses in prominent personnel much better, not solely because leadership is less well-developed and decision-making is more democratic, but also because people have the option of aligning themselves with kinsmen in other similarly structured camps.[26] Moreover, while lack of leadership and direction may engender socioeconomic independence, it may also contribute to a lack of prosperity, and thus a more nomadic tendency. Finally, because egalitarian relations are impossible to maintain in societies where social status is dependent upon economic success, contradictions may arise which generate tensions that require residential moves to resolve.

The Umanaqjuarmiut and Qikirtarmiut exhibit what I believe to be the two main options available to most Central Inuit groups for reproducing productive forces and social relationships. In their most rudimentary form, and after Damas (1963, 1964), these structural tendencies may be best described as *Naalaqtuq* and *Ungayuq* principles of group formation or strategies of affiliation. While the former provides 'order' to society, the latter provides the 'glue' that holds it together. Contradictions arise when social relationships are dominated by either hierarchical or egalitarian relationships, for as Graburn (1964) so correctly observed, all local groups are in the process of being and becoming. In this regard, variability in socioeconomic organization among most Central Inuit groups may stem not just from which option governed interpersonal relationships among co-resident kinsmen, but how regional populations coped with systemic contradictions inherent within each structural tendency. From these perspectives we will reconsider Cumberland Sound prehistory in the next chapter, and explore variability in social organization among other Central Inuit groups in the following two chapters.

# Chapter 7

## Cumberland Sound Inuit Prehistory Revisited

Given that two distinguishable social formations existed in Cumberland Sound historically, a reexamination of the region's prehistory and a consideration of the environmental basis of these two structures are warranted. What is important in the study of Inuit social organization is not the search for ultimate causes, but the discovery of how society and environment articulate to form a coherent structure of social reproduction, and how changes in one might affect the other. Only by abandoning the search for 'ultimate causes' will we be able to move beyond that level of explanation currently limiting our understanding of Inuit social organization.

As the Cumberland Sound Inuit apparently did not undergo a transformation in social organization as a consequence of contact with the Qallunaat, we might expect the Talirpingmiut and Kinguamiut of the late prehistoric/early contact period to exhibit organizational differences of roughly the same character and order as those recorded between the Umanaqjuarmiut and Qikirtarmiut of the contact-traditional period. Although the quantity and quality of information from each time period obviously differ, this, for the most part, is what we find.

### THE TALIRPINGMIUT AND KINGUAMIUT

#### Leadership
Leadership at contact appears to have been better developed among the Kinguamiut than the Talirpingmiut. The headman of Anarnitung was an elderly man with considerable influence, whereas Kingmiksoo's leader was a much younger individual with substantially less status. Not only was the latter of the same generation as Eenoolooapik, who was no more than 20 at the time, but M'Donald (1841: 101) could find little difference between this man and other Kingmiksormiut. Conversely, M'Donald's descriptions of the antics of Anniapik, Anarnitung's leader, leaves little doubt that this man possessed substantial influence, at least in his own mind and those of his followers—Penny's men were less impressed.

## Local Group Size

There were also apparent differences in local group size between the Kinguamiut and Talirpingmiut. At Anarnitung and Kingmiksoo, the principal settlements of each division, M'Donald found 40 and 60 people, respectively, in September of 1840. However, a not insignificant number of families from each village appear to have been elsewhere at the time. The facts that Sutherland (1856) counted 111 people living in 16 dwellings at Kingmiksoo in the fall of 1846, and that Kingmiksoo's population in 1839 was estimated to be about the same (Chapter 2), are perhaps more instructive. Information contained on Eenoolooapik's map of the Sound adds further support to differences in local group size. Whereas the average size of eight settlements associated with the Talirpingmiut was determined to be 46.8, the average size of 13 sites in Kingua Fiord was estimated to be 33.4. Differences in the variability of the sizes of local groups also conform to expectations. Relying once again on information provided on Eenoolooapik's map, the mean number and standard deviation of dwellings per settlement among the Talirpingmiut were estimated to be 6.75 +/- 4.06, while the average Kinguamiut settlement contained 5.85 +/- 1.28 dwellings. Employing the coefficient of variation (CV), we find that Talirpingmiut sites are almost three times as variable in size as Kinguamiut settlements, $CV_{Talirp}$= 60.15%, $CV_{Kingua}$= 21.88%.

Little information exists about residential stability or individual mobility during the early contact period. However, what there is seems to suggest that the Talirpingmiut were not only able to assemble in very large groups—e.g., some 270 Inuit apparently gathered at Kingmiksoo during the fall of 1853 (Penny 1854a)—but they were able to form and dissolve aggregations with considerable ease. For example, during the 1850s, in response to the movements of the whalers, Inuit on the southwest shore of the Sound moved often between Naujateling, Kingmiksoo, and Nuvujen (e.g., Barron 1895: 43, Hantzsch 1977: 38–9; cf. Ross 1985). Although speculative, it seems doubtful whether such large groups and frequent changes in residence would have been as common among the Kinguamiut.

## Core Group Structure

An analysis of Sutherland's (1856) 1846 census of Kingmiksoo revealed that same generation relationships outnumbered adjacent generation ties among married couples within the same household, 3 to 2 (see Table 2). While this finding supports the proposition that the Talirpingmiut were governed largely by horizontal relationships, comparative data from Anarnitung or other major Kinguamiut sites are not available. Nevertheless, we can compare the number of single, double, and triple platform late precontact/early contact sod houses at Kingmiksoo and Anarnitung in order to determine the probable occurrence of vertically-structured vs horizontally-structured multi-family households at each site. The independent nature of most Oqomiut nuclear families predicts

TABLE 21. Occurrence of single, double, and triple platform/room dwellings at Anarnitung and Kingmiksoo. Data from Schledermann (1975: 39) and Gardner (1979: 382).

|            | Single | Double | Triple | Totals |
|------------|--------|--------|--------|--------|
| Anarnitung | 9      | 4      | 3      | 16     |
| Kingmiksoo | 10     | 14     | 1      | 25     |
| Totals     | 19     | 18     | 4      | 41     |

that extended family households, whether held together by same- or adjacent-generation ties, would have had more than one sleeping/working platform. However, if horizontal relationships were more prevalent within households at either site, then we would expect a greater occurrence of twin platform, as opposed to single or triple platform, houses. The reasons why this might be so are clear. Reciprocal rights and obligations, not to mention cooperative socio-economic living arrangements, are easier to maintain when only two principals, in this case two related nuclear families, are involved. The old adage, 'two's company, three's a crowd', is perhaps not irrelevant here. Using Schledermann's (1975) map of Anarnitung and Gardner's (1979) map of Kingmiksoo (see Figures 10 and 11), the frequency of single, double, and triple platform dwellings is presented in Table 21. If we consider the occurrence of bilobate vs other types of sod houses, we find that Kingmiksoo demonstrates far more double-room dwellings than Anarnitung ($X^2=6.7$, df=1)—a result significant at the .01 level of confidence.

In addition, only one out of the four bilobate houses at Anarnitung demonstrates the dual symmetrical shape one would expect two-family houses to exhibit where egalitarian relationships predominate. Conversely, 11 of the 14 bilobate dwellings at Kingmiksoo are proportionate in outline. While the Fisher's Exact Test indicates that the probability of this result being random is one in 6.4 ($p=.078$), differences in the way each archaeologist illustrated double-room sod houses may account for some of this result. Although we cannot demonstrate with absolute certainty that Kingmiksoo's contact communal households were more horizontally structured than those at Anarnitung, archaeological evidence certainly suggests that they were more predisposed to such arrangements.

It was suggested in Chapter 2 that the emergence of communal, multi-room autumn houses during the late prehistoric period was related to a greater emphasis on bowhead whaling, whereby such dwellings helped to coordinate and maintain cooperative activity among related adult males for the fall whale hunt. However, given that the Umanaqjuarmiut placed considerably more emphasis on same-generation relationships, and especially female sibling and cross-sex sibling ties, than the Qikirtarmiut, it seems plausible that the predominance of bilobate dwellings at Kingmiksoo may be indicative of another productive activity not directly associated with men or whaling at all, i.e., the making of caribou skin clothing by related same-generation females (e.g., sisters, cousins, and sisters-in-law) prior to the community moving out onto the sea ice after freeze-up to hunt seals at breathing holes.

## Productive Activity and Relationships

Differences in the nature of productive activity also appear to have distinguished the Talirpingmiut from the Kinguamiut. Not only were these differences congruent with others described above, but they may have also supported the underlying structural tendency of each regional subdivision. Specifically, while Inuit on the southwest shore of the Sound appear to have hunted whales aboriginally from kayaks, those at the head of the Sound hunted whales predominantly from *umiat*. Although we know from Sutherland's census that the Kingmiksormiut possessed *umiat*, they were apparently not used for whaling, at least not to the extent they were at the head of the Sound. As was related to, and by, Jim Kilabuk, while 'some Inuit may have hunted whales from umiaks (*sic*), we (the Umanaqjuarmiut) hunted them from kayaks':

> A man in a kayak . . . is no threat to a whale. The kayak is silent, moves quickly and is much better to handle than any umiak. . . . When we saw whales we could move among them and they were not afraid of our little kayaks. There was no fear of trying to kill a great whale if you know how to do it. My father was such a man. He was the one who knew the right place to stick in the spear. He would paddle beside the whale, carefully looking at her body. There is a place below her spine where you can see a movement. . . . Thats (*sic*) where the kidney is, and that's the only place where it is safe to stick the spear.
>
> This was done carefully and quietly, and you may be surprised to know that the whale did not even know that she was being killed. There was no fight. She kept swimming on and began to bleed to death. We would follow her sometimes for a very long time until she died. As soon as she was dead, we would come to her side and fasten lines to her body . . . and together we paddled towards the shore. There was much hard work and much rejoicing because she gave us food and oil and everything else that we needed in the making of things. . . . (Hallendy 1985: 127–8)

Conversely, the ancestors of the Qikirtarmiut appear to have hunted whales aboriginally from *umiat*. Even the vital organ that was targeted, the heart, which is located behind and under the flipper (Etuangat Aksayuk, personal communication, 1984), suggests the use of a spear longer and larger than that which could be handled from a kayak. In this connection, it is instructive that all the specialized whaling artifacts recovered from archaeological sites in Cumberland Sound come from the north and east sides of the Sound—even *niutang*, after which the Kingnaimiut settlement is named, refers to the drogue

FIGURE 67. Whaling scene on ivory bow drill recovered from LlDj-1, near Imigen Island.

or sea anchor used in aboriginal whaling. While such weaponry includes large walrus tusk harpoons (LlDj-1), heavy slate endblades (LlDj-1), and large socketed lance heads (MbDj-1), most instructive of all is a prehistoric whaling scene engraved on an ivory bow drill recovered from a site (LlDj-1) near Imigen Island (Figure 67). This scene clearly depicts the harpooning of a bowhead whale from an *umiaq* manned by a harpooner and five other crew members. Large, heavy-shafted harpoons (*sakurpang*), whaling lances (*kalugiang*), and drogues (*niutang*) are thought to be associated exclusively with whaling from *umiat* as these items are simply too bulky or heavy to be used from a kayak (Taylor 1979).

Occasions whereby the Kinguamiut used kayaks and the Talirpingmiut employed *umiat* for aboriginal whaling might be anticipated. Nonetheless, each technique had different implications for the organization of the groups employing them. Specifically, the use of *umiat* and specialized weaponry presupposes a ranking and division of labour not associated with kayak whaling. *Umiaq* whaling requires three specialized positions, each with its own level of prestige—in ascending order, rowers, a harpooner (*sivutik*), and a boatsteerer (*aggutik*). The latter was usually the boat owner (*umialiqtak*) and the one responsible for lancing the whale's heart after it had been harpooned. Similarities between this system and that of the whalers were noted previously. Yet, while some component of this division of labour may have been adopted from the commercial whalers, several legends recorded by Boas (1907, 1964) refer to the existence of harpooners and boatsteerers in the context of aboriginal whaling.

Every time a whale was caught, the rank and social status of each crew member, particularly that of the *aggutik/umialiqtak*, would be validated, thus reaffirming his position in society. Conversely, kayak whaling was carried out by groups of men performing essentially the same task, i.e., piercing the whale with a hunting knife attached to a kayak paddle (Hallendy 1985). No one individual killed the whale for it died on its own through loss of blood as a result of sustained cooperative effort, though we might expect the man to draw first blood to have claimed some recognition, such as was the case among the Aivilingmiut (Parry 1969: 509–10). Clearly, these different aboriginal whaling techniques served to facilitate and promote the egalitarian and hierarchical tendencies of the groups that practised them.

Comparisons between the Talirpingmiut and Kinguamiut suffer from an obvious lack of data. Nonetheless, what information can be brought to bear on the subject indicates that these two subregional groups exhibited roughly the same structural tendencies that distinguished the Umanaqjuarmiut and Qikirtarmiut a century later. While we may infer from these comparisons that the Talirpingmiut were *ungayuq*-structured and the Kinguamiut were *naalaqtuq*-structured, we may also conclude, with more conviction than ever, that the Cumberland Sound Inuit did not undergo a significant transformation in social organization during the historic period. In fact, because of the mixing of groups that occurred within the context of contact with Euroamericans, we might expect differences in group structure to have been even more pronounced during the precontact era than those observed for the contact-traditional period. However, where does this leave the Kingnaimiut, the other major regional subdivision known to have occupied Cumberland Sound prior to the arrival of Qallunaat?

## THE KINGNAIMIUT: BIG GROUPS, BIG MEN, BIG PROBLEMS?

### Local Group Size

If Boas' (1907, 1964) descriptions of aboriginal Cumberland Sound Inuit society are accurate, well-developed leadership was characteristic of both the Kinguamiut and Kingnaimiut. However, the Kingnaimiut appear to have lived in larger groups, at least during the fall. Referring once again to Eenoolooapik's map, we find that the Kingnaimiut lived in five settlements at contact, with an average of 7.4 +/- .55 dwellings in each. Employing the average number of people occupying dwellings at Kingmiksoo in 1846 (111/16) and Anarnitung in 1840 (40/7), we find that the population of a typical Kingnaimiut fall settlement ranged from 42.3 to 51.4, which is considerably larger than the average Kinguamiut site. Perhaps more importantly, variation in local group size among the Kingnaimiut, as reflected in the number of dwellings per site, is three times smaller than that recorded for the Kinguamiut, $CV_{Kingnait}=7.43\%$. Although we are dealing with a population of only five sites, this low CV would seem to suggest that Kingnaimiut principles of group formation, as reflected in the number of dwellings per site, were significantly more invariant than even the Kinguamiut.

### Conflict and Contradiction in Kingnaimiut Society?

Kingnaimiut settlements at contact, then, may have been characterized by strong leadership and large, invariant local group size. However, as we have seen, big men and big groups represent a contradiction in Oqomiut society; authoritarian tendencies conflict with the fiercely independent nature of the productive family unit. This may be especially so if both features are embedded in the social consciousness of each local group. Do the Kingnaimiut represent a new, heretofore undiscovered type of social formation in Cumberland Sound? Such a scenario might be reasonable if antagonisms did not manifest themselves through overt behaviour. If big men and big groups among the Kingnaimiut did engender incompatibilities in regard to the maintenance and perpetuation of productive forces and relationships, we would anticipate solutions to have been found to mitigate these conflicts. This would be expected especially in the arena of recruitment, where new rules of marriage and post-nuptial residence might be invoked to maintain the structural integrity of the local group under such circumstances. Yet, there is no evidence that the Kingnaimiut differed significantly from the Talirpingmiut or the Kinguamiut in these regards. Where systemic contradictions may have manifested themselves, however, is in interpersonal and intergroup relations. And, in this respect, Boas (1907, 1964) supplies incontrovertible evidence that the Kingnaimiut were the most hostile and aggressive of any Oqomiut subdivision (see Chapter 2). The Kingnaimiut not only murdered Inuit from other subregional groups, they killed each other, even members of the same local group. We cannot be sure whether Kingnaimiut hostility was solely or even directly a result of antagonisms arising from attempts to recruit and maintain individuals within big groups led by big men. However, we may conclude that a lack of social reciprocity between local groups introduces contradictions into Central Inuit society which must be resolved eventually through alteration of existing social norms. The fact that the Kingnaimiut still 'played by the same rules' as other Oqomiut suggests that the association of big men and big groups in this area of the Sound may have been a relatively recent development.

**Productive Forces and Relationships**
Why the Kingnaimiut possibly endeavoured to maintain both large groups and well-developed leadership is another question altogether. However, we may assume that this proclivity was intimately bound up in the nature of productive activity in the vicinity of Kingnait Fiord, particularly during the fall (i.e., that time of year depicted on Eenoolooapik's map) before land-based settlements were abandoned for winter villages on the sea-ice. With an increasing emphasis on bowhead whaling during the late prehistoric period, a conflict in resource scheduling may have arisen in the Kingnait Fiord area between the hunting of bowhead whales and caribou, wherein there was not sufficient labour or organization to harvest both species effectively. In this context, the development of large groups with well-developed leadership seems plausible. Most bays and inlets in Kingnait Fiord, especially along its southern shore, have always been known as productive caribou range, and in this connection, Schledermann's (1975) excavations of Niutang clearly established the importance of caribou relative to seal in the village economy. The development of big men and big groups during the fall among the Kingnaimiut may be akin to the

Iglulingmiut, whereby older men hunted large marine mammals along the coast, while younger men ventured inland after caribou (Damas 1963).

The conclusions reached concerning Kingnaimiut, Kinguamiut, and Talirpingmiut social organization must remain speculative in the extreme insofar as no systematic study has yet been undertaken to substantiate them nor the major source upon which they were based, Eenoolooapik's map. Nonetheless, what little evidence is available, e.g., archaeological site maps provided in Schledermann (1975) and Gardner (1979), indicates that the amount and perhaps even configuration of dots on Eenoolooapik's map represent the number and possibly arrangement of occupied houses at various Cumberland Sound locations in the fall of 1839.

## TASAIJU: 'THE ODD SITE OUT'

Although the Kingnaimiut may have been the most hostile of any subregional group of Oqomiut, this does not account for the apparent uniqueness of Tasaiju's burial population relative to Niutang and other Central Inuit groups. Inter- and intraregional comparisons led Salter (1984: 291) to conclude that Tasaiju, with its discrete skeletal traits, of which large size was the predominant characteristic, 'was the odd site out'. This, she attributed to the relative isolation of the Kinguamiut *vis-à-vis* the Kingnaimiut and Talirpingmiut. However, given the strong exogamous tendency observed among the Qikirtarmiut (formerly the Kinguamiut and Kingnaimiut) during the contact-traditional period, the discreteness of Tasaiju's burial data may have been simply a local development, unique to only one local group of Kinguamiut over a span of several generations.

### Communal Houses: a Unique Occurrence?

Two lines of evidence appear to warrant this conclusion. The first involves the occurrence of triple platform communal dwellings. Of the six house foundations recorded at Tasaiju by Schledermann, four are clearly trilobate in shape. Another two, obviously older, less distinct features also appear to demonstrate triple room configurations. While such structures occur with some frequency at other archaeological sites in the Sound, there are significantly more trilobate dwellings at Tasaiju than anywhere else in Cumberland Sound (Table 22)—a difference significant beyond the .0005 level of confidence, where $X^2=17.28$, df=1.

The question that emerges from this finding is, of course: Is the discreteness of Tasaiju's burial population related to the unique predominance of triple room communal dwellings at this site? And, if so, why? If symmetrical, double room sod houses are associated with co-resident families held together by horizontal kinship ties, then triple room communal houses might be expected to be characteristic of vertically-structured extended families. Support for this

TABLE 22. Occurrence of triple platform dwellings at Tasaiju and three other major Kinguamiut sites, Imigen (LiDj-1), Anarnitung (MbDj-1), and Kekertelung (MbDi-1). Data from Schledermann (1975).

|  | TRIPLE-PLATFORM DWELLINGS | OTHER DWELLINGS | TOTALS |
|---|---|---|---|
| Tasaiju | 6 | 2 | 8 |
| Other Kinguamiut sites | 5 | 41 | 46 |
| Totals | 11 | 43 | 54 |

inference comes to us from Iglulik (Mathiassen, personal communication to Damas 1963: 103). During the 1920s trilobate snowhouses in this region were apparently occupied mostly by 'parents and children'. As the rear platforms of triple room communal sod houses in Cumberland Sound are almost invariably larger than the two adjoining side platforms, we may infer that such features were most probably occupied by extended families headed by a married couple and their adult married children. Under the ideal of patrilocality, we might expect that most dwellings were occupied by a man and his adult sons. However, as powerful and prosperous leaders might be expected to retain both married sons and daughters within the productive unit, any combination thereof might be anticipated. Just as the number of burials at Tasaiju (n>80) indicates a measure of social stability and economic security found at few other sites in Cumberland Sound, the retention of married children within the local group presupposes the development of a certain degree of social insularity and economic autonomy. Local and kin group exogamy may still be practised under such circumstances, but a level must be reached where resources can no longer support growth in local group size. The probability that no more than five or six of Tasaiju's dwellings were likely occupied at the same time suggests that this limit was reached. This leaves us to consider the question: Did Tasaiju's extended families marry amongst themselves, thus restricting genetic exchange with other local groups? Even though we cannot presently evaluate this proposition, the discreteness of Tasaiju's burial population and the predominance of triple platform communal dwellings, in which parent-child relationships likely characterized social arrangements, strongly suggest this possibility.

### History Repeats Itself?
The theory that local group and, ultimately, some form of kin endogamy accounts for the discreteness of Tasaiju's burial population cannot be tested empirically at present. Nonetheless, it is interesting to observe that the development of a biologically distinct population in the vicinity of Nunaata Island was not restricted to the late prehistoric period, but appears to have also occurred during the contact-traditional period. Just as Tasaiju's skeletal traits are unique because of their large size, so too are the descendants of Keenainak, who are considerably larger in stature than most other Cumberland Sound

Inuit. Like his father and uncles before him, Elija Keenainak is a large man, a fact he often laments for his size hinders his productive capabilities, especially in cold weather (personal communication 1989). Is the large physical stature of Keenainak and his descendants an inherited characteristic from the late prehistoric inhabitants of Tasaiju, which is located on the mainland just across from Nunaata Island? While we know that Inuit from both Kekerten and Umanaqjuaq returned to their former camps and hunting grounds with some regularity during the late nineteenth century, perhaps the productive nature of the environment around Nunaata, particularly the occurrence of a well known *sarbuq*, encouraged the development of increased physical stature.

Although we must conclude that, in the absence of testable data, some combination of the two accounts for the phenetic distinctiveness of Keenainak and his descendants, it is noteworthy that Nunaata during the 1950s and 1960s also differed from nearby camps in terms of its social organization. As will be recalled, leadership was not well-developed, male sibling ties dominated core group relationships, and local and *quasi*-kin group endogamy where multiple ties of affinity were established between consanguines, were not uncommon—features more in keeping with *ungayuq*- rather than *naalaqtuq*-structured settlements. Could such social conditions have led eventually to the discreteness of Tasaiju's burial population? Perhaps not exclusively so. However, I do not think it a coincidence that, when local groups demonstrate social tendencies atypical of other camps around them—and here I am referring specifically to Nunaata and Kipisa during the contact-traditional period—they also tend to exhibit more internal marriage arrangements, and thus a more closed structure.

## Environment and Society in Cumberland Sound

Little mention has been made of the relationship between the structural differences exhibited by Cumberland Sound's regional subdivisions and the local environments in which they lived. However, it seems clear that there are important environmental differences between the head and southwest shore of the Sound where the Qikirtarmiut and Umanaqjuarmiut lived, respectively. Specifically, while *sarbut* are numerous in Kingua Fiord, the number and duration of tidal rips during the winter between the southern shore of Nettilling Fiord and Umanaqjuaq are substantially less. Thus, while both subregional groups hunted seals at *aglu* and the *sina* during the winter, the Qikirtarmiut hunted seals principally at *sarbut*. Conversely, only a few Umanaqjuarmiut had enough resources (i.e., dogs or ammunition) to travel to and remain at such places (e.g., Nettilling Fiord) for any length of time (Hantzsch 1977).

The abundance of *sarbut* at the head of the Sound undoubtedly contributed to the small size and restricted nature of most Qikirtarmiut winter hunting territories relative to those of the Umanaqjuarmiut (Figure 66). Conversely, the many small islands between Keemee and Kingmiksoo encourage the formation of fast ice which is prime ringed seal winter habitat and excellent for breathing-hole sealing. However, as a single seal may maintain numerous breathing holes over an area of several square kilometres, specific locales were often quickly depleted, requiring hunters to range farther in search of new sealing grounds. The occurrence and distribution of seals during the winter in each respective area of the Sound, then, may have influenced not only the size of local group hunting territories but also the extent to which they overlapped.

Is it possible that different social formations evolved in the wake of these different ecological circumstances in order to maintain the required mode of production in each area? For example, did the low rate of residential stability and high rate of individual mobility observed among the Umanaqjuarmiut evolve directly out of the need to hunt seals over large areas? Did well-developed leadership among the Qikirtarmiut emerge as a way to maintain territorial boundaries? Did the higher rate of local and *quasi*-kin group endogamy among the Umanaqjuarmiut develop as a means to foster the necessary co-operative relationships required to hunt seals at breathing holes?

We have come too far to dismiss the many differences observed between the Umanaqjuarmiut and Qikirtarmiut simply and solely on the environment. Fortunately, a number of lines of evidence and reasoning can be marshalled to suggest that social and historical factors played just as important a role in determining the nature of Cumberland Sound Inuit social organization. Given the diversity of local environments, not only between the head and southwest shore of the Sound, but also within each area—e.g., Sauniqtuajuq's immediate environment differs from that of Idlungajung's in having no *sarbut* nearby—one would expect a far greater degree of diversity in local group organization than documented if environmental circumstances were the ultimate architect of social structure. Not only do we find far less diversity in local group organization than this hypothesis would lead us to suspect, but most contact-traditional camps are characterized by only one of two structural patterns—one Qikirtarmiut, the other Umanaqjuarmiut. There were instances, to be sure, where some aggregations belonging to one subregional division adopted the lifestyle of the other after it moved into the other's area. Consider, for example, Keemee during the mid-1960s. This camp was occupied principally by an opposite-sex sibling core from Idlungajung. Leadership was poorly developed among this aggregation and its exploitive zone overlapped considerably with those of other Umanaqjuarmiut camps. Alternatively, there were cases whereby some camps moved into the other subregional group's territory, but still maintained most of their traditional organizational characteristics, and thus distinctiveness. Recall Ussualung during the 1920s, where native leadership was lacking and caching exhibited a typically Umanaqjuarmiut pattern. Similarly, the extent to which leadership and territoriality was demonstrated by Akpalialuk's kin group at Kingmiksoo during the 1950s and 1960s suggests that this local group did not adopt a typically Umanaqjuarmiut style of living after it moved from Ussualung.

If the environment were the ultimate source of structural variation in Cumberland Sound Inuit society, we would find a considerably greater variety of social formations, if not mixture of non-complementary social features, among local groups. We do not. Rather we find two rather distinct structural tendencies, which, in the final analysis, may be no more than antitheses of each other. Is not *ungayuq* the antithesis of *naalaqtuq*; hierarchy the antithesis of egalitarianism? Does not the wholesale negation of one structural tendency and its attendant social features predispose groups towards acceptance of the other? There, I believe, are few other options available to Central Inuit groups once the prevailing social structure, whatever it might be, is rejected.

The conclusion that may be drawn from these arguments is simply this: Neither social nor environmental factors are determinant in shaping Central Inuit society; both play complementary roles. A hypothetical example should suffice to make this clear. Imagine, 250 years ago somewhere on south Baffin Island, there were two brothers who headed two small camps at the mouths of adjacent fiords. Seal was the primary staple in their diet, and while they occasionally got together to hunt seals at breathing holes over the winter, the camps were, for the most part, economically independent. However, for the past several years, bowhead whales have been seen entering their fiords in ever increasing numbers during the early fall. A joint decision is then taken to form a single productive unit in order to create a large enough labour force to form an *umiaq* whaling crew. Co-operative relationships predominate and *naalaqtuq* directives take a 'back seat' to *ungayuq* behaviours. This arrangement works well for several years as the camp prospers. However, contradictions implicit in their kinship tie on the one hand, and economic arrangement on the other, begin to emerge. The older brother claims what is rightly his in terms of prestige and social recognition, particularly if he is the boat owner and/or his sons are beginning to reach productive age. The relationship between the two brothers resumes a hierarchical format and, inevitably, social tension forces the camp to split apart. Bowhead whales are just as numerous as before, but neither extended family is exploiting them. Instead, both groups return to a dependence upon seals as their dietary mainstay.

In this fanciful, though informed, reconstruction we see how the environment can shape the social organization of a group. At the same time, we also see how the structure of productive relationships, because, in this instance, of contradictions within the kinship system, may also determine the use of the environment. There was no attendant transformation in the nature of the resource base, the environment did not change. However, the size and social organization of the group(s) did.

A more proximate example of the interplay of environment and society entails the high rate of adult male death and the predominant use of kayaks among the Talirpingmiut/Umanaqjuarmiut. Sutherland's (1856) census of Kingmiksoo, as well as other evidence (e.g., Ross 1985c: 235) suggests that there was a shortage of productive adult males in this settlement. We do not know for certain whether the same situation existed among the Kinguamiut/Qikirtarmiut. Yet, what little evidence can be brought to bear on the issue indicates that males at the head of the Sound had a longer life span. We also know from informants that hunting seals and whales in open water from kayaks was common among the Umanaqjuarmiut relative to the Qikirtarmiut. Whereas almost every Umanaqjuarmiut hunter during the contact-traditional period owned a kayak, only a few Qikirtarmiut in each camp possessed this craft. While the Qikirtarmiut used kayaks predominantly at the *sina*, they preferred to hunt seals and whales from whaleboats. Although sealing from kayaks is a relatively more independent activity than whaling from kayaks, both are dangerous pursuits which expose the individual to the elements much more so than the hunting of marine mammals from larger craft.

If kayak hunting was, in fact, primarily responsible for the high rate of adult male death among the Talirpingmiut/Umanaqjuarmiut, the question that surfaces is this: Why didn't they abandon such practices in favour of safer pursuits or the use of larger boats? Clearly, the latter would have been more adaptive from the standpoint of individual survival and reproduction. Differential availability of resources appears not to have been a factor in this decision, for, while the Qikirtarmiut may have had more opportunities during the contact-traditional period to acquire whaleboats and ammunition—a prerequisite for sealing from boats—wood as a boat-building material was more accessible to the Talirpingmiut during the late prehistoric period. Alternatively, perhaps the dependence of the Talirpingmiut/Umanaqjuarmiut on kayaks is somehow related to the possibility that the nature of resource distributions on the southwest shore of the Sound favoured the use of such craft over boats? We know from Haller (1967: 57–60) that ringed seals generally tend to migrate out of the Sound in the summer, thus forcing hunters to range further afield in search of seals. While the summer scarcity of seals throughout the Sound exposed kayak sealers even more to the mercy of the elements, the annual migration of harp seals to the mouths of Kangiloo and Issortuqjuaq Fiords each summer may have alleviated the need to travel great distances in search of game near the head of the Sound. Certainly, the migration of beluga to Milurialik created less dependence upon ringed seal among the Kinguamiut each summer.

Differential environmental opportunities and constraints, thus, may have contributed to the predominance of kayak hunting among the Talirpingmiut/Umanaqjuarmiut and of *umiaq*/whaleboat hunting among the Kinguamiut/Qikirtarmiut. However, as discussed previously in reference to aboriginal bowhead whaling, each hunting method has different implications for the social structure of the groups employing them. In this connection, despite its drawbacks (i.e., high adult male death) kayak hunting may have been favoured over the use of boats simply because it supported and maintained the prevailing social structure and ideology of egalitarianism. Conversely, the use of *umiat* and, later, whaleboats presupposes a ranking and a division of labour which undermines this structural tendency. The question whether society or the environment is determinant in this context is rendered irrelevant; both play a role in shaping Inuit social organization and relations of production.

The study of Umanaqjuarmiut and Qikirtarmiut organizational differences could have been approached from a materialist perspective such as that employed by Lewis Binford (1980). After all, does not each regional subdivision exhibit characteristics that Binford associates with hunter-gatherers employing foraging and collecting strategies? Indeed, numerous social phenomena distinguishing these subregional groups (e.g., strong vs weak leadership, exogamy vs endogamy, exclusive vs overlapping hunting territories) might be interpreted as correlates of each subsistence/settlement system. However, classifying the Qikirtarmiut as collectors and the Umanaqjuarmiut as foragers is unrewarding, not to mention intellectually suffocating, for it fails to consider the role of historical factors and social structure, including behavioural directives implicit within kinship systems, in shaping social organization. Any attempt that rejects the environment as the pre-eminent factor in structuring human organization in Arctic and Subarctic regions, such as Ives' (1990) analysis of variability in northern Athapascan kinship, must be applauded. However, we must be careful not to replace the 'vulgar materialism' of Binford and others with 'vulgar kinshipism'. Rather, we must seek to understand how both social structure and the environment influence each other to produce particular socioeconomic arrangements, and how changes in one might affect changes in the other. Only through such endeavours will we go beyond the search for ultimate causes to appreciating fully the intricate and complex interplay of structural, historical, and ecological forces in shaping Inuit and northern hunter-gatherer socioeconomies. And, only from employing such perspectives will we be able to begin to understand and appreciate the operation and reproduction of societies different from our own.

# Part Four

# Central Inuit
# Social Structure

# Chapter 8

# Iglulingmiut and Netsilingmiut Social Organization

Can the foregoing structural analysis of Cumberland Sound Inuit social organization inform our understanding of other Central Inuit groups? More specifically, can the diversity and systemic relationships of various social features observed among the most thoroughly documented of all Central Inuit groups, the Iglulingmiut, Netsilingmiut, and Copper Inuit be explained with reference to the two structural tendencies delineated above? The preceding chapters have laid the groundwork for exploring variability in Central Inuit social organization employing the concepts of *naalaqtuq* and *ungayuq*. Structural, historical, and environmental factors were acknowledged to be the primary architects of Central Inuit social organization. However, some component of the diversity observed within and between regional socioeconomies may stem not just from which structural tendency governed interpersonal relationships, but also how local groups coped with systemic contradictions inherent within each tendency. It is from these perspectives that we will re-examine variability in Iglulik and Netsilik social organization in this chapter, and Copper Inuit social organization in the next, and consider new explanations for the fundamental differences that distinguish these Central Inuit groups.[1]

## THE IGLULINGMIUT: COPING WITH *NAALAQTUQ*

### History of Contact

The Iglulingmiut were composed traditionally of three regional subdivisions, the Tununirmiut of Pond and Admiralty Inlets, the Aivilingmiut of the west coast of Melville Peninsula and Repulse Bay, and the Iglulingmiut of northern Foxe Basin and the east coast of Melville Peninsula (Damas 1963, Mathiassen 1928, Rasmussen 1929) (Figure 68). No marked differences in culture or dialect appear to have distinguished these subregional groups (Rasmussen 1929: 10), and intermarriages among Inuit in all three were common. Among the Iglulingmiut proper, walrus was the most important species, followed by the ringed seal, and then caribou. Major settlements, however, were associated primarily with the occurrence of the former species. Whales, both large and small,

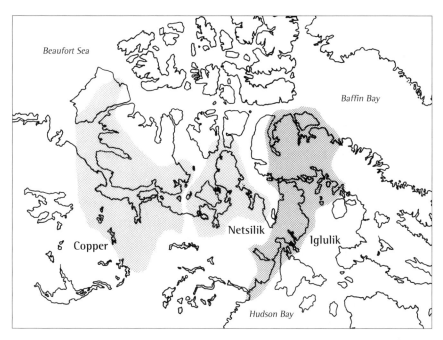

FIGURE 68. Distributions of Iglulik, Netsilik, and Copper Inuit (after Damas 1969b: 41).

attained greater importance among the Tununirmiut. The Aivilingmiut of Repulse Bay also apparently hunted bowhead, though caribou and fish were of greater significance, particularly to the more inland oriented local groups.

The Iglulingmiut and Aivilingmiut were visited by Parry in the early 1820s (1969) when he over-wintered among more than 200 Iglulingmiut, describing various aspects of their culture. Here, he found the ratio of men to women to be approximately equal, while adults outnumbered children two to one (1969: 492). Iglulik Point was the most important rendezvous and most Iglulingmiut gathered there in September to hunt walrus. In addition to several smaller sites on Iglulik Island, Lyon (1824) observed another three major winter habitations at Amitsuq, the Uglit Islands, and Pingirqalik.[2] During the fall and early winter the Iglulingmiut inhabited *qammat* of sod, rock, and whalebone, while living off stores of walrus built up during the summer and early fall. By midwinter, families moved into snowhouses on the sea ice, where they pursued seals at breathing holes and, when wind conditions allowed, walrus at the *sina*. Interestingly, the most common type of snowhouse constructed during the winter was the cruciform or trilobate feature referred to in the previous chapter.[3] As spring approached the large winter sealing aggregations split into smaller basking seal hunting camps headed by pairs of men. Summer was the period of maximum dispersion, with younger men venturing inland after caribou, while older men remained along the shore pursuing walrus and seal from kayaks.

Parry (1969: 528) recorded 12 cases of polygyny among the Iglulingmiut, with an average age difference of five or six years between co-wives. The disparity in age between a man and his spouse(s), however, was frequently considerable, with the man often being older by 20 or more years. Parry also noted instances whereby fathers and sons had married sisters. Although spousal exchange was observed, this custom was infrequent and practised without formality. The authority of the husband, while considerable, was not so great as to preclude his wife from decision-making, especially in matters pertaining to the domestic sphere. Adoption was common, and apparently always arranged between the fathers (Parry 1969: 532–3). Servants were known but generally rare in Iglulingmiut society. The picture that emerges from Parry's and Lyon's brief descriptions of the Iglulingmiut is one of a Central Eskimo culture modified by ... 'large numbers of walrus, which made possible a somewhat higher standard of living than was available for groups farther west' (Damas 1963: 20–1).[4]

Although commercial bowhead whaling never entered the Iglulik area, beginning in 1820 the Tununirmiut occasionally encountered British whalers and/or wreckage of their ships off Pond Inlet and Lancaster Sound. Even so, contact was largely fortuitous and of short duration (Ross 1979a: 251). Wood and metal also reached the Iglulik area via indirect contact with the HBC trading sloop at Fullerton Harbour (Damas 1963: 20). Forty years later, American whaling activity in Roes Welcome Sound attracted the Aivilingmiut. Whaling and trading activities along the west coast of Hudson Bay during the late nineteenth and early twentieth centuries also resulted in the relocation of the Netsilingmiut (Ross 1975). While the Aivilingmiut migrated southward along the west coast of Hudson Bay into Roes Welcome Sound, and eventually onto Southampton Island after disease wiped out the Sadlirmiut, the Netsilingmiut also pushed southward into areas formerly occupied by the Aivilingmiut.

Additional economic and, to a lesser extent, social changes resulting from Aivilingmiut/American whaler interaction have been illuminated by Ross (1975) and Robinson (1973). In summary, these include (a) intensification of the caribou and whale hunts, as well as increased productivity in the seal and walrus hunts—both made possible by the adoption of the rifle and whaleboat; (b) loss of certain items of traditional material culture (e.g, bows), with increased dependence upon the outside world for supply of goods, particularly ammunition; (c) changes in subsistence patterns, most notably a shift from an inland orientation to a coastal adaptation; (d) gradual erosion of the winter aggregation/summer dispersion cycle to a more semi-sedentary, centralized existence associated with specific whaling harbours or ships; (e) relaxation of old tribal boundaries, but greater territoriality of local groups; (f) creation of new leadership roles and collaborative activities through employment and the acquisition of whaleboats; and (g) individualization of terrestrial hunting through the introduction of rifles. Despite these changes, both Ross and Robinson saw no significant overall alteration in the structure of productive relationships among the Aivilingmiut. Indeed, the arrival of the RCMP and HBC during the early twentieth century, and the subsequent adoption of fur trapping and dispersal of the population, were acknowledged to counter some of the acculturative trends initiated during the whaling period (Ross 1975: 137).

Contact between Inuit and commercial whalers in the north Baffin region was never as intensive as that in Cumberland Sound or Roes Welcome Sound/Repulse Bay. It was not until the free traders moved into Pond Inlet in the early 1900s that Tununirmiut labour and produce became attractive to outsiders. Even though the nature of Inuit-whaler interaction in the Pond Inlet region is poorly known, we turn our attention exclusively to the Iglulik area. Not only do we know, largely through Damas' (1963, 1964) work, more about the Iglulingmiut, but Iglulik appears to have been the nexus of interaction for all three regional subdivisions, especially during the contact-traditional period.

In the late 1860s the Iglulik region was visited briefly by C.F. Hall. Here, he observed villages at Uglit, the Tern Islands, and at Iglulik Point (Damas 1963: 21). Throughout the 1860s and the remaining decades of the nineteenth century, goods entered the Iglulik region via trade with the Aivilingmiut, who interacted directly with American whalers wintering at Repulse Bay and Depot Island. As a consequence of this trade, there appears to have been an increase in individual mobility and interaction among Inuit living at Pond Inlet, Iglulik, and Repulse Bay (Mathiassen 1928: 21). Interregional mixing continued well into the first decades of this century after trading posts were established at Fullerton Harbour, Pond Inlet, Chesterfield Inlet, and Repulse Bay. The Iglulik area, however, remained on the periphery of this trade. Although the adoption of the whaleboat greatly increased productivity in the Pond Inlet and Repulse Bay areas, this craft was not introduced into the Iglulik region until 1930.[5] Nonetheless, both walrus and caribou hunting became more productive enterprises after the introduction of the rifle. Yet, as the Iglulingmiut had no direct access to ammunition, traditional implements and methods of hunting continued to play important roles in the economy. As in Cumberland Sound, trapping took a back seat to sea mammal hunting.

The possibility that Inuit in the Iglulik region were not as affected by Euroamerican contact as other Central Inuit populations is evidenced by the fact that the size and structure of the population remained virtually the same between Parry's visit in 1822 and the Fifth Thule expedition one hundred years later (Damas 1963: 23, Parry 1969, Mathiassen 1928).[6] At that time, Mathiassen (1928: 15–21) recorded five winter camps in the Iglulik region, including Pingirqalik, Amitsuq, Itibjiriaq, Manirtuq, and Iglulik which was the largest village with 74 people. Damas' (1963) reconstructions of the composition of these camps identified several general characteristics, including tendencies towards extended family organization based on bilateral ties, virilocality, local and kin group exogamy, as well as the prevalence of F-S cores and, to a lesser degree, B-B cores.

**Culture Change and Continuity**
The introduction of Christianity and the Peterhead whaleboat after 1930 had important impacts on Iglulingmiut socioeconomy. While the work of Anglican and Catholic missionaries appears to have resulted in a considerably faster replacement of the old religious ideology than was the case in Cumberland Sound, the impact of the whaleboat was similar to that experienced in the latter region. Specifically, the whaleboat served to maintain the co-operative structure of the extended family and the authority of its leader at a time when the rifle individualized productive activities. Moreover, the whaleboat engendered increased productivity in the walrus hunt, and thus sedentism throughout the winter. Yet, the method of acquisition differed from that in Cumberland Sound. While prominent Oqomiut hunters were given whaleboats, first for their service in the whaling industry, and then later for moving to better fox trapping areas, whaleboats among the Iglulingmiut appear to have been purchased jointly from the HBC by several men (Damas 1963). As polar bears were scarce and as profits from fox trapping were always small, even after the establishment of a HBC trading post at Ikpiakjuk on Iglulik Island in 1939, the process of purchasing a whaleboat often took years (Damas 1963: 24). Thus, the acquisition of the whaleboat served to reinforce both adult male ties and virilocal tendencies within the local group. Although only three or four of these craft had been obtained by the early 1930s, the Iglulingmiut were concentrated into two large villages, Aqungniq and Abadjaq. Notwithstanding the centralization of the population, and despite the introduction of the rifle, whaleboat, and steel trap, the annual cycle appears to have remained relatively unchanged from earlier times—the addition of fox trapping during the winter and a gradual transition from winter breathing-hole sealing to hunting seals at the *sina* with rifles constituted the major changes.

By 1940, the population had doubled owing not just to the disappearance of walrus from Repulse Bay and subsequent immigration of those left in that area to Iglulik in 1936, but also to the high level of food production that characterized Iglulingmiut economy during the 1930s (Damas 1963). The whaleboat and the continued occurrence of walrus were largely responsible for this development. Indeed, the average number of dogs per team in the Iglulik region during this time was estimated to be 15 or more (Manning 1943: 101–2). The trading post played only a minor role in this new level of economic prosperity as ice conditions frequently prevented the arrival of its supply ship.

After 1940, further acquisitions of whaleboats, continuing emigration from Repulse Bay and Arctic Bay, as well as encouragement from the local trader and police to disperse and relocate to better trapping areas, resulted in the fragmentation of the population and an increase in the number of camps. By 1949 there were two large villages, Iglulik (n=68) and Qarmat (n=83), and nine smaller camps in the Iglulik region. While immigration contributed to a significant increase in the overall population of the area, the indigenous population also doubled during the same period owing to an increase in infant survival—children outnumbered adults in the 1949 census. Most of the social characteristics noted above for the 1921–22 Iglulingmiut continue to be reflected in the 1949 aggregations (Damas 1963: 69–71). However, while male sibling groups appear even less numerous than before, kin group endogamy had become more prevalent, particularly among the descendants of Ituksarjuak—the most influential leader in the Iglulik region during the twentieth century. Local group and kin group exogamy, however, remained both the ideal and norm.

In 1948 family allowances, which helped to alleviate the impact of an Arctic-wide reduction in fox fur prices, were instituted. After 1950 even more whaleboats, now equipped with engines, entered the region and supported the trend begun earlier towards the socioeconomic independence of the extended family (Damas 1963: 27). Permanent wage labour opportunities became available in the Iglulik region in 1956 after a DEW line site was built at Hall Point. However, fewer than ten full-time jobs were available to locals, and seasonal employment during late summer sea-lifts remained the largest source of cash income from wage labour. Beginning in 1955, children were flown to Chesterfield Inlet for schooling. However, by 1960 a school house was erected at Ikpiakjuk (Igloolik). At the same time, Ikpiakjuk began to emerge as the major village in the region—the presence of a trading post, an Anglican church and rectory, as well as the availability of seasonal wage labour and plywood from the DEW line site contributed to the attractiveness of the site (Damas 1963). A year later, Damas conducted his extensive ethnography and study of Iglulingmiut society.

Damas recorded the composition of 14 camps in the Iglulik region. While the population of the region had grown steadily, this did not manifest itself in the occurrence of larger aggregations (1963: 97). Rather, the size and composition of local groups, and by extension group formation principles, remained the same. While Ikpiakjuk (Igloolik) remained the largest village in the area, another four campsites demonstrated populations of between 40 and 64 people. In terms of group composition, kin and local group solidarity was greatest only when the father was living; the death of the latter usually resulted in the fissioning of the local group along male sibling lines. Virilocality was still predominant, as was local group exogamy, which appears to have been strengthened over the years with the increased permanency of winter villages. Although religious endogamy—created by the presence of both Catholic and Anglican denominations—represented a new and strictly enforced development, marriage between relatives increased from only one in 1922 to six in 1961 with an additional three 'border-line' marriages between adoptive relatives (1963: 98). The seasonal economic cycle of the Iglulingmiut also appears to have been little altered from earlier times—increased sedentism brought about by more effective modes of hunting and transportation represented the most notable modification. In summary, up to Damas' study, the processes of cultural modification in the Iglulik area followed a gradual course with no major adjustments.[7] On the contrary, 'rather than being responsible for a gradual abandonment of the hunting economy . . . culture contact made possible a new level of prosperity' (1963: 32) in the Iglulik region.

### Enduring Features of Iglulingmiut Social Organization

Damas (1963: 98) has clearly demonstrated that a great deal of continuity prevailed in Iglulingmiut local group composition from the early 1820s to the early 1960s:

> Though population increase, wage labour situations, and a major florescence of the hunting economy with attendant increase of sedentariness have characterized recent years at Iglulik . . . village size and the sorts of alignments that one finds within have remained very much the same.

Although the essential features of Iglulingmiut social organization were described and contrasted with those of the Netsilik and Copper Inuit in Chapter 1, they are briefly summarized here with specific reference to the kinship system.

The economic cycle of the contact-traditional Iglulingmiut was divided into four seasons: *mauliqtuq* sealing in winter, *uttuq* sealing in spring, open water hunting of walrus throughout the summer and early fall, and caribou hunting in the autumn, during which time a division of labour existed between younger and older men. The household, which was the primary unit of production, was based on the extended family with parents and married children at the core. On occasions a single large extended family would constitute a local group. At other times, particularly during the winter when large aggregations of people assembled on the ice to hunt seals at breathing holes, two or more related extended families would form a village. In all seasons, band composition tended to be bilateral with a prominence of male-relevant ties.

Leadership among the Iglulingmiut was well-developed with the eldest resident hunter usually assuming the role of *isumataq*. The authority of the *isumataq* reached beyond his own extended family to social and economic matters affecting the residential group. He also frequently regulated game sharing and distribution. Yet, food sharing and eating patterns in Iglulingmiut society were highly variable with individuals, nuclear families, extended families, and whole villages forming commensal units on different occasions. While marriage with relatives became more common in recent times among Ituksarjuak's kin, both kin and local group exogamy were still rigidly practised. In addition, there appears to have been a preference for subregional group exogamy, especially in later times (Damas 1963: 69), and perhaps earlier times as well.[8] For the male, marriage was often accompanied by a year's bride-service, after which the couple usually returned to the husband's camp. Under the ideal of patrilocality, matrilocality was more a luxury than a preferred living arrangement, though very strong bonds of affection still prevailed between a woman and her parents.

*Hierarchy*
Perhaps the most striking features of Iglulingmiut social organization are its emphases on super-subordinate relationships and the solidarity of the extended family. In regard to the former, in-marrying males are subordinate to males born into the kin group regardless of age or generational relationships: sister is subordinate to brother, younger brother to older brother, Ego's generation to parent's generation, etc. This structure, as noted in Chapter 1, is acutely reflected in the kinship system whereby age, generation, gender, and consanguineal affiliation determine one's place in the social hierarchy. As Damas (1963: 201) observed, Inuit whose personal relationships are governed largely by *naalaqtuq* directives are continually inquiring about one's age, purpose, and connection to others in the community to see where you fit into the status hierarchy.

Hierarchy is also reflected in other aspects of Iglulingmiut ideology. For example, in forms of reference and address, personal names were rarely used. Rather, kin terms and the behavioural content that their use presupposes—deference and respect—were employed. This was most notable among opposite-sex in-laws, where the greatest avoidance and respect relations prevailed; those in this category were called *illiyuariik*, 'those who could not mention the names of each other' (Damas 1963: 47). While the sea goddess was recognized to be the supreme deity, as she was among other coastal-oriented Central Inuit, there was a well-developed hierarchy of higher/lower and malevolent/benevolent spirits in Iglulingmiut mythology, each serving a higher deity, and ultimately the sea goddess (Rasmussen 1929: 62, 70). In the *qaggi*, dominance-subordination was symbolically reflected in ritual performance whereby the drummer/singer stood alone in the middle of the house, surrounded by an inner circle of men and an outer circle of women. The latter never handled the drum, but sang in support of their husbands as the men took turns in the centre (e.g., Rasmussen 1929: 228).[9]

Closely related to the hierarchical structure of kinship relationships was a spirit of friendly rivalry or competitiveness that served to establish dominance-subordination between non-kin or distantly related males. As Rasmussen (1929: 227, 231) noted:

> Underlying all the games is the dominant passion of rivalry, always seeking to show who is the best . . . the swiftest, the strongest, the cleverest (etc.).

The same spirit of rivalry ... is also found in the song contests which are held in the feasting house.... (However) two ... opponents in song contests must be the very best of friends; they call themselves, indeed, *iglorek*, which means 'song cousins', and must endeavor not only in their verses, but in all manner of sport, each to outdo the other; when they meet, they must exchange costly gifts, here also endeavoring each to surpass the other in extravagant generosity.

As noted previously, *naalaqtuq* and *ungayuq* are strongest between parents and children. This upward channelling of affectional and respect-obedience behaviours is congruent with many features of Iglulingmiut social organization, including local and kin group exogamy, the greater frequency of adoption by grandparents (Damas 1975c: 21), and strong leadership. The authority of the eldest resident hunter with the largest circle of kinsmen extended to all members of the local group, and sometimes even to other camps. As in Cumberland Sound, the strongest leaders were also associated with the best hunting grounds (Damas 1963: 87). Although the authority of the *isumataq* in later times was undermined by individualistic tendencies engendered by the rifle, this was offset by the whaleboat and the fur trade whereby the camp leader had to co-ordinate a greater diversity of activities and serve as middleman in the exhange and distribution of goods. Leadership of local groups was also sometimes inherited, not just achieved. Boas (1907: 115) provides an example of this, as does Damas (1963: 178–9). With respect to the latter, a son of Ituksarjuak assumed leadership of his father-in-law's camp, which was incongruent with kinship directives. Hypergamous tendencies were observed to be a corollary of leadership in Cumberland Sound. And, not surprisingly, Damas describes this propensity among the Iglulingmiut. In fact, so strong was this tendency among Ituksarjuak's descendants that it appears to have led to kin group endogamy and the development of an élite or 'royal family' (Damas 1963: 56).[10]

*Solidarity*
In contrast to the Netsilik, Copper, and even the Cumberland Sound Inuit— indeed most Central Inuit populations—affines are not equated with consanguines in Iglulingmiut society (Table 23). For the male, in-marrying males (*ningaut*) and females (*aiit*) in Ego's and the adjacent generations are strictly separated from blood relatives. For example, MZH is not lumped with MB nor is FBW equated with FZ. Alternatively, collateral relatives in the first ascending, Ego's, and first descending generations show relatively greater and lesser father- and mother-like, brother- and sister-like, and son- and daughter-like roles, respectively (Damas 1963: 53).

One of the most unique features of Iglulingmiut kinship reckoning is its three-cousin terminology. Opposite-sex cousins are often identified with

TABLE 23. Iglulingmiut, Netsilingmiut, and Copper kinship terminology for Ego's, first ascending, and first descending generations, male speaking. Based on Damas (1975c).

| Gen. | Relationship | Iglulingmiut | Netsilingmiut | Copper Inuit |
|---|---|---|---|---|
| First Ascending Generation | F | ataata | apak (ataata) | angut (aappak) |
| | M | anaana | amaama | arnaq (amaama) |
| | FB /male cousin | akka | akka | pangnaaryuk |
| | MB /male cousin | angak | angak | angak |
| | FZ /female cousin | attak | attak | attak |
| | MZ /female cousin | aiyak | arnarvik | arnarvik |
| | FBW, MBW | ai | attak | arnaaryuk |
| | FZH | ningauk | akka | pangnaaryuk |
| | MZH | ningauk | angak | angak |
| Ego's Generation | W | nuliaq | nuliaq | nuliaq |
| | B (older) | angayuk | angayuk | angayuk |
| | B (younger) | nukaq | nukaq | nukaq |
| | Z (older) | nayak | nayak | aliga |
| | Z (younger) | nayak | nayak | nayak |
| | FBS | angutiqat | illuq | angutiqat |
| | FBD | nayak | nayak | angutiqat |
| | MBS, FZS | illuq | illuq | arnaqat |
| | MBD, FZD, MZD | nayak | nayak | arnaqat |
| | MZS | arnaqat | illuq | arnaqat |
| | WB | sakiaq | hakiaq | hakiaq |
| | WZ, BW | ai | ai | ai |
| | ZH | ningauk | ningauk | ningauk |
| | WZH (older) | angayuunnguq | angayuunnguq | angayuunnguq |
| | WZH (younger) | nukaunngaq | nukaunngaq | nukaunngaq |
| First Descending Generation | S | irniq | irniq | irniq |
| | D | panik | panik | panik |
| | BS, BD | qangiaq | qangiaq | qangiaq |
| | ZS, ZD | uyuruk | uyuruk | uyuruk |
| | SW, BSW, ZSW | ukuaq | ukuaq (ukuvak) | ukuaq |
| | DH, BDH, ZDH | ningauk | ningauk | ningauk |

opposite-sex siblings, *anik(saq)* for males and *nayak(saq)* for females, and thus exhibit a typically 'Hawaiian' terminology. However, for male Ego, FBS (*angutiqat*) is differentiated from MZS (*arnaqat*) and the cross-cousins, MBS and FZS (*illuq*). The behavioural directives of this system, though discussed in Chapter 1, warrants summary here. FBS is closest to male Ego, followed by FZS and MBS, and finally MZS. The exact complementary pattern is obtained for female Ego, whereby MZD is the most sibling-like, followed by MBD and FZD, and then FBD. Avoidance characterized relationships between both opposite-sex siblings and cousins after childhood.

Damas (1963, 1975c) considers the Iglulingmiut to be the most 'internally adaptive' or 'integrated' of any Central Inuit society. Indeed, he felt that the correlation between Iglulingmiut kinship terminology and behaviour was 'truly amazing in its ingenuity and internal consistency' (1975c: 19). And in this regard, the great emphasis placed on gender solidarity, the autonomy of the consanguineal unit, and affinal-consanguineal separation within the context of local and kin group exogamy points to the operation of a social system once based on groups of brothers marrying groups of sisters. Ives (1990), following Asch (personal communication), has noted that such systems are designed to 'pump people' out of local aggregations, extend alliances beyond the local group, and place emphasis on the regional group. Although this interpretation is supported by levirate and sororate tendencies (Damas 1963), male sibling cores are not enduring foundations upon which Iglulingmiut groups were constructed in recorded times. While Parry provides some anecdotal evidence to suggest that male sibling ties may have been stronger in the past, the male sibling group represents the major structural weakness in Iglulingmiut society.

### Structural Problems and Solutions in Iglulingmiut Socioeconomy

Like the Qikirtarmiut/Kinguamiut of Cumberland Sound, the Iglulingmiut appear to have placed so much emphasis on *naalaqtuq* that it undermined the reproduction of local groups from one generation to the next. As noted above, male sibling groups tended to split apart after the death of the father, and 'pronouncedly so after the death of both parents' (Damas 1975c: 25). Such splitting of male sibling groups is a direct function of super-subordinate directives implicit in their relationship. Damas (1963: 106) noted that disagreements between brothers revolved around matters of authority and decision-making on the horizontal level, and the term *isumakattiginituk*, or 'they disagree', was used frequently to explain the separate residential locations of brothers. Such splits also appear to have conspired against the emergence of large permanent groups of kindred and strong unilineal tendencies (Damas 1963: 105). In the life cycle of Iglulingmiut male siblings, most splits occurred after the age of forty, subsequent to the passing of the father and the emergence of their own productive extended family units. As Damas observed,

> the disappearance of the cohesive influence of the parental generation and the development of economic and emotional independence through maturity and through the emergence of the new extended family organization make the time ripe for a break, whether the motive may be economic or personal. (1963: 106)

Irrespective of the reasons cited for the fissioning of male sibling groups, the underlying cause was structural. Simply, *ungayuq* directives and co-operative behaviours between brothers were not sufficiently well-developed to allow the local group to maintain its structure and organization after the death of the father. In particular, the new dominance-subordination relationship that pertained between male siblings conspired against the perpetuation of the group and the cohesive nature of its former productive relationships. It is with reference to this structural weakness that most unique features of Iglulingmiut society can be explained.

The splitting of male sibling groups probably did not present too serious a problem in areas that were continuously productive from one year to the next. However, such environments are rare in the Canadian Arctic. While the Iglulik area is considered to be far more productive than most Arctic regions (Damas 1963, 1969b), it is still subject to unpredictable resource fluctuations. In this regard, the maintenance of same-generation adult male ties, and male sibling groups especially, would be advantageous. This is particularly so around Iglulik Island where walrus constituted the foundation of the economy. As walrus may be even more dangerous to hunt from kayaks than bowhead whales (Parry 1969: 509–10), to hunt this animal effectively in open water requires groups of several adult men co-operating in a highly organized manner. Winter breathing-hole sealing and fall caribou hunting by traditional means also necessitated sustained co-operative male effort, as did spring basking seal hunting, though to a somewhat lesser degree. While the Iglulingmiut regarded male sibling cores to be a preferred living arrangement (Damas 1963), second only to F-S cores, they evolved a variety of other mechanisms to promote male solidarity among both kin and non-kin in the formation and maintenance of productive relationships.

Foremost among these was the three-cousin terminology. In this system FBS for male Ego assumed sibling-like status, particularly with respect to the *ungayuq* axis. Yet behavioural directives between same-sex parallel cousins were often less intense than those between siblings (Damas 1963, 1968a). Cross-cousins were also governed by *ungayuq* directives; indeed, a roughhouse joking relationship (*illuriik*) often characterized this dyad (Damas 1975c: 20). As *naalaqtuq* directives were less well-developed between cousins than brothers, *ungayuq* behaviours often characterized the bond. Nonetheless, because *naalaqtuq* directives were less structured between parallel cousins (i.e., open to other considerations), they were even less viable foundations than *nukariik* cores upon which to build local groups after passing of the parental generation. Even so, the three-cousin system of the Iglulingmiut can be viewed as an attempt to override deficiencies inherent within male sibling cores by isolating collaterals in terms of affectional closeness and promoting another dimension of male solidarity in productive activity. The fact that a complementary system existed for female Ego underscores the emphasis placed on kin group exogamy and gender solidarity.

The many voluntary or non-kinship alliances that once characterized Iglulingmiut society can also be viewed from the same perspective. Damas (1972b, 1975c) has noted that the Iglulingmiut possessed the most complete system of alliances not founded on kinship of any Central Inuit population. These included spouse exchange, child betrothal, adoption, name-sharing, and age-sharing relationships as well as singing/dancing, trading, and rough joking partnerships. In fact, the only voluntary alliance missing from the Iglulingmiut inventory was the well known seal-sharing partnerships of the Netsilik and Copper Inuit (see below). Alternatively, not only did the Copper Inuit attach comparatively little significance to kinship as a means of organizing productive relationships, but they also displayed the most incomplete system of non-kinship alliances. Damas (1972b) was, at first, puzzled by these findings. After all, should not 'Eskimo' groups who place little emphasis on kinship have relied upon a host of voluntary alliances to organize society, and vice versa? However, he found just the opposite. Iglulingmiut society was organized largely on kinship directives, but it also demonstrated the most complete inventory of non-kinship alliances, many of which played important roles in establishing socioeconomic relationships. Employing an eco-evolutionary perspective, Damas (1972b) suggested that the Iglulingmiut were the most 'internally integrated' of the three Central Inuit populations he analyzed. But by concentrating on 'internal integration' in his comparative analyses Damas obscured important differences among these regional groups, and missed the central purpose of voluntary alliances in Iglulingmiut socioeconomy. Many voluntary partnerships in Iglulingmiut society supported the kinship system, to be sure, but others appear to have overcome its deficiencies, especially its emphases on social hierarchy and kin group solidarity, by providing alternative means for forming productive relationships and extending alliances beyond the boundary of local and kin groups.

Summarized succinctly, the diversity of voluntary alliances in Iglulingmiut society and its unique three-cousin system were means that evolved to overcome structural weaknesses inherent in its emphases on *naalaqtuq*. By taking the 'edge' off *naalaqtuq*, these features, like exogamy, avoidance of opposite-sex consanguines, and leadership, served complementary, and yes, 'integrative' functions.[11]

### Iglulingmiut/Qikirtarmiut Comparisons

This brief treatment of Iglulingmiut social organization raises questions about the Qikirtarmiut of Cumberland Sound, the subregional group that the Iglulingmiut perhaps most closely resemble. As concluded in Chapter 6, Qikirtarmiut/Kinguamiut socioeconomy was governed largely by *naalaqtuq* directives, perhaps just as much as the Iglulingmiut—recall that brothers hardly ever formed the basis of productive activity. Yet, the Qikirtarmiut appear not to have developed any means by which to mitigate the effects of *naalaqtuq* in the formation of their productive relationships.[12] While missionary and other contact influences may have rendered some voluntary alliances obsolete, it seems clear that such partnerships in Cumberland Sound at the time of Boas' study were far less numerous and important than they were in the Iglulik region 80 years later.

The kinship terminology of the Cumberland Sound Inuit also does not favour or place any special emphasis on the formation of productive relationships beyond the vertically-structured extended family. On the contrary, the fact that all cousins are lumped under the same category (*illuq*) and differentiated from siblings reaffirms the importance of the conjugal family. Aunt and uncle terms also show no favoured connection to either the mother's or father's line—all parents' siblings assume equal rank. Both the single cousin terminology, which derives from the same root as the word for house (*iglu*) and can be roughly translated as 'house mate', and the affinal-including aunt terminology of the early historic Cumberland Sound Inuit recall a system of cross-cousin marriage. Yet, they are largely exogamous.[13] While the lack of differentiation in terminology and behaviour among collaterals in Ego's and the adjacent generations may be viewed as logical correlates of exogamy in this context, they also suggest no particular preference beyond the nuclear family in the formation of productive relationships; all possess equal status, thus increasing the potential number of alliances that can be formed. Such a system would surely seem advantageous in areas of unpredictable and variable resource distributions. However, Kingua Fiord appears to have been at least as productive as the Iglulik area. Thus, it is not surprising that *irniriik* (F-S) cores formed the basis of most local groups. *Panniriik* (F-D, M-D) cores were also common, as were *ningaugiik* (F-DH) cores, as sons more often than not went to reside with their wife's parents rather than their brothers after the death of the father. Conversely, *nukariik* (B-B, Z-Z), *nayagiik* (B-Z), and other same-generation cores rarely constituted viable foundations for group membership among the Qikirtarmiut, as they did among the Umanaqjuarmiut. The environment may not have been the ultimate architect of Qikirtarmiut/Kinguamiut social organization. Nonetheless, it certainly allowed certain aspects of Central Inuit social structure, i.e., its *naalaqtuq* tendencies, to develop and flourish.

THE NETSILINGMIUT: *UNGAYUQ* INTENSIFIED

If the Iglulingmiut and Qikirtarmiut embody the essence of *naalaqtuq* social structure, then the Netsilingmiut appear to have carried the alternative structural principle of *ungayuq* to a similar extreme. Here a brief summary of Netsilingmiut culture history and change is warranted.

**History of Contact**
The Netsilingmiut lived on the Arctic coast between Boothia Gulf and Queen Maud Gulf (Figure 67). The Netsilingmiut were first contacted in 1830–33 when John Ross wintered in Lord Mayor Bay. Although little interaction occurred, the loss of one of Ross' ships provided local groups with a windfall of wood and metal for years. In 1833 George Back approached Netsilingmiut country from the south, reaching the mouth of the river that now bears his

name, while Dease and Simpson penetrated the region from the east. Fifteen years later, John Franklin's expedition abandoned their ships near King William Island, eventually succumbing to disease and starvation. Subsequently, a number of Franklin search parties entered the region. However, like their predecessors, they recorded very little of Netsilingmiut social life. Lacking proper training, they stressed the sensational aspects of Inuit social and material culture, while neglecting the relations that existed among the people in everyday life (Balikci 1970: 93).

Between 1880 and 1920 whaling and trading activity at Chesterfield Inlet and Repulse Bay attracted up to 150 Netsilingmiut, or *c.* 40 per cent of the population (Rasmussen 1930: 84–8). Most of these individuals apparently came from the less abundant western Netsilingmiut district around King William Island, not the more productive eastern area around Pelly Bay (Remie 1985: 72). Roald Amundsen, who wintered on the south coast of King William Island in 1903–5, was the first to record Netsilingmiut culture in any detail. However, Knud Rasmussen's descriptions of 20 years later are more informative, despite the fact that he documented few details on social organization.

At the time of Rasmussen's observations the Netsilingmiut numbered 260 people in five local groups with a ratio of about 1.5 males to every female (1931: 84).[14] Considering the high incidence of adult male death, selective female infanticide was clearly indicated, a feature that Rasmussen sought to document extensively during his eight-month stay among the Netsilingmiut. In comparison with the Iglulingmiut, Netsilingmiut socioeconomy was based much more on fish and caribou than marine mammals—the lack of an open water hunting technology suggests an inland origin.[15] During the summer and fall small extended family units dispersed inland for caribou and fish, whereas during the winter and early spring they aggregated in larger villages to pursue seals at breathing holes.

**Culture Change and Continuity**
Any reconstruction of Netsilingmiut society must rely on Rasmussen's descriptions. Yet, by 1923, firearms obtained from the Aivilingmiut had begun to make significant impacts on aboriginal society, perhaps more so than any other Central Inuit population. Although many features of aboriginal Netsilingmiut socioeconomy remained intact up to and during Rasmussen's study, after the acquisition of the repeating rifle, the co-operative hunting of caribou at summer water crossings was soon abandoned in favour of pursuing caribou in smaller groups throughout the year. Not only did the rifle simplify, intensify, and individualize the caribou hunt (Balikci 1960), but it also resulted in the abandonment of the kayak. At the same time, firearms allowed seals to be more easily killed at the *sina*. Gradually, breathing-hole sealing as well as its rigid meat-sharing system began to disappear. Concomitantly, extended family organization as well as religious ideology began to erode:

Now we can shoot caribou everywhere with our guns, and the result is that we have lived ourselves out of the old customs. We forget our magic words, and we scarcely use any amulets now. We forget what we no longer have any use for. . . . We remember them no more. . . . (extracted from Rasmussen 1931: 500)

Yet as food became more plentiful, dog teams became larger and mobility increased while territorial boundaries decreased. Whereas the pattern of aggregating in large winter sealing villages on the sea ice and dispersing inland during the summer and fall to hunt and fish continued into the mid-1920s, regular trading was also carried on with the Repulse Bay trading post. Soon after, trapping was adopted, as were fish nets and canoes. The latter resulted in the adoption of summer sealing with rifles, and maintenance of productive relationships, at least temporarily, within the extended family.

The establishment of a mission on Pelly Bay in 1935 accelerated acculturative trends begun with the introduction of the rifle. As the missionary operated a trading store and supplied medicines, most of the population began to gather permanently around the mission. By 1950 the Netsilingmiut were concentrated at three settlements, Gjoa Haven, Spence Bay, and Kugardjuk, where they were subject to the combined acculturative influences of missionaries, traders, and police. By 1960 no communal hunting nor widespread sharing of caribou or seal meat took place, the latter being restricted to close kinsmen only (Balikci 1960). Likewise, trapping remained an individualistic activity and fur money was never shared, even between the closest of relatives. As Balikci (1960, 1964) has observed, the individualization of hunting practices and restriction of sharing produced by the rifle, together with trapping and individual ownership of imported goods, resulted in the emergence of the nuclear family as the primary socioeconomic unit.

Balikci (1960: 151) felt that the rifle was responsible for the interruption of relations among neighbouring groups and the consequent stabilization and isolation of these groups. This led, in turn, to the search for mates within the group, and ultimately preferential cousin marriage. There is little doubt that the rifle had a profound impact on Netsilingmiut socioeconomy. However, many of the unique customs that anthropologists have come to associate with the Netsilingmiut appear to have been integral, well-developed institutions long before Rasmussen's study and, probably, the introduction of the rifle (e.g., see Amundsen 1908). Such features include cousin marriage, female infanticide, mutual suspicion and hostility between groups, and formal seal-sharing partnerships, in addition to Netsilingmiut kinship terminology and behaviour. Together, these features define a logically coherent structure designed to maintain affectional ties and productive relationships within the group. Damas (1972b, 1975c) contends that *ungayuq* directives did not apply to the Netsilingmiut, and that their society was less 'integrated' than that of the Iglulingmiut.

Although the Netsilingmiut may have never explicitly used or articulated the former concept, aboriginal Netsilingmiut society represents an extreme example of a Central Inuit society attempting to accentuate and maintain *ungayuq* behaviours, perhaps even to the ultimate detriment of its own reproductive forces. While the reasons why they stumbled upon this path may never be known (but see Chapter 9), it will be shown below that Netsilingmiut society represents an integrated system of institutions designed to keep kinsmen within the residential unit and others who were not, out.

## Traditional Features of Netsilingmiut Social Organization
### Closed Groups

In the early twentieth century the Netsilingmiut apparently consisted of six territorially defined groups. These include the Arviligjuarmiut of Pelly Bay, Netsilingmiut of Spence Bay and Lord Mayor Bay, Arvertormiut of the northern end of Boothia Peninsula, Qeqertarmiut of King William Island, and the Ilivilermiut of Adelaide Peninsula. A sixth group, the Ukjulingmiut of Boothia Isthmus, had become all but extinct through starvation, co-residence, and intermarriage with the Utkuhikjalingmiut of the Back River (Caribou Inuit) prior to Rasmussen's expedition to the region in 1923 (Balikci 1970: xx-xxi). One notable aspect of Netsilingmiut socioeconomy was that there was no recognition of an overall 'tribal' identity as there was, for example, among the Iglulingmiut or Qikirtarmiut. Indeed, Balikci (1970: xx) viewed the Netsilingmiut on the whole as 'a divided and unstable group of people'. Prior to the twentieth century, sustained co-operative interaction between local groups was not a feature of Netsilingmiut social organization (e.g., Rasmussen 1931: 202). As one of Balikci's (1964: 71) informants stated 'in the past the *ilagiits* (sic) didn't mix; now they are all mixed up'. In fact, mutual suspicion and hostility characterized relations between most local groups.

Rasmussen (1931) describes at length the aggressive nature of Netsilingmiut intergroup interaction. As he recounted, 'in the old days a tribe was really at war with all others outside of its own hunting grounds, and many are the tales that have been handed down of strife, murder, in fact massacre' (1931: 202). The genesis of much murder and feuding was wife stealing resulting from a shortage of women created, most certainly, by the practice of female infanticide. Indeed, it was not unusual for women to have had a succession of husbands, each killed by the next in line (1931: 206). Both polygyny and polyandry—which were far more and far less common, respectively, than the shortage of marriageable females would lead us to suspect—only exacerbated feelings of jealousy and hatred, and ultimately murder (1931: 54, 74, 77, 78). While lying, thievery, and wife stealing were expected between strangers, such behaviours were 'strongly condemned within the tribe' (1931: 200). Yet, mutual fear and suspicion were not restricted to outsiders. As an old informant of Rasmussen's (1931: 203) recalled, whenever travelling 'a man would carry a harpoon and snowknife for fear of being attacked by his companions'. Moreover,

> A man in the procession could not stop to make water without great risk, for the man who walked in front might easily get the idea that the man for some reason or other would strike him down from behind, and the suspicion alone might be sufficient cause of bloodshed.[16]

Balikci (1964) identifies two main socioeconomic units beyond the nuclear family among the Arviligjuarmiut, the extended *ilagit* and the restricted *ilagit*. The former constituted the widest circle of relatives recognized among the Netsilingmiut, and was composed predominantly of consanguines. It was, however, not a residential, ceremonial, political, or economic unit, but an 'ego-based kindred' which formed the largest sphere of security for individuals (Balikci 1964: 25–9). The extended *ilagit* was also the domain beyond which mates were not actively sought. Indeed, the establishment of affinal ties between extended *ilagit* was not valued, and therefore discouraged. Had it not been for wife stealing and the exorbitantly high prices paid for women (Rasmussen 1931: 32)—methods of obtaining women without being obliged to reciprocate in kind—extended *ilagit* might have otherwise functioned as marriage isolates. Although extended *ilagit* often assembled in winter sealing villages, rarely did this unit form a residential group, the essential characteristic of the restricted *ilagit*.

The latter socioeconomic unit, which may be equated with the local group or *nunatakatigiit*, was the more important and primary unit of production in Netsilingmiut society. Although local group composition was sometimes bilateral, *irniriik*, *nukariik*, and male cousin cores formed the basis of most local groups. Whereas the eldest active hunter was usually regarded as the leader or *ihumataq*, super-subordinate relationships appear to have been secondary to the maintenance of co-operative and affectionate relations among local group members. As the woman always went to live with her husband's family (Rasmussen 1931), there was a heavy patrilocal slant to most groups. At the same time, male dominance and solidarity within local groups found expression in the separation of men and women at meal times, the considerable affectional and joking relationships that prevailed between male cousins (Balikci 1970: 121–2), and the practice of female infanticide, which was the man's prerogative (Freeman 1970, 1971). Marriage with members of the same kin group, preferably between first cousins, was the ideal arrangement, whereas unions between non-resident kinsmen were less desirable. In addition, adoption was carried out predominantly within the local group. As both these customs served to multiply and intensify relationships within the group, in perhaps no other Central Inuit group did the *nunatakatigit* more closely approximate the *ilagit* than it did among the Netsilingmiut.

The inward focusing of relationships and the maintenance of the local group at the expense of the regional group is also reflected in Netsilingmiut mythology. For example, the myth of the 'salmon (char) and sea sculpin' warns of the dangers of leaving one's group and associating with others who are not your kind (Rasmussen 1931: 397–8). Similarly, 'the raven who married a snow-goose' describes the fate of those males who leave their group for that of their spouses—the snowgoose wife and her brothers eventually conspire to kill the outsider (1931: 400–1). The myth of the sea goddess (Nuliajuk) also underscores the importance of kinship ties within the group. In the 'Sedna' myth of the Iglulingmiut, Oqomiut, and other Baffinlanders, 'Sedna' (Nuliajuk) is thrown into the water by her father, where her fingers are dismembered and she drowns, ultimately as a consequence of denying appropriate suitors and/or marrying a dog, i.e., rejecting social norms or conventions. In the Netsilingmiut version (1931: 225–6) Nuliajuk meets a similar fate not because she repudiates society, but because society rejects her, i.e., she was a stranger: 'no one cared about her, no one was related to her, and so they (the boys and girls she was playing with) threw her into the water.'[17] The myth of the origin of 'thunder and lightning' repeats the same theme—misfortune befalls those who are without close relatives (1931: 210). Rasmussen recorded a number of variations of the same myth among the Netsilingmiut. Some of these discrepancies, however, may be the result of differential contact influences, both with Euroamericans and/or other Inuit groups. Nonetheless, among a people exhibiting such a closed structure, it is perhaps not surprising that there would be considerable variation in customs and beliefs among local groups.

## *Cousin Marriage*

Netsilingmiut kinship terminology is similar to that recorded by Morgan (1870) from Cumberland Sound insofar as there is an assimilation of affinal terms in the first ascending generation. However, Netsilingmiut terms are much more complete in this regard (Table 23). The extension of FZ not only to FBW, but to MBW, moreover suggests an emphasis on the male line. While same-sex cousins were *illuq*, regardless of whether they were cross or parallel, sibling terms were extended to opposite-sex cousins. Thus, for male Ego, *nayak* was used for sisters as well as female cousins. Yet, as first cousin marriage was the preferred marriage arrangement, individuals grew up to marry their classificatory siblings. Damas (1975c: 17) correctly regards this practice as a transferral of the appropriate sentiments from a consanguineal to an affinal context. During childhood cousins of both sexes are second only to siblings in terms of affectional closeness to Ego. However, unlike the Iglulingmiut and most Central Inuit groups, where avoidance begins to characterize cross-sex cousin relations at puberty, affectional behaviours are not abolished among the Netsilingmiut. Rather, upon reaching adulthood these sentiments are transferred to prospective spouses. In this way, bonds of affection established in childhood were maintained into adulthood, thus preserving deep feelings of trust between local group members and the solidarity of the local group.

Rasmussen (1931: 191–2) provides ample evidence of this transference. Husbands and wives never addressed each other by their real names, but by pet-names or terms of endearment appropriate to their consanguineal relationship, particularly, it seems, if their fathers were related. Thus, Itiqilik called his wife 'my cousin's daughter', while she addressed him as 'uncle' (*akka*, FB). Tarrajuk called his wife 'my dear little younger sister', whereas his wife called him 'my dear little elder brother'. Qaqortingneq's name for his wife, Quertilik, was 'my big, younger sister', while hers for him was 'my big, elder brother'. The latter apparently called each other by these names because they wanted to emphasize the fact that their fathers were brothers.

Damas has stated that *ungayuq* does not apply to the Netsilingmiut. However, it would be difficult to conceive of a better demonstration of this principle at work. In this regard, there is no contradiction between Netsilingmiut behaviour and kinship terminology, as Damas (1975c) suggests. The Netsilik do not represent a case whereby kin behaviour changed before the kinship system. On the contrary, Netsilingmiut kinship terminology and marriage practices are in perfect harmony.

*Female Infanticide*
In the anthropological literature on female infanticide the Netsilingmiut occupy a special position. In comparison with most Inuit, the Netsilingmiut appear to have practised this custom to an extreme. Remie (1985, 1988), however, has questioned the reliability of Rasmussen's statistics. While Rasmussen (1931: 140–1) reported that 38 infant girls were killed out of the 96 births he recorded in the western Netsilik region, in actual fact there were only 34 cases of female infanticide (Remie 1985: 73). Re-examination of Rasmussen's figures also indicate that there were 114 live births, with additional, but an undetermined number of, cases of female infanticide. However, these latter data also appear suspect (Remie 1985: 69–70). Even so, Remie concluded that the rate of female infanticide in the less productive western Netsilik district was nearly 60 per cent (34/(114/2)), or almost twice that of the eastern district (18/(114/2)). Remie (1985) suggests that these differences were the result of the greater ecological productivity of the Pelly Bay area and the fact that Netsilik groups in the western district were severely depleted by emigration, thus restricting the circle of relatives from whom a marriage or adoption partner could be found. While Remie's thesis may have some validity, the pattern of selective female infanticide among the Netsilingmiut is undeniable: the culling of infant girls was some 30 times that of infant boys— Rasmussen recorded only one instance of an infant boy being put down. Moreover, Remie does not address why female infanticide was practised so extensively among the Netsilingmiut in the first place.

Numerous explanations have been offered for the high rate of Netsilingmiut female infanticide. Rasmussen (1931) advanced the popular notion that extreme environmental pressures necessitated the elimination of unproductive members of society. As resources were scarce and as boys were more valued than girls (Rasmussen 1931: 140–2), the latter were the first to be sacrificed for the good of the community. Alternatively, Balikci (1967, 1970) has suggested that female infanticide arose as a means to correct the imbalance created in later life by high adult male mortality. However, this practice, as Freeman (1971: 1015) noted, would seem to contribute to, rather than alleviate, any numerical imbalance between the sexes. While these views have found acceptance in the anthropological literature, Freeman (1970, 1971) considers that female infanticide may possess complementary functions, one ecological, the other social. Specifically, whereas female infanticide might have enabled a higher proportion of available energy to be channelled to mature individuals, who use energy more productively and were less likely to perish than children, this practice was more readily an expression of male dominance within the extended family household—recall that males took the decision to extinguish infant girls. It must be noted that, while Rasmussen's (1931: 141–2) informants supported the widespread and strongly expressed sentiment in favour of boys, both Freeman (1971: 1014) and Steenhoven (1962: 50) observe that they did not rationalize female infanticide in ecological or any other terms. Rather, they accepted it as customary behaviour, as they did first cousin marriage and other cultural practices.

While the above hypotheses may have varying degrees of merit, the high rate of female infanticide among the Netsilingmiut may be explained most parsimoniously as means of maintaining the closed structure of local groups by restricting or controlling access to women. As noted above, Netsilingmiut groups were heavily patrilocal and constituted on the basis of male relevant ties. However, by eliminating excess females who could not be betrothed to or adopted by relatives within the local group, access to women was denied to non-kin and non-local group members. In effect, female infanticide served to alleviate the obligation to form reciprocal exchange relationships outside the local group. At the same time, as with preferential first cousin marriage, female infanticide functioned to keep and focus affectional bonds within the group.

Although Riches (1974: 357) broached the 'closed group' explanation of Netsilingmiut female infanticide when he investigated its relationship to cousin marriage and infant betrothal, he attributed the development of these institutions to competition for scarce resources and extreme environmental pressure. The latter 'so exacerbated competitiveness for live game that preferential kindred endogamy (and female infanticide) developed as (strategies) to deny the reproductive services of females to groups of unrelated and rival Eskimos.' The well engrained custom of betrothing children before birth to close kinsmen (Rasmussen 1931: 194) was simply a corollary of these institutions insofar as it too functioned to preserve the closed structure of Netsilingmiut groups. While Riches' and my explanations for closed group structure among the Netsilingmiut differ, we are in general agreement that female infanticide, kin endogamy, and child betrothal served this end well.

Given the closed structure of local groups, their heavy patrilocal tendency, strong preference for boys, as well as the betrothal of children before the gender of the infant was known, the principle of reciprocity between co-resident families may have also been involved in the practice of female infanticide. In this regard, it is perhaps little wonder that Rasmussen (1931: 140) observed that girls are usually 'killed or given away at birth until enough boys have been born'. Under these tendencies, female infanticide makes some sense, especially when a gender balance among producers is desired within the local group. And, in this regard, it is in adulthood that the imbalanced sex ratio created by female infanticide begins to correct itself through the higher frequency of adult male deaths. Thus, we see in both Rasmussen's western Netsilik and Remie's eastern Netsilik infanticide data a latent tendency to maintain two to three times as many boys as girls.[18]

Female infanticide not only served to focus ties inward while restricting relationships with other groups, but it also reinforced male unity and solidarity within the group. Freeman (1970, 1971) comes close to this explanation when he suggests that female infanticide was an assertion of male dominance within the household. However, more proximately, female infanticide, by restricting access to women, reduces the potential for the formation of exchange relationships with outsiders, thus preserving the solidarity of the kin group, and, in particular, adult male relationships.[19]

## Closed Groups and Ecological Necessity

Many of the unique customs associated with Netsilingmiut society appear to have developed as means to maintain the closed structure of kin and residential groups. However, there were occasions, particularly in the winter, when two or more restricted *ilagiit*, and often extended *ilagiit*, came together for economic reasons. As ringed seals were more dispersed in the Netsilingmiut district than most regions of the central and eastern Arctic, it was absolutely essential in the winter that several extended families live and work together in the interest of the common good (Balikci 1970: 139). In this context, the mutual suspicion, jealousy, and hostility that were directed to people outside the extended family took a 'back seat' to the formation of co-operative relationships between different groups, the most common of which was the seal-sharing partnership, or *niqaiturasuaktut*. This system of sharing, which was established only among non-kin or distantly related individuals (i.e., people outside the residential unit), depended upon an interlocking set of sharing partnerships involving precise and inflexible rules (Van de Velde 1956). Ideally, every man had 12 partners, each named after the part of the seal (e.g., head, right front side and

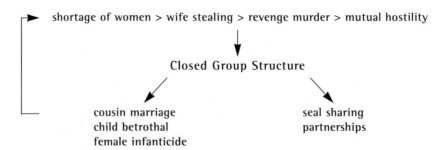

FIGURE 69. Schematic of interrelationships among Netsilingmiut socioeconomic features.

flipper, etc.) shared between them. Such partnerships, or *nangminariit*, were usually chosen by a boy's parents in early childhood or inherited from the father. So rigid was this system that, if a man was unable to fulfil his role as a partner because of death or absence, a substitute (related in some way or another to the absentee) was found. This inflexible and formal system of sharing worked to overcome latent hostilities between non-relatives, thus giving the winter aggregation social cohesion and economic viability (Balikci 1970: 138). Yet, it was quickly 'mothballed' in times of resource abundance. Joking, song, spouse exchange, and other partnerships may have also served to alleviate suspicion and fear between members of different *ilagiit*. However, they were less formal and more ambiguous in content, and thus just as likely to engender hostility as to alleviate it.

### Contradiction and Integration in Netsilingmiut Society

If Netsilingmiut society is viewed as an extreme development of *ungayuq* directives and closed group structure, we can see, *contra* Damas, that it is not any less integrated than Iglulingmiut society. Rather, the above institutions form a systemically related, if somewhat tautological, structure that can be expressed in graphic form (Figure 69). No contradiction is apparent in this structure so long as a balance is maintained among its constituent parts. However, such a scenario exists only in theory. Netsilingmiut society, because of the tenuous interrelationships of its component behaviours, was always subject to dysfunction arising from both internal and external forces. For example, increased attempts to maintain closed group structure might trigger higher rates of female infanticide, cousin marriage, wife stealing, and intergroup hostilities—such as might have occurred in the western Netsilik district during the late nineteenth century—ultimately leading to the extinction of some groups.

The Netsilingmiut represent an extreme development of *ungayuq* behaviours. However, they also demonstrate the inherent deficiency of too great an emphasis on establishing affectional ties and co-operative relationships only with close kinsmen within the local group—a weakness that the Copper Inuit appear to have avoided.

# Chapter 9

# The Copper Inuit:
# Antithesis of Central Inuit Social Structure?

The Copper Inuit represent an entirely different solution to the problems of material and social reproduction in the Canadian Arctic than either the Iglulingmiut or the Netsilingmiut. Consequently, and because the Copper Inuit play a critical role in theories of Inuit social organization, considerable attention is devoted to possible mechanisms of culture change. Knowing how groups reacted to and were influenced by culture contact will allow us to better understand the structure of aboriginal Copper Inuit social organization.

## History of Contact

### Nineteenth Century Contact and Culture Change

The Copper Inuit, so named for their extensive use of copper and the cultural uniformity of those groups in possession of this metal (Jenness 1922: 42, Stefansson 1919: 33), occupied the coastal plains of Victoria Island and the Arctic mainland between Amundsen Gulf and Queen Maud Gulf (Figure 65). The Copper Inuit were first contacted in 1771 when Samuel Hearne's Dene guides encountered and massacred 20 Inuit at Bloody Falls on the Coppermine River. Fifty years later, the British Navy's search for the Northwest Passage brought Sir John Franklin to the region (Franklin 1828). However, Franklin's 1821 and 1826 expeditions made little contact with the Copper Inuit—the latter were dispersed inland during the summer—and recorded virtually nothing of their habits or culture. Under the auspices of the HBC, Dease and Simpson travelled down the Coppermine River to its mouth in 1838 and 1839 where they explored the coastline east of the Kent Peninsula. At Bloody Falls, Cape Turnagain, and the mouth of the Richardson River they encountered small groups of Copper Inuit, the largest of which numbered 30 people.

The next and most intensive phase of European activity began ten years later when the British Navy entered the region in search of Franklin. In 1848 Rae explored the south and west coasts of Victoria Island, where at Cape Hamilton he encountered 13 Inuit families, 'all . . . very fat, having an abundance of seal's fat and flesh' (Jenness 1922: 30). Several winters later, M'Clure wintered in Prince of Wales Strait, where he met five families at Berkeley Point. Both Rae and M'Clure remarked on the curious habit of the Inuit they met of repaying gifts given to them. According to Jenness, while Cape Hamilton natives found it difficult to understand that 'no return was expected for ... the presents made them', the Berkeley Point natives were 'incapable of supposing that anyone would give them an article without expecting an equivalent' in return.[1] In 1851–52 Collinson wintered in Walker Bay on Victoria Island and then proceeded to Cambridge Bay where he spent the following winter among 200 to 300 Inuit. About the same time, M'Clure almost circumnavigated Banks Island before abandoning his ship, the *Investigator*, at Mercy Bay. For the next 50 years the Copper Inuit remained virtually undisturbed by outsiders. Yet Euroamerican materials had entered the country via trade with the Caribou Inuit and Netsilingmiut for at least a century.

The greatest influx of Euroamerican materials into Copper Inuit society, however, came from the abandonment of the *Investigator* and its cache of supplies at Mercy Bay. Until about 1890, when musk-ox depletion may have forced Inuit to return to their homelands on Victoria Island, the Mercy Bay depot was exploited for its copper, iron, and wood. For about two dozen years, tons of valuable materials and manufactured goods entered the economic system of the Copper Inuit (Hickey 1984: 19). While some view the infusion of materials from the Mercy Bay cache and other sources as contributing to the development of many of the unique features of Copper Inuit culture (Damas 1969b; Hickey 1984, 1992; McGhee 1972), the possibility that most of these characteristics may have been integral features of Copper Inuit existence well before 1850 is considered below. Interestingly, materials from this cache were not hoarded by those few groups who hunted on Banks Island in order to gain a more secure livelihood for themselves or some measure of socioeconomic advantage over groups with whom they had contact. Rather, its materials were dispersed widely throughout Copper Inuit society, particularly during the winter when large numbers of people met on the sea ice (Hickey 1984, 1992). As implied above, the Copper practised a system of mutual and generalized reciprocity, wherein, except for close kinsmen or formal exchange partners, gift-giving had to be met with an immediate repayment of equal value (Hickey 1984: 25; see also Jenness 1922: 30).

How then did the Copper Inuit cope with this infusion of wealth? Hickey (1984: 25) postulates that exchange parity was maintained by procuring extra quantities of traditionally valuable goods not found in abundance in those areas exploited by the discoverers of the Mercy Bay depot, the Kanghiryuarmiut and Kanghiryuachiakmiut. For example, the former group obtained composite wood bows from the Haneragmiut of the Rae River drainage southeast of Coronation Gulf in exchange for iron goods and copper from Mercy Bay. In other words, 'by intensifying production and putting more goods into circulation the Copper Inuit altered a number of highly visible, material aspects of their society in order to maintain others that they perceived to be at the core of their own self-identity and ideological self-expression' (Hickey 1984: 24). Among other things, this may account for the demise of basking seal hunting at about the same time the *Investigator* cache was discovered. Stefansson (1919, 1964) in 1910 found very few old men who had ever practised this type of sealing, despite the annual occurrence of hundreds of seals on the ice near spring settlements. Hickey (1984, 1992) attributes this development to the fact that, with the need to procure more items of traditional value to Banks Islanders (e.g., wood and caribou skins) and these being found only on land, southern groups began to leave the sea ice earlier in the spring. This resulted in a change in seasonal subsistence patterns whereby more time was spent inland to acquire land products and to prepare them for exchange.[2]

Hickey (1992) has elaborated further on the effects of *Investigator* materials on the Copper Inuit. With the need to procure terrestrial resources to maintain exchange parity, a greater ideological separation of the traditional land-sea dichotomy developed. Socioterritoriality may have also become more pronounced so 'that subgroups could protect (exclusive) access to appropriate exchange resources that were disjunctive in location'. At the same time, there developed a more elaborate and ceremonialized context for the enhanced trading and circulation of both exotic and local goods, especially British Navy copper and iron, as well as native copper. In turn, there emerged a strong preoccupation with individualistic trade and the overshadowing of kinship ties in order to maintain egalitarianism on a personal basis through economic and specific (i.e., neither kin-based nor generalized) reciprocal material exchange. Concomitantly, the development of an 'entrepreneurial' ideology tempered by the need to maintain egalitarian ideals, led to a period of intense social friction and high rates of homicide as well as the abandonment of an 'irrelevant' mythology based on other, older values.

Hickey's ideas are not only elegant in their formulation and plausibility, but go far beyond any attempt thus far to account for the distinctiveness of traditional Copper Inuit society. Nevertheless, many key features of Copper Inuit culture may have been present well before the abandonment of the *Investigator*. Whatever impact the *Investigator* materials may have had on the Copper Inuit, it could not have been as significant a harbinger of cultural change as some suggest. While certain aspects of the economy were altered to maintain others, the basic ideology and foundations of Copper Inuit society remained unchanged well into the first quarter of this century, when traders, missionaries, police, and ethnographers entered the region.

**The Contact-Traditional Period and Culture Change**
During the first decade of this century Qallunaat came not as an explorer, but as an adventurer and trader. Several individuals passed through the area between 1900 and 1905, but none stayed for long and contact was restricted to

a few casual encounters. As part of a natural eastward expansion of the Beaufort whale fishery, C. Klengenberg and 'Billy' Mogg wintered on the west coast of Victoria Island in 1905 and 1907, respectively. Both men returned with stories of 'blond Eskimos',[3] a discovery which brought V. Stefansson to the region in 1910. Here, he spent the summer and the following spring, recording Copper Inuit culture in considerable detail (Stefansson 1919). With the demise of commercial whaling around 1910, the few commercial whalers who remained on the Arctic coast turned to fox trapping and general trading. After 1910, largely as a result of Stefansson's trip to Great Bear Lake, the Copper Inuit also began to trade with the Hare Dene with whom they formerly had hostile relations (Stefansson 1919). This new trade and the need to convert the 'recently discovered' Copper Inuit ultimately resulted in the murder of two Catholic priests near Bloody Falls in 1913 at the hands of the infamous Copper Inuit *angaquk*, Uluksak, and his companion.

Between 1914 and 1916 the southern party of the Canadian Arctic Expedition was stationed at Bernard Harbour in Dolphin and Union Strait. Many records of this expedition, especially Jenness' (1922) excellent study of Copper Inuit culture, remain to this day definitive works on the region. Both Jenness and Stefansson (1919), and to a lesser extent Rasmussen (1932), documented many details about Copper Inuit social life, and these form the basis of our understanding of their society.

Anxious to convert and to trade with the Copper Inuit, the Anglican Church and HBC arrived at Bernard Harbour in 1916. Prior to this date, trade was carried out in a desultory manner by free traders. However, trading now became more entrenched in the annual round. A year later, two trading posts were established near Bathurst Inlet (Usher 1965: 50). The impact of this trade, particularly the acquisition of rifles, fox traps, and fish nets, was quickly felt. While only a few men owned more than one dog in 1910, five years later Jenness (1922: 89) noted that four or five dogs per man was common. By 1920 there were four posts in Copper Inuit territory and nearly every man possessed a rifle (Jenness 1922: 249). In addition, the 'old copper culture' had given way to iron, caribou had become the main dietary staple, and most families were leaving their sealing grounds two months early to trap fox and hunt caribou inland (Jenness 1922: 248–9). In order to maintain traplines, hunters had to travel farther over shorter periods of time.[4] This placed a premium on dogs, which in turn, put greater pressure on local caribou and fish resources. Even though the rifle and fish net may have allowed the hunter to trap fox and maintain larger dog teams, the former also led to local shortages of caribou through over-hunting and the abandonment of former ranges (e.g., Rasmussen 1932: 62; Usher 1965: 135–7).[5]

While the inland orientation of the Copper Inuit may have been fostered, in part, by the infusion of materials from the *Investigator*, this tendency became even more pronounced after the introduction of the rifle and trapping.

By the mid-1920s many Copper Inuit families were leaving their sealing camps as early as mid-March to intercept caribou long before this animal reached its calving grounds on the coastal plains (Usher 1965: 137). In addition, many groups remained dispersed in the interior from early spring to late fall, establishing temporary hunting camps at fishing locales. With the failure of migrating caribou to reach Victoria Island, summer subsistence in this district turned to net fishing and sealing from wooden boats during the summer, with winter settlements becoming centred at points of land on the coast (Damas 1988: 112). Eventually, the large winter sealing camps of the past began to give way to alternative configurations. Still, many families continued to gather in large sealing villages during the winter. For instance, four winter camps of 14 to 25 families were observed in Coronation Gulf over the winter of 1928–29 (Damas 1988: 115).[6] However, winter aggregations became more variable in size during the 1930s.[7]

The rifle, fox trap, fish net, wooden boat, seal hook, and other technological introductions, facilitated by local caribou shortages, led to a diversity of adaptations not experienced before in Copper Inuit country. In consequence the homogeneity of Copper Inuit culture, and the strict ideology of egalitarianism that drove it, began to break down. Local adjustments in subsistence and settlement patterns served, for example, to undermine the foundation of reciprocal rights and obligations between former geographically distant and proximate families. As the basis for such arrangements were formerly instituted through the formation of dance/song, spousal exchange, and trading partnerships during winter sealing aggregations, it follows that the latter features, and reciprocity in general, would begin to erode as well. These changes, in turn, were compounded by the attraction of local trading posts, whereby at least some of the year in various parts of the district was spent in close contact with contact agents. Yet, interaction led to an increase in the incidence of disease, and by the late 1920s many Copper Inuit had lost their lives to tuberculosis and influenza.

As ill-equipped as the Copper Inuit were to cope with these disruptions, they were even less prepared to deal with the system of credit introduced by the traders. To trap, the hunter had to be supplied with an outfit. Having no means to pay for his outfit, he went in debt to the trader and settled his account the following spring by trading his catch of furs (Usher 1965: 62). Yet, because of the fluctuating availability of fox and its variable market value, debts were not always paid off. This indebtedness not only prevailed for 30 years until other sources of cash income became available, but it undermined the Copper Inuit unique system of mutual and direct reciprocity. Indeed, as Jenness (1922: 241) noted, 'the custom of offering an equivalent for everything received' was quickly dropped after the coming of the white man. By incorporating the trader into Copper Inuit socioeconomy the average hunter/trapper was often placed in an inferior position. This altered the traditional basis of exchange away from maintaining mutual reciprocity with one's countrymen, while fostering the accumulation of personal wealth. Combined with the disruption of aboriginal exchange relationships through the diversification of subsistence and settlement patterns, individuals such as Uluksak began to attain positions of considerable power and authority in Copper Inuit society. Even though Uluksak's stature appears to have been achieved more out of coercion and fear than mutual respect or admiration, the fact that he could rise to a position of such power was surely influenced by contact, and specifically the erosion of traditional egalitarian relationships and the introduction of hierarchical relations.

The fur trade, the rifle, and the fish net not only undermined traditional reciprocal rights and obligations, but they began to turn families into self-sufficient production units. Thus, the traditional economic foundation for sharing began to erode and socioeconomic relationships became more inwardly focused. At the same time, the abolition of spousal exchange and ritual dance/song partnerships by Christian authorities destroyed the traditional means by which co-operative activity and reciprocity were established. Ultimately, the option of forming and dissolving productive egalitarian relationships at will became less viable, if not less attractive. As group membership became more stable and prolonged within the context of producing for the fur trade, kinship directives implicit in age and gender relationships within the nuclear family began to replace and assume the roles of voluntary partnerships of the past. Eventually, as adult children of both sexes remained in the group, extended family relationships became more important in the formation of productive units, despite the fact that the rifle tended to individualize hunting practices.[8] While the influx of goods and materials from the *Investigator* undoubtedly affected Copper Inuit society in some of the ways that Hickey has postulated, even to the point of underscoring the operation of already established social features, it was a combination of more direct influences and proximate impacts during the contact-traditional period that wrought the most change.

Since the deaths of two priests at the hands of Uluksak in 1913 there had been a perceived need by the Canadian Government to establish law and order in the region, and the first RCMP post was built at Tree River in 1919. However, the murder of this post's corporal and HBC trader in 1922 resulted in increased efforts to bring law enforcement to the area and a detachment was soon established at Bernard Harbour. Fur traders and the RCMP were not the only ones interested in the area. A dozen years after the Anglican Church established a mission at Bernard Harbour, the Catholic Church arrived at Coppermine, resulting in a competition for souls that was as fierce as that for furs among the traders.

The long-term effect of the Euroamerican presence during the contact-traditional period in the Coronation Gulf region was to concentrate services and eventually people in a few central locations (Usher 1965: 58), while disrupting aboriginal social and exchange relationships and their socioeconomic underpinnings. Trading posts attracted very few Inuit at first as traders often carried goods to camps in mid-winter when furs could be bought at low prices and Inuit families suffered no want. However, the trading post began to assume greater importance in the annual round of most Copper Inuit after this practice was outlawed in 1927 (Usher 1965: 58–9). While the Copper Inuit family was still nomadic, it began to depend increasingly on the mission/ trading post to satisfy its needs. Gradually, time spent at traditional meeting places on the sea ice declined, until snowhouse villages were abandoned altogether in favour of congregating at trading posts/mission stations for Christmas and Easter (Usher 1965: 59). The introduction of family allowances in the late 1940s served to further increase the dependence of the Inuit on Qallunaat. Trapping during the 1930s and early 1940s brought fair returns. However, with the post-war decline in fur prices and continuing scarcity of caribou herds owing to a combination of over-hunting, population dynamics, and changes in feeding ranges, most Inuit suffered economic hardships. While this was mitigated somewhat by the introduction of federal health and education services around 1950 at Coppermine, alternative sources of cash income to trapping did not become available until the establishment of DEW line stations along the Arctic coast in 1955. The Copper Inuit hunter/trapper may have no longer lived in large winter settlements on the sea ice or engaged in breathing-hole sealing. However, his indebtedness to the trader served, ironically, to maintain his nomadic lifestyle well into the mid-1950s.[9] DEW line and other government activities during the late 1950s created permanent wage labour positions as well as seasonal employment in construction. This resulted in increased centralization of the population, while initiating a shift away from trapping/hunting towards a wage labour economy.[10] Still, traditional land-based activities are as valued and important as they ever were.

## FEATURES OF TRADITIONAL COPPER INUIT SOCIETY

### Economy

Setting aside temporarily the question of culture change during the late nineteenth century, the salient features of traditional Copper Inuit socioeconomy can be summarized. From spring to fall most Copper Inuit families were dispersed in small groups hunting caribou and fishing at numerous locations in the interior. As freeze-up approached they descended the rivers to the coast to make clothing and wait for the ice to freeze. When the ocean froze and the sewing was complete, they moved into large snowhouse villages; settlements of over 50 snowhouses or 150 people were not uncommon (Jenness 1922, Stefansson 1919). Yet, as with summer life, changes in winter residence and composition occurred frequently. While *mauliqtuq* sealing dominated the economic sphere, trading, ceremonial dancing, and song contests characterized social life. By the end of April most families had dispersed along the coast to hunt caribou and/or bearded seals (Jenness 1922: 120), though the latter pursuit was abandoned in the late nineteenth century. By June, these families began to ascend the rivers to fish and hunt caribou in the interior. Despite the rigid division of their seasonal round, the social organization and religious life of the Copper Inuit remained unchanged during all seasons (Jenness 1922: 143), as did the size, composition, and fiercely independent nature of the nuclear family.

**Kinship and the Nuclear Family**

Perhaps the most intriguing feature of Copper Inuit social organization is its pronounced lack of emphasis on the extended family, *naalaqtuq*, and kinship in the organization of society. In fact, among all Central Inuit populations, the Copper Inuit alone represent the classic nuclear type of family organization formerly ascribed to 'Eskimo' society. The independence of the nuclear family was absolute in all seasons, whether dispersed inland in small groups or assembled in large aggregations on the sea ice. Indeed, adult sons rarely occupied the same dwellings as their fathers (see Jenness 1922); three-generation families account for less than 7 per cent of those families recorded by Rasmussen (1932: 78–84). Although communal sharing and eating of food was widely practised, the eldest productive male of a family had no inherent authority over other local group members. So great was the emphasis on egalitarianism that, outside the nuclear family, there were no positions demarcating certain individuals as standing above or apart from others (Hickey 1984: 20):

> Every man ... has the same rights and ... privileges as every other man in the community. One may be a better hunter, or a more skilful dancer, or have greater control over the spiritual world, but this does not make him more than one member of the group in which all are free and theoretically equal. (Jenness 1922: 94)

While a man, because of his ability or character, may have attained a position of some influence, as his powers faded so too did his prestige and authority (1922: 93). Even women outside the domestic sphere enjoyed equal status with that of men in decision-making (1922: 162). At the same time, the great emphasis on individualism made it virtually impossible for communal action: 'there is no common council wherein the will of the people can find a voice, no spokesman to give it public expression, and no leader to translate it into action' (1922: 94). Subsequently, murders and other transgressions against society often went unpunished, and there was no more respect given to elders or to people with superior wisdom or skills than to anyone else (1922: 169). Simply, the Copper Inuit were intolerant of social inequality,[11] a factor which certainly contributed to their legendary mercurial temper (Jenness 1922, Rasmussen 1932). While jealousy and vendetta fuelled aggression, murder was frequently the outcome of its release. Indeed, Rasmussen (1932: 17–18) encountered a village of 15 families wherein all of its men had been 'involved in a killing in some way or the other'.

Just as intriguing is the indifference shown to kinship beyond the nuclear family in the formation of productive relationships. That kinship was simply an unimportant means by which to organize society is apparent in the frequent occurrence of agamous marriages, wherein marriages between blood relations, whether of the same or adjacent generations, were neither actively encouraged nor discouraged. Even so, the affinal-including terminology in the first ascending generation and the separation of cross-sex siblings from cross-sex cousins is consistent with the practice of cousin marriage (Table 23). Paradoxically, however, the Copper Inuit possess the most extensive system of kinship reckoning of any Central Inuit population. Compared to the Oqomiut and Iglulingmiut, who usually did not recognize blood relationships beyond second cousins, the Copper Inuit could trace kin ties to the third or fourth degree of collaterality, especially if same-generation kinsmen were involved (Damas 1975c: 13). Yet no behavioural distinctions accompanied these relationships. In fact, as Damas (1975c: 12) has observed, for the researcher seeking to correlate kin terminology with behaviour, work among the Copper is bound to be frustrating. Indeed, male Ego's nepotics (*gangiaq* and *uyuruk*) are 'all the same', as are first cousins, although FBch (*angutiqat*) are, in fact, separated terminologically from all other cousins (*arnaqat*) (Table 23). Different sets of avuncular relatives are not differentiated. Nor is there an over-riding of generations among consanguines; each generation has its own discrete terms. Some sets of relatives, however, have behavioural directives inherent in their classification. For example, opposite-sex in-laws should avoid each other, while male Ego should obey his father, uncle(s), father-in-law, and older brother. However, enforcement of such codes was weak, and egalitarianism was stressed at all times.

Consistent with the vague and infrequently observed correspondence between kinship and behaviour is the lack of use of kin terms in forms of address. For example, outside the nuclear family, personal names were preferred, even between elders and children, and between members of the opposite sex. This stands in marked contrast to the Oqomiut, Iglulingmiut, and most other Central Inuit groups, where kin terms were used almost exclusively. In fact, taboos often existed against using personal names in various contexts. While the Copper Inuit practised the universal Inuit custom of naming children after deceased relatives, there was no transfer of the latter's kin term nor its behavioural concomitant to the newborn (Jenness 1922: 239), as among other Inuit groups.

**Voluntary Alliances**

If kinship was simply unimportant in organizing society, and if the extended family was not the major unit for food getting, sharing, or commensuration, on what basis were socioeconomic relationships established in Copper Inuit society? More than any other Inuit society, Copper Inuit social integration was based on voluntary reciprocal exchange relationships. Foremost amongst these were spousal exchange and singing/dancing associates. As Jenness (1922: 87) noted, 'it is by ... wife exchange and association in dancing that the Copper Eskimo establishes friendships wherever he goes and travels from group to group without danger.' Ideally, these partnerships were created between individuals who could trace no kinship connection (Damas 1975c: 14). In reality, however, such partnerships found acceptance so long as there was no close blood tie between the exchanging parties. While song/dance partnerships found expression in the dance house, multi-family households were founded on the basis of spousal exchange. Indeed, two families engaged in spousal exchange relationships normally lived together in a double house with separate living/cooking areas (Stefansson 1919: 65, 293; Jenness 1922: 74). Such dwellings also became the foundations for dance houses (Jenness 1922: 71). While Jenness (1922: 74–6) found that double snowhouses accounted for $c.$ 25 to 50 per cent of all winter dwellings he encountered, he also recorded settlements consisting entirely of these features (1922: 35, 110). However, spousal exchange arrangements rarely endured for any length of time for, as settlements were abandoned and people moved to more favourable sealing grounds, old partnerships dissolved and new ones formed: 'It was rare for two families that lived together in one settlement to stay together in the next, apparently because they had tired of each other's company and were anxious for a change' (1922: 74). Jenness and Stefansson noted the common occurrence of bilobate dwellings among the Copper Inuit, and features of this type are commonly found in archaeological context throughout Copper Inuit territory from northern Banks Island (Hickey, personal communication 1991) to the headwaters of the Rae River (Stevenson 1992).

**Individual Mobility and Group Membership**
Group membership in all seasons, but especially the winter, was volatile as individual families were always changing residence from one group to another (Jenness 1922: 32). Despite the facts that local groups were territorially defined with a fixed name and possessed exclusive hunting and fishing rights—strangers had no rights of access unless they shared food and conformed to the customs of the resident group (1922: 91)—aggregations usually consisted of members of many different bands. For example, in 1915 Jenness (1922: 32) encountered a winter camp which was composed of people from 11 different bands. Such dynamics are reflected in the existence of 20 or so territorially defined groups among a population of about 750 (Jenness 1922: 42). These figures contrast markedly with the Cumberland Sound Inuit and Iglulingmiut, where there were four and three regional subdivisions, respectively, for populations of approximately 1000 and 550 people. Comparisons with other Inuit groups,

however, may not be appropriate; Jenness (1922: 39, 120–2) and Rasmussen (1932, 1942) provide evidence to suggest that groups were splitting and forming all the time—a fact which undoubtedly accounts for differences in the number of groups recorded by ethnographers. Rasmussen (1942: 37–42), in fact, recorded several groups so new in their formation that they apparently did not include any children. Clearly, these characteristics reflect the fiercely independent nature of the individual and nuclear family in Copper Inuit society.

### Mythology, Ideology, and Diametric Dualism

Compared to the rich mythology and religious beliefs of most Inuit groups, the Copper Inuit possessed only vague and indefinite religious notions, which left open much room for individual interpretation (Jenness 1922: 174, 184). Whatever rules existed, particularly those supporting the land and sea dichotomy, were frequently broken. As Jenness (1922: 184–6) observed, 'there seems to be a tendency . . . to limit restrictions and taboos as far as possible; they say themselves that they do many things now that were forbidden in former days.' Moreover, there was no belief in features so fundamental to other Central Inuit such as segregation of women during birth or menses, reincarnation of the soul, fear of death or the dying, or a conception of an afterlife (Jenness 1922: 169). The Copper Inuit, in fact, lacked many of the myths and legends that underpinned Inuit social structure and ideology to the east. As Jenness (1924: 1) noted, 'a man may live to old age and die without ever learning more than half a dozen of the tales that have been handed down by his forefathers. . . .' There was simply no interest in old traditions, and whatever stories ethnographers were able to collect are even more crude and disjointed than their English translations would indicate.[12]

Although the Copper Inuit may have been the poorest storytellers of all Central Inuit groups, they were regarded by early ethnographers as the most creative dancers, songwriters, and poets (Jenness 1924, Rasmussen 1932). In these respects, the Copper Inuit appear to have explicitly denied the lessons of the past while focusing on the present. Indeed, 'every notable incident, every important experience or emotion in daily life (was) recorded in dance (and) song . . .' (Roberts and Jenness 1925: 9). They also attached great importance to ritual and ceremony, as evidenced, for example, by the fact they possessed two kinds of dance, the *aton* and *pisik*, as well as elaborate festive parkas and headdresses expressly for use on such occasions (Hickey 1984: 21). The dance house was, in fact, the centre of Copper Inuit social life (Roberts and Jenness 1925: 9).

A reciprocal egalitarianism permeated many aspects of Copper Inuit society from the custom of immediately repaying gifts given with goods of equal value (Jenness 1922: 89–90) to the configuration of houses shared by spousal exchange partners. In the dance house, where both men and women performed, the drummer sang his/her song in two sections while encircling his/her dancing partner (Rasmussen 1932: 140, Roberts and Jenness 1925,

Jenness 1922: 223–5). Upon completion, the drum was passed to the dancing associate, with this being continued until all partners of each gender had sung. While dances and feasts were always held for visitors, the latter were obliged to repay their hosts on the following day (1922: 55, 87, 224–5).

The mythology of the Copper Inuit, impoverished though it might be, also reflects a pervasive dualism (Jenness 1924: 70–85). In fact, a number of tales appear to dwell on the theme of abrogation of reciprocity. An inventory of Inuit string figures compiled by Jenness (1924) reflects both the singularity and dualistic structure of Copper Inuit ideology. Of the 25 string figures unique to the Copper Inuit, 16 or 64 per cent were perfectly symmetrical as opposed to asymmetrical. Of these, six contained diametrically opposed figures (two lemmings, two kidneys, etc.). Similarly, Rasmussen (1932: 271–88) collected several dozen Copper Inuit string figures, 53 per cent of which were symmetrical, with over half of these representing perfect oppositions, indicating once again that the Copper possessed significantly more diametric figures in this medium than other Central Inuit groups.

**Female Infanticide and Seal-Sharing**
Jenness documented the occurrence of female infanticide among the Copper Inuit. However, this practice in 1915 was not nearly as prevalent as it was among the Netsilingmiut. For example, Jenness (1922: 42) recorded 67 males for 60 females at Bernard Harbour in 1915, with the ratio of male to female children being about the same as adults. Moreover, the three instances of female infanticide that Jenness reported were considered to be excessive (1922: 166). This example, along with the fact that there was no child betrothal among the Copper Inuit (1922: 159), leads us to suspect that female infanticide was practised for reasons different than among the Netsilingmiut. In this regard, Rasmussen's documentation of a number of groups composed entirely of young couples and the case whereby a young couple committed female infanticide twice within a one- or two-year period are instructive (1922: 166). Specifically, they suggest that female infanticide was practised not as a means to maintain the closed structure of local groups, but as a way to maintain the independence of young couples and their ability to form life-long relationships with other couples in other groups. Young couples, by delaying reproduction and parenthood, thereby escaping the demands of childrearing and obviating claims upon them by close kinsmen, were granted the social mobility and freedom to form productive relationships that would serve them well in later life. As boys were preferred over girls, and as the Copper Inuit exhibited virilocal tendencies (Damas 1975c), infant girls were put down before infant boys.

Despite an overall increase in individual productivity, Rasmussen's (1932: 70) census reflects a considerably higher frequency of female infanticide than Jenness' statistics. While this indicates that environmental factors had little to do with this practice among the Copper Inuit, it also suggests that contact may have led temporarily to higher rates of female infanticide whereby even more value was placed on the freedom, mobility, and autonomy of young couples.[13] Given the uncertainty and acculturative forces at work during the early 1920s (e.g., disruptions to traditional relationships and lifestyles, and individualization engendered by the rifle), an increase in this practice is understandable.

As among the Netsilingmiut, seal-sharing partnerships were common among the Copper Inuit. However, unlike the Netsilingmiut, seal-sharing partnerships were established with both kin and non-kin. Of the 107 pairings documented by Damas (1972a: 224), 74 per cent were constituted between kinsmen—a figure undoubtedly influenced by the broad range of kinship reckoning among the Copper Inuit. Of these, nearly half were cousins, a not entirely unexpected finding since we might expect most partners to be men of the same generation. However, the largest category were *angutikattigiit*, i.e., the children of two brothers, suggesting an inherent tendency to form male sibling ties—a feature reflected in kinship terminology, but not generally other behaviours, as all cousins were 'the same'. Nonetheless, the fact that seal-sharing partnerships were formed with both kin and non-kin suggests that such dyadic pairings served not to bring otherwise hostile groups together, as among the Netsilingmiut, but to integrate individual families into productive relationships. The possibility that the basis of seal-sharing partnerships was different in each region is suggested by the fact that sharing among Copper Inuit families was not abandoned in times of abundance (Jenness 1922: 89), as it was among the Netsilingmiut and all other Central Inuit societies.

## Breaking the Rules: A Rejection of Central Inuit Social Structure and Ideology

The Copper Inuit are so fundamentally different in social structure and ideology from other Central Inuit groups, including the Netsilingmiut, as to suggest that they represent an entirely different type of social formation. They are what might be termed the 'fierce egalitarians'. In essence, Copper Inuit society constitutes a rejection of Central Inuit social structure and ideology. In fact, it would be difficult to conceive of a better antithesis. While the Netsilingmiut and Iglulingmiut, respectively, represent embellishments of *ungayuq* and *naalaqtuq* directives implicit in social relations, the Copper Inuit have abandoned kinship as an organizational principle altogether. Rather, Copper Inuit social structure and ideology appear to be based on a pervasive dualism, whereby social relations were governed by individualistic, egalitarian ideals that left no room for hierarchical expression. Under such an ideology, social relationships attained a kind of symmetrical equality or duality which rarely endured for any length of time. Not only were individual families continually changing affiliations, but band membership was constantly in a state of flux, with named groups being formed and disbanded all the time. In essence, the

Copper Inuit rejected both the affectional and the obedience directives explicit within their kinship relations and terminology, in favour of creating symmetrical ties with individuals who may or may not have been linked through blood or marriage.

### Reconciling Theories

There remains a problem with this view of Copper Inuit society, insofar as it exhibits little evidence of dialectical materialism. While this in itself argues for a relatively recent origin, it is curious that Copper Inuit society represents an almost perfect antithesis of Central Inuit social structure and ideology. There is virtually nothing in traditional Copper Inuit society, save perhaps for a single ritualistic dance involving both sexes recorded by Jenness (1922: 224–5), that even remotely resembles hierarchical expression. It is possible that those local groups who eventually became the Copper Inuit did so recently (e.g., during the late eighteenth century), and did not have the time to develop into a synthesis of former and newly constituted social structures and ideologies. Alternatively, something may have retarded this dialectic, while serving to reinforce the strict egalitarianism of prevailing social norms, attitudes, and behaviours. In this regard, we must reconsider Hickey's (1984, 1992) views on the development of historic Copper Inuit society. Hickey believes that the many unique features of Copper Inuit culture which appear antithetical to traditional Central Inuit society, emerged directly as a means of coping with the influx of European materials from the *Investigator* cache. However, he also feels that strict reciprocity and egalitarianism may have been present before this event. In this regard, what the impact of the *Investigator* materials may have served to accomplish was to reinforce the predominant social structure and ideology, thus preventing a synthesis of the old and new social orders from taking place prior to documentation.[14]

### Environment and Society in Copper Inuit Country

We have not yet addressed the role of environment in traditional Copper Inuit socioeconomy. The region inhabited by the Copper Inuit is generally regarded to be a marginal, if not harsh and unproductive, environment, at least compared to most areas of the Canadian Arctic (e.g., Damas 1969b, McGhee 1972, Morrison 1983). And social complexity and 'integration' did not evolve to levels experienced in other regions because of these limitations (Damas 1969b). These views stand in contrast to those of Rasmussen who considered Dolphin and Union Strait to be among the best winter sealing grounds he had ever seen. Indeed, in some locations as many seals were caught in a day as in a month among Netsilingmiut villages of the same size (Rasmussen 1932: 75). Stefansson (1919: 52) and Jenness (1922) also speak favourably of Coronation Gulf's productivity. Moreover, it is well-documented that Copper Inuit groups often left for the interior when hundreds of seals were basking within the vicinity of their spring camps. Hard times, of course, were not unknown to the Copper Inuit; many people appear to have perished around the Tree River during the mid-nineteenth century. But famine is also known to occur with surprising regularity in supposedly more productive Arctic environments such as southeast Baffin Island.

What is perhaps unusual about the Copper Inuit region is that caribou and fish, while not locally abundant, are distributed fairly evenly across the landscape from spring to fall. Almost every small lake that drains into Coronation, Amundsen, and Queen Maud gulfs contains fish, whereas throughout the summer caribou are dispersed evenly across their calving grounds, where they forage for food. Indeed, one can walk south in a straight line in almost any direction away from Amundson or Coronation Gulf and be guaranteed of intercepting caribou or a good fish lake within a day. This fact may account for the remarkable inability, noted by Jenness, of Copper Inuit hunters to reproduce maps of their environments. While Jenness (1922: 229) lamented the fact that 'not a single native, save for Uloksak (*sic*), had the slightest conception of a map . . . (or could) reproduce his own topographic knowledge', a similar albeit less pronounced lack of terrestrial cognition has been noted among contemporary males in Coppermine (Stevenson 1992).[15] This finding is in marked contrast with the Iglulingmiut, Oqomiut, and other Central Inuit hunters who could reproduce topographic knowledge with stunning detail and accuracy (e.g., Boas 1964). Although Jenness (1922: 229) obtained some tolerably accurate maps from a few Bathurst Inlet Inuit and southwest Victoria Islanders, the fact that only Uluksak could reproduce topographic knowledge in any detail is instructive. First, it suggests that Copper Inuit hunters could, if so inclined, acquire and reproduce topographic knowledge, but chose not to. Evenly, albeit thinly, distributed interior resources undoubtedly contributed to this proclivity as the accumulation and storage of such information in such environments may have simply been irrelevant. Additionally, the autonomy, mobility, and transitory nature of relationships experienced by the nuclear family may have undermined the need to gather and exchange ecological knowledge and other information, especially between adjacent generations (e.g., from father to son). However, just as importantly, the reluctance to gather, store, and overtly disseminate knowledge concerning resource distributions may be related to the maintenance of individual independence and strict egalitarian relationships in Copper Inuit society. In this context, it is perhaps not surprising that only those individuals who chose to reject such relationships and place themselves above others, e.g., Uluksak, would be concerned with the acquisition and demonstration of topographic knowledge. While the environment may not have been the original or main architect of Copper Inuit social structure and ideology, it certainly did not present a liability to their operation.

## CONCLUSION

Traditional Copper Inuit society exhibits a very different kind of social organization than that demonstrated by the Iglulingmiut and Netsilingmiut. Even so, most features unique to Copper Inuit society can be understood with reference to the principles of *naalaqtuq* and *ungayuq*. Specifically, while the Iglulingmiut and Netsilingmiut exemplify exaggerations of these directives in the formation of productive relationships, Copper Inuit social structure and ideology represent a rejection of these behavioural dimensions. In this respect, Copper Inuit socioeconomy is not any less integrated than that of the Iglulingmiut or the Netsilingmiut. While the reasons why the Copper Inuit chose this particular path are explored in the next and final chapter, the possibility that they may have possessed a different type of organization in the past is suggested by their kinship terminology.

Comparing and theorizing about the social organization of various Central Inuit populations is fraught with difficulties. As Damas and other researchers have discovered, it is difficult to avoid sophistry and eliminate contradiction within a single regional group, let alone between regions. Nonetheless, the theory advanced here has attempted to minimize contradictions while offering a fuller, more complete understanding of Central Inuit social organization than heretofore has been presented in the literature. Based on a detailed analysis of Cumberland Sound Inuit social organization we saw how the principles of *naalaqtuq* and *ungayuq* can be invoked to explain many of the dynamic and integral features of Iglulingmiut, Netsilingmiut, and Copper Inuit socioeconomy.

While space does not permit us to explore the utility of these concepts for explaining variability within and between other regional groups, preliminary study indicates that populations demonstrating features explainable with reference to these two structural tendencies occur elsewhere in the Canadian Arctic, and often in close proximity. For example, in the Clyde Inlet district of Baffin Island, Stevenson (1972) has recorded the existence of two rather distinct regional subdivisions that differ with respect to kinship terminology and behaviour, as well as marriage and residence patterns. While northern groups exhibit a typically Iglulingmiut pattern, southern groups demonstrate a more closed formation, complete with the acceptance of cousin marriage. Similarly, prior to Guemple's fieldwork in the Belcher Islands, Freeman (1967) distinguished two groups of settlements in the islands that differed with respect to numerous social features. Specifically, the northern group exhibited well-developed leadership, stronger kinship linkages, less variable group composition, and decreased mobility and interaction with other local groups. Inuit living in these camps were also less acculturated than Inuit living in the southern camps. In the Sugluk region, Graburn (1969) recorded a preference among local groups for either kin endogamy or exogamy, but not both. Susan Rowley (personal communication, 1989), based upon the published and unpublished journals of Charles F. Hall, has documented the occurrence of two regional subgroups in Frobisher Bay which demonstrate differences in leadership, group size, as well as other features explainable with reference to the two structural tendencies observed in Cumberland Sound. Although too early to state with any degree of confidence, it seems as if the co-existence of *naalaqtuq* and *ungayuq* tendencies among adjacent formations may constitute an integrated, binary system that lies at the heart of Central Inuit social structure.

As intriguing as the detailed exploration of these and other Central Inuit populations might be, they are left for the future. Having determined that Central Inuit social organization does indeed possess a coherent structure, we now address the archeological, anthropological, and political implications of this fact.

# Part Five

## In Consideration of
## Central Inuit Social Structure

# Chapter 10

## Canadian Arctic Prehistory Reconsidered

The search for structure in Inuit social organization has been an elusive quest for anthropologists. The inability to explain variability within and between regional groups has led to *post hoc* accommodative arguments that hold that Inuit society is somehow less structured than other preliterate societies, or that some combination of environment and historical accident is the ultimate architect of Central Inuit society. While composite groups and marginal resources are indisputable facts of life in the Canadian Arctic, and without denying the roles of historical and other factors, we have seen that Central Inuit social organization does indeed possess a coherent and identifiable structure.

Differences between the two major historical groups in Cumberland Sound, the Qikirtarmiut and Umanaqjuarmiut, can be explained with reference to two structural dimensions inherent within most Central Inuit groups. Whereas the Qikirtarmiut were governed largely by *naalaqtuq* directives, productive relationships among the Umanaqjuarmiut were constituted more on *ungayuq* behaviours. Although both structural tendencies possess contradictions that challenge social and material reproduction from one generation to the next, their elucidation permitted a detailed re-examination of the late prehistory of the Sound. Just as importantly, they allowed us to undertake a closer exploration of Iglulingmiut, Netsilingmiut, and Copper Inuit social organization. While the former two regional populations were found to be embellishments, respectively, of *naalaqtuq* and *ungayuq*, the latter was seen to be a rejection of Central Inuit social structure and ideology.

What implications do the above findings have for interpreting Canadian Arctic prehistory? More specifically, in what ways can we profitably apply what we have learned about Central Inuit social structure to our understanding of human occupation of the Canadian Arctic prior to direct contact with Euroamericans? There are a number of issues pertaining to Canadian Arctic prehistory to explore: (a) the Thule Inuit expansion, (b) Copper, Netsilik, and Caribou Inuit origins, and (c) Paleoeskimo social structure. What follows should be regarded less as a series of definitive statements than as a body of informed speculations advanced to guide future discussion and research.

## THULE INUIT OUT OF ALASKA

Around A.D. 900–1000 there occurred an eastward migration of Inuit out of northwestern Alaska bearing distinctive traits and elements known as Thule culture (Mathiassen 1927). Considered to be the direct ancestors of Canada's Inuit, the Thule Inuit possessed a specialized maritime hunting economy. While bowhead whaling was originally thought to have been its foundation (Mathiassen 1927, McGhee 1970), Thule socioeconomy was also well adapted to the exploitation of seals, smaller whales, walrus, and other Arctic species such as caribou and char (Taylor 1966).

Numerous theories have been advanced for the causes of this migration, the most popular of which is the climatic warming model. McGhee (1970) originally speculated that Thule culture developed from a terminal phase of the Birnirk culture in north Alaska around A.D. 900, and spread rapidly eastward within a century or so during a climatic warming trend known as the Neoatlantic period (A.D. 900–1200).[1] With milder temperatures and less ice cover, both spatially and temporally, Alaskan bowheads and Thule Inuit groups expanded into the Canadian Arctic. Once entrenched in their new homeland, these Inuit colonized different regions, adapting to varying ecological conditions and developing distinct socioeconomies in the process. The importance of bowhead whaling in Thule economy, however, has been questioned by some researchers (e.g., Freeman 1979, Yorga 1979). Indeed, Stanford (1976) suggests that, because sealing was far more important in Birnirk economy than whaling, the impetus for the Thule migration may have been the search for better sealing grounds (i.e., greater expanses of fast ice) under deteriorating (i.e., warming) climatic conditions.

An alternative view holds that the early Thule culture in Arctic Canada demonstrates greater affinities with the Punuk culture of the Bering Sea region than with Birnirk of north Alaska. Motivated by population pressure and warring factions on the shores of the Bering Sea, bearers of the Punuk culture migrated north and eastward during the second half of the tenth century (Maxwell 1985: 252–3), bringing with them their walrus and whale hunting traditions and mixing with Birnirk people along the way. In this context, Canadian Thule in its initial development is viewed as a composite of both western and northern Alaskan traits. Concomitantly, Thule culture's arrival in Arctic Canada was seen to be the outcome of a number of factors including warming climates, reduced fast ice for ringed seal populations, and population pressure in the west.

Analyses of harpoon head styles, and to a lesser extent radiocarbon dates, indicate that the earliest Thule migration veered north across the Arctic islands to the east coast of Ellesmere Island and ultimately northwest Greenland. The main features of this phase include a distinctive square-to-roundish house style, west as opposed to north Alaskan-related artifact traits, and the occurrence of Norse materials (Schledermann and McCullough 1980)—the latter indicating yet another possible reason for the initial Thule expansion: trade. While this migration appears to have been a relatively rapid event which occurred during the eleventh century A.D., other 'classic' Thule sites from the central and eastern Canadian Arctic date to the thirteenth century or later. Although most of these assemblages suggest a direct relationship with earlier, northern Thule groups (McGhee 1982: 71), there appears to have been a separate movement of people along the Arctic coast to as far east as King William Island (Morrison 1983, 1990). Archaeological assemblages from the latter area resemble those from north Alaska more so than assemblages from other regions of Arctic Canada and, as such, appear to represent a Thule Inuit population distinct from that which initially colonized and expanded throughout the eastern and central Arctic (McGhee 1982: 71). While the relationship between these population movements is not adequately understood, most Arctic archaeologists believe that there was an initial high Arctic migration of Thule Inuit out of Alaska around A.D. 1000 and a less extensive, low Arctic migration several generations later.

In recent years there has been an increasing tendency to view the Thule expansion as the result not just of changing climatic conditions, but of other phenomena such as social pressures arising from population growth or the introduction of bow-and-arrow warfare (McGhee 1982, Maxwell 1985, Morrison 1990). Regrettably, other than a recognition of the possibility that some combination of social, cultural, and environmental factors influenced Thule movements into various regions of the Canadian Arctic (Yorga 1979), few researchers have attempted to explore these and other explanations to the extent they deserve. In the following pages I will investigate the propositions that (a) the initial Thule Inuit migration was predominantly a rejection of social processes occurring in western Alaska around A.D. 1000, (b) the causes of the second Thule Inuit expansion were not unrelated to the first, and (c) a later migration of Inuit from the west during late prehistoric times was more a function of the maintenance of the prevailing social order in Alaska than of its systemic rejection. Combined with specific adaptations to local environmental conditions and varying Dorset Inuit influences, the first two migrations set the stage for the emergence of socioeconomic variability in Central Inuit regions.

## Speculations on the Initial Thule Expansion

Befu's (1964) analysis of Inuit kinship (see Chapter 1) indicated that systems in west Alaska and the central/eastern Canadian Arctic resembled each other far more than either approximated the north Alaskan system. Specifically, and most importantly, both west Alaskan and Canadian systems possessed bifurcate collateral terms for parents' siblings, while bifurcating nepotics on the basis of sibling's sex (1964: 95). In these and other respects, north Alaskan systems manifested opposite patterns. For example, the sex of the connecting relative, whether parent or sibling, was not used to separate parents' siblings or siblings' children; first-degree collaterals in each adjacent generation possessed single terms. While this provides empirical support for the predominantly west Alaskan derivation of the initial Thule expansion, all kinship systems in the central and eastern Arctic, with the exception of the Iglulingmiut, differ markedly from those in both west and north Alaska in one fundamental way: they do not distinguish cross and parallel cousins! Even the north Alaskans, who did not differentiate between their aunts or uncles, separated cross from parallel cousins. This pattern is the exact opposite of most central/eastern systems whereby bifurcate collateral aunt-uncle terms are maintained but the cross/parallel cousin distinction is not. As observed previously, the Iglulingmiut and Port Harrison Inuit represent exceptions to these patterns. Only the west Alaskans bifurcate collateral aunt-uncle terms and distinguish cross and parallel cousins (Befu 1964), a system that is consistent with cross-cousin marriage. And in this regard, exogamous clan organization appears to have been most common throughout west Alaska and the Bering Strait region (e.g., Bandi 1995, Fienup-Riordan 1983, Giddings 1952, Harritt 1995, Hughes 1958, Heinrich 1960, Thalbitzer 1941).

Perhaps not too surprisingly there are vestigial remnants of the former importance of cross-cousin marriage in Central Inuit society. As noted before, with the exception of single cousin terms, most Central Inuit kinship schedules possess features that recall a system of cousin, and specifically cross-cousin, marriage. As noted before, in populations practising dual exogamy or restricted exchange, the term for FZ is often extended to FBW, and MB to MZH, as first ascending generation affines would have assumed the roles of one's parents' cross-sex siblings who likely lived elsewhere. Central Inuit society exhibits other latent features consistent with dual organization, not least of which is that Inuktitut, unlike English, has a dual plural form (e.g., Inuk=one person; Inuuk=two persons; Inuit=three or more people). Additionally, in ritual performance during the 'Sedna' ceremony and communal recreational activities throughout the rest of the year the Iglulingmiut, Oqomiut, and other Baffinlanders divided into two totemic groups, the ducks and the ptarmigans. While membership was life-long and depended on the season of birth (summer or winter), no marriage nor other reciprocal rights or social obligations were observed outside the 'Sedna' ritual (Boas 1907, 1964). The Iglulingmiut do not marry close relatives. Yet, their cousin terminology recalls a system of dual exogamy insofar as FBS for male Ego and MZD for female Ego are supposed to maintain sibling-like relationships, while cross-cousins are lumped together (Levi-Strauss 1969). Avoidance relationships between affines and consanguines, the levirate and sororate, spousal exchange, as well as other features may also be survivals of former dual clan organization (see Fainberg 1967). Even the *qaggi* may be a reflection of former dual exogamy as it reunited husbands and brothers-in-law in ritual and political collaboration (Levi-

Strauss 1963: 118), resolving the conflict between wife-givers and wife-takers (see also Fienup-Riordan 1983).

While these features perhaps indicate that the forebears of the Central Inuit may have once been organized into moieties, there is more overwhelming evidence to suggest that the direct ancestors of Thule Inuit society formerly possessed an asymmetrical or generalized exchange marriage system. This is not to deny that moiety systems possessing asymmetrical exchange are not possible—Levi-Strauss (1963, 1969) has demonstrated that they are. However, generalized exchange systems, as exemplified by the Kachin of highland Burma and the Gilyak of northeast Siberia, were apparently once wide-spread throughout eastern Asia (Levi-Strauss 1969). And in west Alaska emphasis on one or the other side of descent occurs frequently. In terms of parallel cousins, FB children are often accorded special sibling-like status, as they are in Central Inuit society. More proximately, while FZch are sometimes separated from the other cousins, MBch are more normally singled out for special treatment. The importance formerly attached to MB in Central Inuit society is evidenced in the use of the root *angak* to denote positions of power, authority, and influence. Thus, *angayuk* = older brother or sister, *angaquk* = shaman, *angajuqqaq* = leader or person to be obeyed, etc. In some western Alaska groups, *angakuk* (Fainberg 1967) was used to refer simultaneously to MB and 'chief'. Alternatively, in other groups male Ego's sister's children (*uyuruk*) are denoted special status, while both male and female Ego's Bch and female Ego's Zch are terminologically lumped with Ego's children (Fainberg 1967). Finally, the common Central Inuit customs of bride-price, bride-service, and the deep respect and obedience *ningaut* show their fathers-in-law may be symptomatic of asymmetrical exchange inasmuch as they would have been superfluous in systems practising direct exchange (Levi-Strass 1969).

This evidence would seem to point to the former and frequent occurrence in west Alaska of marriage with the matrilateral cross-cousin (male Ego), the distinctive characteristic of Levi-Strauss' (1969) generalized or asymmetrical exchange among patrilineal, patrilocal groups. Such systems, because of their emphasis on this relation, preclude the operation of direct reciprocity or dual exogamy. Consequently, men and women must marry into different groups. In its simplest form, this system of exchange tends to circular, closed, and therefore egalitarian (i.e., A->B->C-> . . . A->, etc.). However, generalized exchange also has the potential to engender social tension, ambiguity, and conflict. Having negated the practice of endogamy and direct exchange, generalized exchange systems leave open the possibilities of either maintaining old alliances to former wife-givers and wife-takers, or establishing new ones. In the former case, groups are linked in one or more circles of wife-givers and wife-takers (Friedman 1984: 169), whereby mechanisms must be developed in order to (a) mitigate the delayed reciprocity inherent within this system,[2] and (b) maintain the same alliance networks over the generations. Such a system works well as long as there is closure to the circles and obligations are fulfilled. However, it is also fraught with difficulties and, in particular, social tension arising from the failure to maintain reciprocity owing to demographic imbalances, economic inequities, and other factors. At the same time, the inability to fulfil kinship obligations instituted by such networks leads to the creation of new alliances and the formation of secondary circles, thereby abrogating former exchange arrangements, while promoting social inequality, the need to gain social, political, and material advantage, the emergence of exploitive relationships, etc.[3] From this perspective, both the maintenance and the expansion of generalized exchange—the renewing of former alliances and the establishment of new ones—have the potential to create significant social conflict. Indeed, such generalized or open exchange systems, as opposed to restricted or closed exchange systems, are closely correlated with political, economic, and thus territorial expansion, whereby the population included in the circle(s) expands and a multiplication of new lineage segments results (Friedman 1984: 169). As Levi-Strauss (1969: 238) observed:

> The longer the cycle of exchange tends to become, the more frequently it will happen, at all stages, that an exchange unit, not being immediately bound to furnish a counterpart to the group to which it is directly in debt, will seek to gain advantage either by accumulating women or by laying claims to women of an unduly high status.

Not surprisingly, hypergamy, polygyny, and the birth-order ranking of children are features common to most Inuit societies. The basic intrasystemic contradiction of generalized exchange, then, is that its operation presupposes equality, but is itself a source of inequality (Friedman 1974, 1984; Levi-Strauss 1969: 266).

In this connection I believe it is no coincidence that, during the last centuries of the first millennium A.D., the Bering Strait region was experiencing a period of population pressure, territorial expansion, and warfare between rival factions (e.g., Maxwell 1985: 252-3). This conclusion is supported by the archaeological record of the region, whereby the appearance of formalized burial grounds, armoured vests of bone, fearsome spears, and other weaponry (Bandi 1995, Harritt 1995, Maxwell 1985) attests to considerable conflict in Punuk society. This, in turn, leads us to consider the possibility that the causes of the initial Thule migration were primarily social, not environmental. In effect, those Inuit who participated in this migration, by 'voting with their feet', represent an explicit rejection of the social forces, structures, and ideologies that bound people into productive (and not so productive) relationships in west Alaska. Indeed, the act of migration often is itself a rejection of social forces governing society. Social conflict appears to have been the motivating force behind the Copper Inuit migration to the Arctic coast (see below) as well as other, more recent Inuit migrations.[4] And, just as the Copper Inuit constitute an explicit repudiation of Central Inuit social structure and ideology, so too may the earliest Thule Inuit in Arctic Canada represent a renouncement of social and political forces prevailing in west Alaska during the tenth century A.D.

The possibility that intrasystemic contradiction inherent within, and dysfunction arising from, generalized exchange may have been the driving forces behind this rejection finds support in the emergence of an entirely new cousin terminology across the Canadian Arctic that is consistent with neither generalized nor restricted exchange systems.[5] Rather, this terminology is reminiscent of complex marriage systems insofar as there are no positive marriage rules, unilineal descent, nor, especially, cross/parallel distinctions. Here, I am referring specifically to the term *illuq* to denote both cross and parallel cousins. While there are slight variations in the application of this term, its distribution is congruent with the initial Thule migration into Arctic Canada and subsequent expansion into northern Hudson Bay, Baffin Island, west and east Greenland, and Ungava Bay. While the Iglulingmiut employ a three-cousin terminology, recalling the St Lawrence Island system, they also use the term *illuq* for cross-cousins (Damas 1963). In addition, the Netsilingmiut employ this term to refer to all same-sex cousins, though some families in contact with the Caribou Inuit apparently also used the more common western cousin terms, *angutiqat* and *arnaqat* (Sperry 1952: 14). The Copper and Caribou Inuit also utilize the latter terms, though for different categories of cousins—the former use *arnaqat* for all cousins except FBch (*angutiqat*), while the latter separate cousins on the maternal side (*arnaqat*) from those on the paternal side (*angutiqat*). Only the Iglulingmiut employed these terms to distinguish parallel cousins. With the exception of the latter, the terms *angutiqat* and *arnaqat* are associated with those regions where the second Thule expansion occurred. While the Iglulingmiut appear to represent an amalgam of both cousin systems, the Netsilingmiut, who share a common heritage with the Caribou and Copper Inuit (see below), appear to have abandoned *angutiqat* and *arnaqat* in favour of the more inclusive eastern term, *illuq*.

While Sperry (1952: 16) believes that Thule culture originally possessed a three-cousin terminology similar to those described above, and that Central Inuit cousin terminologies are relatively recent in their development, Fainberg (1967: 255) attributes the disappearance of matrilineal clans and dual exogamy in the Canadian Arctic to the movement, disintegration, and intermingling of clans in the 'vast unpopulated stretches of the Arctic'. However, following the above discussion, we may suggest, alternatively, that the earliest Thule Inuit groups abandoned cousin terminologies consistent with generalized exchange systems relatively early in their migration as a consequence of their rejection of social forces operating in the Bering Sea region around the tenth century A.D.

In this regard, the origin of the specific term *illuq* begs further examination. Sperry (1952: 15–16) and Thalbitzer (1941: 721) propose that *illuq* replaced earlier cousin terms as a result of a change from single family to communal dwellings. Concomitantly, *illuq* came to denote 'related housemate of my generation'. While this may be a plausible explanation, Thalbitzer believed that the use of *illuq* among the Angmasilik of east Greenland paralleled the emergence of the longhouse, which was supposedly adopted from the Norse around A.D. 1500. However, communal dwellings in Labrador, Cumberland Sound, and other regions in the Canadian Arctic appear in even later contexts (e.g., Schledermann 1976). Alternatively, the earliest Thule Inuit dwellings in most areas of the eastern Arctic from Ellesmere Island to Labrador are small, oval to subrectangular, single family dwellings (*ibid.*)—features that would appear incongruent with the use of the designation *illuq*. Yet, the first Thule Inuit to penetrate Arctic Canada probably abandoned cross/parallel cousin distinctions relatively early in conjunction with the adoption of a single cousin terminology, and specifically the term *illuq*.

While our current knowledge may not be adequate to reconcile these conflicting interpretations, it should be pointed out that as the Thule Inuit expanded eastward and eventually southward they did not enter a land devoid of human occupation. Rather, they encountered Inuit of the Dorset (Tunnit) culture who had recently begun to construct communal longhouses. Damkjar (1987) speculates that this development is related directly to the initial expansion of Thule Inuit into the Canadian Arctic insofar as the emergence of longhouses may have been an attempt on the part of the Dorset Inuit to construct and maintain an ethnic identity in the face of an intrusive and perhaps technologically superior culture. However, such behaviours would not necessarily have restricted interaction between the two ethnic groups, but may actually have served to create an effective setting for intercourse to take place (Stevenson 1989).[6] While the reasons why longhouses emerged just prior to the demise of Dorset Inuit culture may be difficult to ascertain, the context of ethnic group interaction could have served as a stimulus for the emergence of the term *illuq*, especially if intermarriage was common between the two cultures. Whatever the case, having rejected the cross/parallel distinction and the more restrictive marriage systems that this distinction implies, early Thule Inuit groups may have quickly adopted/created a single cousin terminology in an effort ultimately to construct a potentially unlimited, though dispersed, pool of marriage partners and external alliances across the vast expanses of their new homeland.[7]

## A Second Thule Wave

Archaeological investigations in the Amundsen Gulf, Coronation Gulf, and Queen Maud Gulf areas indicate that Thule Inuit culture entered the western Canadian Arctic during the latter half of the twelfth century A.D. Morrison (1983, 1990) has termed this variant of Thule culture in Coronation Gulf, the Clachan phase. Artifacts recovered from sites of this phase suggest considerably greater influences from the west than the east (Morrison 1983). It was this particular expansion that eventually gave rise to the Copper, Netsilik, and Caribou Inuit, though probably in much less direct fashion than McGhee (1972), Morrison (1983), and others might envision (see below). In turn, the facts that these three regional groups possessed bifurcate collateral aunt-uncle terms while making no cross/parallel distinctions in Ego's or the first descending generations suggest that they are more similar structurally to historic Inuit populations in the eastern Arctic than to those in west and north Alaska. If so, we may hypothesize that the causes of the second Thule Inuit expansion were not unrelated to the first. Specifically, like the earlier Thule migration, the second may have been a systemic rejection of social and political processes occurring in west Alaska during the twelfth century A.D.

At this juncture it is important to point out that significant sociopolitical tensions appear to have had a long history in northwest Alaska. Apparently, 'mutual distrust and hostility ... (were) an ancient heritage, and inter-regional feuding and warfare were part of the normal state of affairs' (Burch and Correll 1972: 24). The occurrence of 37 or so regional groups in northwest Alaska during the mid-nineteenth century, many with differing dialects and kinship terminologies suggestive of various forms of exchange, appears to be a correlate of this social conflict. However, while many of these groups were at war with each other, they were also allied in other ways, and principally through marriage and trade (Burch and Correll 1972). In fact, the extensive and formal trading system, known as the Beringian trade network, that emerged in northwest Alaska around the fifteenth century A.D. (Hickey 1979) may have arisen largely as a method of recruiting marriage and trading partners among otherwise autonomous and mutually hostile regional groups. The possibility that intrasystemic contradiction within generalized exchange systems, the latters' resultant rejection, and the exploration of new ways to establish productive forces and relations may ultimately underlie the development of those north Alaskan groups studied by Befu (1964), is supported by the fact that their kinship schedules exhibit opposite patterns to those in west Alaska. While the causes of the Thule Inuit expansion into the western Canadian Arctic after the twelfth century A.D. must remain speculative, given the evidence and theory cited above the possibility that this migration was the outcome of enduring, unresolved structural conflicts within northwest Alaska societies seems plausible. The disappearance of Thule culture from the former area at the onset of the Little Ice Age is another question altogether, and one that we will return to when we consider the origins of the Caribou, Netsilik, and Copper Inuit.

### A Third Expansion

During the late precontact period there appears to have been another expansion of Alaskan Inuit into the Canadian Arctic. Most notably, recent evidence (Stevenson 1992) suggests that Alaskan/Mackenzie Inuit groups spread along the south shore of Amundsen Gulf into Coronation Gulf during late prehistoric times. Though not as extensive as earlier migrations, this occupation was more localized and intensive. Again, while the reason for this expansion remains elusive, forces at work in northwest Alaska during the late prehistoric period suggest that it may be related more to the maintenance of the prevailing social order in the west rather than to its systemic rejection.

Beginning in the fifteenth century in northwest Alaska there emerged an extensive trading system that eventually stretched from northeast Asia to Coronation Gulf. Hickey (1979) has argued that the Beringian trade network was not the result of contact with traders as others have supposed, but was part of a post-expansion Thule Inuit culture phenomenon that evolved in northwest Alaska in the few centuries prior to contact with Russian traders. The emergence of annual trading fairs in which powerful men from hundreds of miles around gathered with extended relations to trade with others of similar socioeconomic ranking was a major characteristic of this network. However, in order to acquire rare, attractive items for exchange, specialization in the procurement and manufacturing of a variety of exotic trade goods developed. This, in turn, required greater movement and expansion of groups into outlying districts containing such resources.[8]

Could the occupation of Amundsen Gulf by north Alaskan Inuit during late precontact times have been the direct result of an attempt to colonize new areas and to exploit rare resources for exchange with other groups further west? Such an explanation seems at least as feasible as the generally accepted model that Mackenzie Inuit groups in the Amundsen Gulf region developed *in situ* out of an earlier Thule Inuit base (Morrison 1983, 1990). The fact that the Copper Inuit knew all Mackenzie Inuit not by the name of the closest historically documented regional division (i.e., the Avvaqmiut), but by the name Kupugmiut, which was one of the most powerful yet most distant Mackenzie Inuit groups, is instructive (Stefansson 1919). This is precisely what might be expected if the Mackenzie Inuit occupation of south Amundsen Gulf was the outcome of a rapid eastward expansion from the Delta area during late prehistoric times. The lack of continuity between early Thule and later Mackenzie Inuit assemblages at sites in east Amundsen Gulf and west Coronation Gulf also favours this interpretation.[9] Although further research will undoubtedly clarify this issue, it is important to bear in mind that, while Mackenzie Inuit and Thule Inuit occupations have been documented in the south Amundsen Gulf region, virtually no evidence connecting these cultures has been found within the same archaeological site. In other words, there seems to be a cultural hiatus between Thule and later Mackenzie Inuit assemblages throughout much of the region.

Although favourable ecological conditions could have drawn Alaskan/Mackenzie Inuit groups to Amundsen and Coronation gulfs,[10] copper and especially soapstone may have been the major attractions. By the early nineteenth century cooking pots and lamps from the Bering Sea to Cape Bathurst at the west end of Amundsen Gulf were apparently made of soapstone from Coronation Gulf (see Morrison 1991: 239 for references). Indeed, local Copper Inuit associated with the Tree River drainage appear to have specialized in the manufacture of soapstone vessels for exchange (Stefansson 1919). As the Mackenzie Inuit appear not to have penetrated Coronation Gulf to any significant extent, they likely served as middlemen in a trade for copper and especially soapstone between Alaskan groups and the Copper Inuit. Along the south shore of Amundsen Gulf the Mackenzie Inuit themselves may have quarried soapstone to manufacture into vessels for exchange, since several outcrops of this material are known locally to occur along this coastline (Tony Green, personal communication, 1991). Evidence of specialized slate manufacture in late eighteenth century *qammat* on the south shore of Amundsen Gulf hints at the possibility that this material was also an item of interregional exchange (Stevenson 1992).

If the Mackenzie Inuit occupation of south Amundsen Gulf existed largely, or even partly, to provision exotic goods for exchange with Delta/Alaskan groups, then any significant perturbations or alterations in this trade network might be expected to have created disruptive and perhaps even significant repercussions throughout the region. In this connection, I would suggest that the abandonment of the south shore of Amundsen Gulf by Mackenzie Inuit groups coincides directly with the massive influx of European trade goods into northwest Alaskan society shortly after 1790. One might expect a sudden increase in the manufacture and exchange of soapstone items for trade in order to maintain exchange parity with west Alaskan groups having direct access to Russian trade goods (cf. Hickey 1984).[11] However, with the introduction of metal pots, soapstone vessels were rendered obsolete almost immediately and the soapstone trade quickly collapsed. Although the south Amundsen Gulf coastline may not have been occupied up to the 1830s as Stefansson (1919) suggested, the recent appearance of the 'western Eskimo' artifacts he encountered indicates that it could not have been abandoned much before. Limited archaeological investigations (Stevenson 1992) also support a late eighteenth- or early nineteenth-century abandonment of the Amundsen Gulf region. Thus, it was not so much the establishment of trading posts on the Mackenzie River during the mid-nineteenth century as the influx of Russian trade goods into Alaska at the end of the eighteenth century A.D. that broke the chain of continuity in coastal trade between Coronation Gulf and the Bering Strait, resulting in the rapid depopulation of the south Amundsen Gulf region.

## COPPER, NETSILIK, AND CARIBOU INUIT ORIGINS

McGhee (1972) and Morrison (1983) suggest that the Copper Inuit of Coronation Gulf developed directly from earlier Thule Inuit groups. However, there is even less evidence for cultural continuity in this area than in south

Amundsen Gulf. Given the lack of evidence for the *in situ* development of historic Mackenzie and Copper Inuit from an earlier Thule base in Amundsen Gulf and Coronation Gulf, we must conclude that 'classic' Thule culture, as represented by the Clachan phase, came to an end around the middle of the fifteenth century. As the disappearance of Thule culture from these shores appears to coincide with the onset of the Little Ice Age, the possibility that changes in local environmental conditions resulted in the abandonment of this region must be considered. Specifically, unable to maintain their maritime-based mode of production under changing ecological conditions, the Thule Inuit left. It is possible that some groups migrated west towards Alaska, exercising social rights and obligations with groups with whom they had been in contact for 250 years. Alternatively, perhaps some groups, particularly those occupying the southern shores of Coronation Gulf and Queen Maud Gulf, moved inland to become the forerunners of the Caribou, Netsilik, and Copper Inuit.

There exist two competing theories for the origins of the Copper Inuit. The oldest of these, which found favour among early ethnographers such as Birket-Smith (1929) and Rasmussen (1932), was advanced by Jenness (1923). This thesis holds that the Copper Inuit were formerly an inland people who migrated to Coronation Gulf from the barrenlands sometime within the last 500 years or so. Jenness saw little similarity between Copper Inuit economy, which lacked an open-water sea mammal hunting technology, and the maritime-based lifestyle of the Thule 'Eskimo'. Rather, the Copper Inuit so closely resembled the Netsilingmiut in material culture that a recent historical connection was implied, a suggestion later supported by Taylor (1963) and Burch (1979). Certain customs and beliefs also linked the Copper to the Netsilik and Caribou Inuit, rather than the Mackenzie Inuit (Jenness 1923: 545). Similarly, linguistic evidence demonstrated close ties to the Caribou Inuit and Netsilingmiut. In fact, later linguistic research (Correll 1968) revealed the language of the Caribou, Netsilik, and Copper Inuit to be a single dialect, distinct from the Iglulingmiut, Oqomiut, and other eastern Inuit groups. On the basis of a number of lines of evidence, then, the ancestors of the Netsilik and Copper Inuit were thought to have come 'to the coast (from inland) only a few centuries before the appearance of (historic) Copper Eskimos in Coronation Gulf' (Jenness 1923: 530).

McGhee (1972) and, more recently, Morrison (1983), however, have advanced the theory that the Copper Inuit evolved directly *in situ* from earlier, western Thule groups in Coronation Gulf, and that this transformation was triggered by climatic changes. Thule economy in this region appears to have been based on the procurement of seals from ice-leads and open-water boats for the accumulation of food surplus to last the winter—bowhead whaling and breathing-hole sealing were practised rarely, if at all. The Little Ice Age strained this economy to the breaking point resulting in starvation and rapid socioeconomic changes, particularly in the size, composition, and location of winter villages (Morrison 1983: 278). In turn, 'an effective breathing-hole hunting strategy, one which made sealing both possible and reasonably productive during the now lengthened winter' was adopted. McGhee (1972), however, sees a less direct transformation from Thule to Copper Inuit. The Little Ice Age eventually forced Thule occupants of the area from their small, land-based winter settlements on the coast into large snowhouse villages on the sea ice. The size of Copper Inuit winter aggregations grew in order to hunt seals effectively beneath the sea ice. Many of the distinctive features of Copper Inuit society (e.g., agamous marriage patterns, lack of kinship directives, elaborate ceremonial complex, etc.) are said to be the result of changing trade relations and the introduction of European technology during the century prior to direct contact (McGhee 1972).

One of the greatest obstacles in accepting either theory is, of course, an absence of archaeological evidence. With respect to the *in situ* development model, the change from land-based to sea ice winter settlement is invoked to account for both the marked differences in Thule and Copper Inuit technology, and the scarcity of late precontact sites in the region (McGhee 1972). Whatever evidence exists from this time period is said to represent an intermediate phase between Thule and Copper. Yet, the features and artifacts provided by McGhee (1972) in support of this phase clearly suggest greater similarities with Mackenzie Inuit than with the historic Copper Inuit. In this regard, McGhee's Intermediate interval sites on southwest Victoria Island probably represent not so much any transitional stage between Thule and Copper, but the extreme northeastward expansion of Mackenzie Inuit during late prehistoric times.

Taylor (1963) and, more recently, Burch (1978, 1979), in addressing the question of historic Caribou Inuit development, have broached the subject of Copper Inuit origins. In contrast to the generally accepted view that the Caribou Inuit were descended from Thule Inuit groups who migrated south along the west coast of Hudson Bay early in the present millennium, Taylor suggested that the Caribou Inuit were descended from Thule folks who migrated overland from the Coronation Gulf/Queen Maud Sea region sometime during the seventeenth century. Burch subscribes to this theory, while providing evidence that the Caribou Inuit did not appear in the district in which they had been documented historically until relatively recently. While around A.D. 1700 there was a sudden change in house form along the west coast of Hudson Bay from semi-subterranean to surface dwellings, Caribou Inuit occupation of the south interior barrenlands remained 'archaeologically thin' until the early nineteenth century, at which time they began to expand inland. In short, Burch (1978: 21) provides compelling evidence in favour of Taylor's hypothesis, even to the extent of eliciting information from his informants that they had come to their present location 'from somewhere to the northwest' via an overland route.

Given the marked similarities among Netsilik, Caribou, and Copper Inuit material culture and dialects, not to mention religious beliefs and other social customs, any attempt to address the origins of one must consider the historical antecedents of the others. Such a model must also endeavour to address the causes of the apparent depopulation of the Arctic coast by Thule Inuit groups during the late fifteenth century and its subsequent reoccupation by the Netsilik and Copper Inuit two or more centuries later. The following model, in attempting to do just this, accounts for archaeological, historical, linguistic, and ethnographic evidence in ways that other models have not.

**Late Precontact Central Arctic Population Movements and Their Causes**
The onset of climatic cooling and increased ice cover at the beginning of the Little Ice Age around A.D. 1450 wrought significant hardship upon the Thule occupants of the Arctic coast. Unable to build up sufficient food reserves to last the winter, and/or unwilling to alter the nature of their productive relationships and activity under deteriorating (cooling) climatic conditions, they abandoned the region. At the same time, decreased snowfall during this cold period may have resulted in an increase in habitat favourable to caribou (C. Hickey, personal communication, 1992). While some Thule Inuit in the south Amundsen Gulf area may have moved west in an effort to maintain their maritime lifestyle, groups on the more protected shores of Coronation Gulf and Queen Maud Gulf expanded the terrestrial component of their economy. Gradually, as they followed the caribou herds south toward their winter ranges in the trees, coastal living became a less attractive and viable existence. Shortly thereafter, the knowledge and technology associated with their former maritime hunting traditions were forgotten. A region incorporating the tree-line and the headwaters of the Back, Thelon, and Coppermine rivers, perhaps centred on the Thelon Woods, soon became the homeland of this proto-Copper-Netsilik-Caribou Inuit culture.

Burch (1978, 1979), after Taylor (1963), sees this migration as occurring during the late seventeenth century at the height of the third and coldest phase of the Little Ice Age. However, given the absence of evidence of Inuit occupation of the Arctic coast between the early sixteenth and early eighteenth centuries, it seems likely that this inland tradition developed soon after the onset of the climatic episode around A.D. 1450. An early abandonment of this coastline would also seem more plausible insofar as, if its Thule residents could have adapted to the disruptive effects of the first cooling episode, there seems little reason why they could not have withstood the effects of the last. For two hundred years or more this inland-oriented culture subsisted principally on caribou, and secondarily on fish and other game. However, as caribou are well known for recurrent fluctuations in their numbers and ranges, existence was marginal and sometimes precarious. Around the beginning of the eighteenth century A.D. there was a massive population movement out of this homeland.

One or more groups descended the Back River to the coast to eventually become the Netsilingmiut, while others descended the Thelon River to a now vacant Hudson Bay coast—according to Burch (1978, 1979) the Thule Inuit had ceased to become a viable cultural entity on this coast by the early seventeenth century. Still other groups descended the Coppermine, Hood, and/or Mara/Burnside rivers where they eventually became the Copper Inuit. The possibility that the Back River Inuit were the first to diverge, while the Copper and Caribou Inuit were the last, is suggested by the closer linguistic similarity of the latter two, and the fact that they apparently had a longer history of trade, which seems to have been centred around Akilinik, or the Beverly Lake/Thelon Woods area (Burch 1978, 1979).

What caused these various migrations to take place has been the subject of speculation by a number of scholars. Although his timing appears to be off by a couple of hundred years, Jenness (1923: 551) suggests that it was pressure from expanding Chipewyan groups that forced the ancestors of the Copper and Netsilik to the Arctic coast. Burch (1978, 1979), on the other hand, believes that those Copper 'Eskimos' who eventually became the Caribou Inuit, migrated to the west coast of Hudson Bay during a particularly cold spell around A.D. 1700 owing possibly to a reduction in game supply. This climatic episode, which lasted until the mid-eighteenth century, was, in fact, the coldest period in the last several thousand years. Here, Caribou Inuit culture developed, until small groups began to expand inland sometime during the early nineteenth century to their historically documented positions. Given the apparent synchronous timing of this cooling trend with major population movements to various coastal settings, an environmental explanation is not beyond the realm of possibility, especially if hard times in the interior were accompanied by an expansion of fast ice across large bays and gulfs along the Arctic coast. The latter would have facilitated the movement of caribou to new grounds, while providing increased ringed seal habitat. It has also been speculated that colder climatic episodes may have favoured an increase in caribou range, and possibly abundance.

While the role of climate change in these migrations is difficult to assess, another possible factor enters the picture during the late seventeenth century. Through sporadic but hostile encounters with the Chipewyan, southern groups of this inland-oriented Inuit culture would have been aware of an European presence along the west coast of Hudson Bay by the 1690s. Yet, it was not until the reopening of York Factory in 1714 and the establishment of a trading fort at the mouth of the Churchill River three years later that a stable basis for trade with native peoples developed in the region. In this connection, it is probably no coincidence that the number of sightings and encounters with Inuit on the west coast of Hudson Bay increased markedly in frequency after 1717. In fact, there appears to be no authenticated sightings of Inuit by traders or explorers on the west coast of Hudson Bay prior to this date (Burch 1978, 1979). While Burch (1978: 28) may not have been entirely correct when he stated that 'by the time Hudson's Bay Company personnel arrived at Churchill (in 1713) ... (this) population would have been firmly ensconced in the centre of its new homeland ... along the coast', his timing for this event was probably not too far off.

If the prospect of new wealth and trading relationships encouraged some southeasterly groups of this inland Inuit culture to descend to the west coast of Hudson Bay, the impact of this migration on groups remaining in the interior would have been felt in a number of ways. First, social relationships and trading partnerships would have been disrupted. Social rights and obligations formerly instituted through aboriginal exchange (marriage, trading, and adoption) relationships would be difficult to maintain and exercise. Concomitantly, an increase in spatial distance between these populations would have been met with an increase in social distance, if not social tension.

Alternatively, if coastal groups were obtaining goods from the HBC, and if trade continued with other groups in the ancestral homeland, the latter would have found it difficult to maintain exchange parity. Following once again the same argument advanced by Hickey (1984, 1992) for the late nineteenth-century Copper Inuit, interior groups may have begun to range farther afield to procure items for exchange, bringing them into more direct and sustained contact with the Arctic coast,[12] and possibly with Mackenzie Inuit groups who were expanding from the west. The time depth of the Akilinik trade network is not known. While Hanbury (1904) believes that it was a relatively recent and insignificant development, both Jenness (1923) and Stefansson (1919) imply that this important institution had a fairly long history. If the latter are correct, the roots of this trade may lie in the migration of a segment of this inland culture to the west coast of Hudson Bay, and their subsequent emergence as middlemen in a trade between other interior groups and the HBC. However, the emergence of these proto-Caribou Inuit as middlemen would have increased economic and social disparity between themselves and other groups in the homeland, resulting in increased inequality, social tension, and hypergamy. Perhaps more than anything else, it was this development that precipitated the migration of proto-Copper and proto-Netsilik groups to the Arctic coast during the mid-eighteenth century.

At least two lines of evidence support the latter theory. First, of the handful of legends handed down from one generation of Copper Inuit to the next, one stands out. This is the belief that a long time ago their ancestors were ruled by very powerful men who lived like 'chiefs' and killed people at will (Rasmussen 1932: 251). Although this story might possibly have been fabricated to help maintain the strict egalitarian ideology and social structure of the Copper Inuit, it is probably not fortuitous that the Netsilingmiut share the exact same legend, while the Caribou Inuit do not. The other line of evidence concerns the social structures and ideologies of the Copper and Netsilingmiut, which differ significantly not just from the Caribou Inuit, but from each other.

It is difficult to accept in this era of 'critical' enlightenment that dramatic sociopolitical transformations, such as those apparently experienced by the Copper Inuit, resulted exclusively from accumulative changes in environmental conditions. On the other hand, social conflict, which is not without economic underpinnings, has long played a role in social change, rebellion, and migration. Thus perceived, the Copper and Netsilik migrations to the coast may have occurred for the same reasons, although not necessarily at the same time—linguistic data seem to suggest otherwise. In other words, the proto-Netsilik and proto-Copper Inuit migrations and subsequent adaptations to the Arctic coast may represent the outcome of a conscious rejection of the prevailing social order among a larger socioeconomic network in the interior. Given the substantial theory and lesser body of empirical evidence presented above, it is suggested that these social revolutions/migrations occurred as a consequence of the emergence of powerful proto-Caribou Inuit middlemen within the context of trade with Europeans along the west coast of Hudson Bay sometime around the middle of the eighteenth century A.D. Although trade was still kept up between the Caribou and Copper, and to a lesser extent between the Caribou and Netsilik, well into the nineteenth century, it was not as intensive as during earlier times.

Not surprisingly, the Netsilingmiut also possess a social system which is very different from that of all other historic Inuit groups in Canada. By way of summary, whereas historic Copper Inuit society appears to be the antithesis of traditional Central Inuit ideology and social structure, the Netsilingmiut have emphasized closeness-affection bonds within kinship relationships (*ungayuq*) at the expense of authority-respect directives (*naalaqtuq*). In so doing, the Netsilik have become the most closed, inwardly-focused society of any in the Canadian Arctic. On the coast, Netsilingmiut groups came into contact with the Iglulingmiut from whom they selectively borrowed certain aspects of their socioeconomy, including numerous religious notions and perhaps the cousin term *illuq*. The Copper Inuit, on the other hand, played by different rules altogether. More specifically, they rejected the respect-obedience and closeness-affection directives explicit within Central Inuit kinship relationships and terminology, in favour of continually creating symmetrically structured socioeconomic ties with individuals who may or may not have been linked through blood or marriage. While Netsilingmiut society could be conceived as a synthesis of traditional Central Inuit social structure (thesis) and its negation (antithesis), the historic Copper Inuit exhibit no such dialectical materialism. While this in itself argues for a later divergence from an inland social network and emigration out of an interior homeland, it has been suggested that the *Investigator* materials may have served to retard this dialectic, while at the same time reinforcing the strict egalitarianism of current social norms and behaviour, thus preventing a synthesis of the old and new social structures from taking place.

## COPPER INUIT AND PALEOESKIMO AFFINITIES

These contemplations on Canadian Arctic prehistory cannot be concluded without referring to the similarities between Copper Inuit and Paleoeskimo material cultures, especially in the shapes of dwellings. Bilobate features are one of the most common forms of Copper Inuit and Paleoeskimo (particularly Independence II and Dorset culture) accommodations. Among the Copper, this style of temporary structure appears, in part, to be a symbolic expression and reaffirmation of the omnipresent symmetrical duality of social relationships and productive activity. As with the Copper Inuit, who possess the most elaborate ritual dance and song complex found anywhere in the Canadian Arctic, Paleoeskimo culture, most notably Dorset Inuit culture, is well known for its ritualistic and artistic expression as represented in portable art. These two fundamental characteristics as well as others—commensality may have been another characteristic of both Dorset and Copper Inuit domestic life (C. Hickey, personal communication, 1990)—leads us to entertain the hypothesis that the underpinnings of later Paleoeskimo and Copper Inuit societies were not dissimilar, particularly with respect to sociopolitical structure. No historical connection is implied here; I am in general agreement with McGhee (1972) that the Copper Inuit evolved out of an earlier Thule base, although in less direct manner than he envisioned. Rather, both social systems may have been originally an explicit rejection of a dominant social order and ideology within a larger cultural milieu in which social inequality was sanctioned. The fact that the Copper Inuit possess a kinship terminology which apparently once placed special emphasis on FBch to the exclusion of all other cousins, is perhaps instructive in this regard. As observed previously, the Copper Inuit represent the antithesis of behavioural directives so common to eastern Central Inuit groups. However, as the Copper constitute only one regional Inuit population at one point in time, it is not unreasonable to expect that, under the right circumstances, there were other disenfranchised Inuit in the past who rejected the prevailing social structure and ideology of a dominant group and set off for a new homeland, altering the foundations of their productive activity and social relationships in the process.

This being the case, once traditional Central Inuit social structure and its underlying ideological foundations, as we have come to know them, are rejected, few alternative social formations may be possible. Copper Inuit society, with its fiercely egalitarian ideology and dualistically structured relationships, represents perhaps one of the few social formations that Central Inuit groups can assume once hierarchy is rejected.

Can the arrival and development of Paleoeskimo culture in the Canadian Arctic be better understood in these terms? Possibly, although Dorset longhouses might be interpreted to represent, for whatever reasons (e.g., accumulation of surplus, increased within-group solidarity, need for defence), a shift back to a more hierarchical expression. But Paleoeskimo culture, or at least its material manifestations, appears to have had a longer history than Thule Inuit culture in the Canadian Arctic. In this regard, the question of which system is characteristic of Canadian Inuit social structure and which is not comes to the fore. Perhaps Canadian Inuit prehistory from its humblest beginnings several thousand years ago to the contact period, and our understanding of it, could best be approached from a perspective which employs as its core the dialectical interplay of these two systemic tendencies.

## Conclusion

The above speculations constitute a more dynamic and complex picture of Canadian Arctic prehistory than has previously been advanced in the literature. And, while these interpretations undoubtedly raise more questions than they attempt to answer, this perspective puts a different, perhaps more human slant on Canadian Inuit history and culture. No longer can we assume that there is, or ever was, a direct link between environment and society in the Canadian Arctic. Indeed, from the foregoing research and analyses, environmental arguments become exceedingly more involved and therefore difficult to sustain—structural and historical factors intervene to obfuscate any direct one-to-one correlation. Even so, the role of environment in shaping Central Inuit socioeconomy has not been ignored. Rather, a more complete picture of Central Inuit society derives from a more thorough understanding of the complex interplay of environment, history, and social structure. Many of the propositions advanced above will be seen to have 'rocked the boat' of Canadian Arctic archaeology. Towards this end, it is hoped that other Arctic researchers will be encouraged to undertake their investigation in more systematic fashion than has been possible here in an effort to guide the discipline away from the rocky shores of environmental determinism and cultural materialism into new waters.

# Chapter 11

# Central Inuit Social Structure and Kinship Theory

Central Inuit social structure at the outset of the Thule migration may have been a rejection of intrasystemic contradictions arising from the operation of asymmetrical exchange systems in the Bering Strait. However, I have not specified how Central Inuit society came to possess negative marriage rules and bilateral descent—the classic features of complex marriage systems. Levi-Strauss (1969) divides social structures into three types: (1) elementary, (2) Crow-Omaha, and (3) complex. These, after Asch (1988: 105–6), can be displayed in graphic form where one axis is labelled descent and the other marriage rules (Figure 70). Each axis can be further subdivided into societies that possess either unilineal or non-lineal descent, and positive or negative marriage rules. Levi-Strauss' *Elementary Structures of Kinship* (1969) sought to explore the structural foundations of societies with unilineal descent and positive marriage rules. He further divided these elementary structures into two types: (1) those with restricted or symmetrical exchange rules, and (2) those with generalized or asymmetrical exchange rules. In societies practising restricted exchange, residence and descent are always incongruous, as a group is either a wife-giver or a wife-taker. Alternatively, in systems of generalized exchange residence and descent are in harmony. Thus, just as every disharmonic regime leads to restricted exchange, every harmonic regime announces generalized exchange (Levi-Strauss 1969: 493).

Although Levi-Strauss' treatment of women as objects of exchange is not without its critics (e.g., see Leacock 1981), his analysis of elementary kinship structures worldwide is exemplary. Nonetheless, he does not deal with the other three types of fundamental social structures to the extent they deserve, i.e., those societies demonstrating (a) unilineal descent with negative marriage rules (i.e., Crow-Omaha systems), (b) non-unilineal descent with positive marriage rules, and (c) non-unilineal descent with negative marriage rules (complex systems). Recently, Asch (1988) has proposed the term 'Bilateral-Dravidianate' for societies with bilateral descent and positive marriage rules, a category that, up until Ives' (1990) and Asch's reflections on Athapaskan socioeconomy, was not described in the ethnographic literature nor even

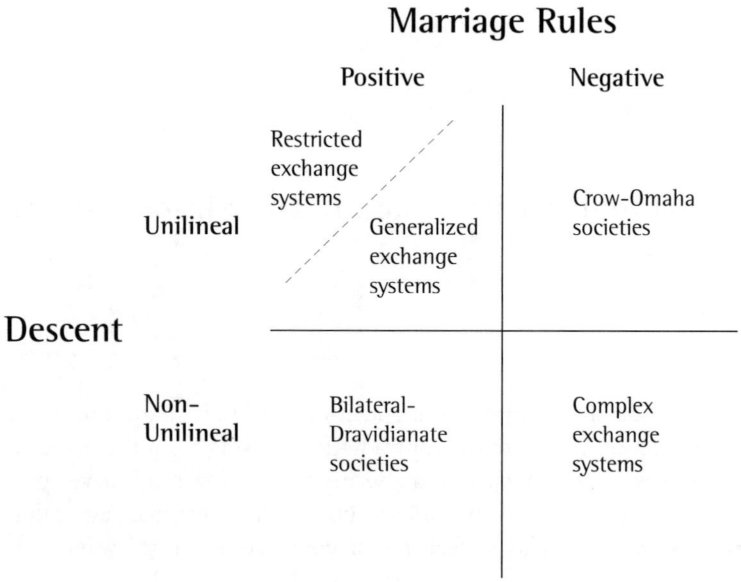

## Marriage Rules

|  | Positive | Negative |
|---|---|---|
| **Unilineal** | Restricted exchange systems / Generalized exchange systems | Crow-Omaha societies |
| **Non-Unilineal** | Bilateral-Dravidianate societies | Complex exchange systems |

**Descent**

FIGURE 70. Four fundamental social structures (from Asch 1988: 105, after Levi-Strauss 1969).

postulated theoretically (Asch 1988: 105). While the origin of Bilateral-Dravidianate social structure is not unrelated to the issues discussed here, our concern is with how complex structures, such as those possessed by the Central Inuit and Euroamerican society, come into being.

## COMPLEX STRUCTURES

In systems of generalized exchange, groups bound by indirect exchange are linked by trust, i.e., speculation that gifts given will eventually be returned and that the circle will close:

> Born as it is out of collective speculation, generalized exchange, by the multiplicity of the combinations which it sanctions, and the desire for safeguards which it arouses, invites the particular and private speculations of the partners. Generalized exchange not only results from chance but invites it, for one can guard oneself doubly against the risk: qualitatively, by multiplying the cycles of exchange in which one participates, and quantitatively, by accumulating securities, i.e., by seeking to corner as many women as possible.... (Levi-Strauss 1969: 265)

And possibly the most parsimonious way of extending exchange relationships and accumulating securities is to widen the circle of affines. One way of doing this is to open up the marriage universe to all cross-cousins, not just the matrilateral cross-cousin.[1] Yet, bilateral cousin marriage may encourage the development of restricted exchange—indeed, such rules may derive from the principle of direct exchange (1969: 445). Alternatively, another, more radical measure is to forbid marriage among all cousins and close relatives while instituting the practice of bride purchase: 'By enforcing, through the gradual augmentation of prohibited degrees, the formation of longer and longer, and theoretically at least unlimited, cycles' (1969: 471), bride-price and a small number of negative marriage prescriptions extend and broaden the formula of generalized exchange. Yet, the substitution of wife-purchase for the right to the cousin allows generalized exchange to break away from its elementary structure by favouring the creation of a growing number of increasingly supple and extended cycles. Ultimately, because of intrasystemic contradiction, either the system collapses before reaching 'critical mass', returning to a more simplified version of its elementary structure and keeping positive marriage rules and unilineal descent intact (e.g., the *gumlao* rebellion among the Kachin), or it abandons these rules altogether. The principles of generalized exchange may still be present under the latter scenario, but the institutions which allow such systems to operate within its own parameters are not.

Levi-Strauss (1969: 474) conjectures that there are two fundamental solutions to the problems of generalized exchange: (1) subdividing into more restricted formations, pairs of which commence to exchange, with local systems of restricted exchange beginning to function within a total system of generalized exchange and gradually replacing it (i.e, reverting back to a more restricted system of exchange); or (2) renouncing a simple form for a more complex form. The latter is precisely the Central Inuit and northern European development.

Intrasystemic contradictions inherent within asymmetrical exchange systems may have set the stage for the emergence of complex exchange systems in Bering Strait over a millennium ago. However, it was the rebellion against these forces and the subsequent migration eastward out of Alaska that allowed Inuit groups to break away, although perhaps not once and for all (see below), from the vexations of generalized exchange. In other words, the impetus for the emergence of complex structure among the Central Inuit was born out of the self-destruction of previous relations of production instituted by generalized exchange and delayed reciprocity. While we may never know if and how other factors (e.g., the technological base) were involved in this process, the merger of dialectical materialism with the structural paradigm accounts for differences and similarities in Central Inuit and Alaskan Inuit kinship and social structure in ways that other theories have not.

### Variations on a Theme
Levi-Strauss (1969: 289) asserts that generalized exchange systems provide an especially favourable formula for not only the integration of ethnically and

geographically remote groups, but also the emergence of cultural diversity, since:

> It is in such a system that they have to renounce the least of their peculiarities. But the system also favours differentiation, even when it does not exist originally, because it reduces exchanges between groups to a minimum, and by reason of its competitive nature, invites the partners to assert themselves.

It is perhaps this aspect of generalized exchange, combined with dialectical materialism, which accounts for the substantial diversity and autonomy of regional groups in Alaska, and possibly even the emergence of such large-scale integrating mechanisms as the Beringian trade network. However, if this integrating, diversity-generating feature is an inherent characteristic of generalized exchange systems, it must be doubly inherent in complex systems. And, in this connection, it is perhaps not surprising that the degree of variation in kinship terminology and socioeconomic organization across the Central Arctic is considerably greater than one would otherwise suppose, given the common heritage and legacy of its inhabitants (Damas 1972c: 23).

It was the Talirpingmiut that provided Morgan (1870) with the most complete terminology upon which 'Eskimo' as a type of kinship system was constructed. The integral feature of this schedule was that it separated one's siblings from one's cousins, while having one term for both cross and parallel cousins. Similarly, the Netsilingmiut had one term for same-sex cousins, but possessed a 'Hawaiian' system for cross-sex cousins, whom they eventually married. Alternatively, the Copper Inuit had one term for all cousins, except the patrilateral parallel cousin. Likewise, they separated FB from all other males in the first ascending generation. On the other hand, the Port Harrison Inuit extended the term for FB (*akka*) to all males in this generation. Different again are the Iglulingmiut, who, perhaps borrowing from the second Thule expansion, perhaps succumbing to forces inherent within complex structures (e.g., the accentuation of consanguineal group solidarity), adopted a three-cousin terminology recalling forms in the Bering Strait region.

Complex exchange systems create seemingly endless possibilities, but they also open up a Pandora's box. For instance, we can anticipate how hypergamy might set into motion other processes, eventually producing alternative configurations. We have hypothesized that, following Levi-Strauss (1969), hypergamy in generalized exchange structures leads to either regressive solutions such as the institution of simpler forms of exchange (i.e., restricted exchange or the egalitarian manifestations of generalized exchange), or the emergence of complex structures. However, unshackled by the chains of restricted and generalized exchange formulae, hypergamy in the context of complex social structure continues to exacerbate significant potential for contradiction and dysfunction. Specifically, hypergamy can lead to either endogamy or 'complete paralysis of the body social' (Levi-Strauss 1969: 475). With respect to the emergence of endogamy, Central and Eastern Inuit populations are not immune. In northwest Greenland during the colonial period, for example, an upper social stratum emerged which sought mates exclusively from within its own class (Rasmussen 1986). This tendency also surfaced among Ituksarjuak's descendants in the Iglulik area, even though it violated the strictest of all social rules, i.e., marriage with blood relatives. Such transformative processes may even underlie the emergence of Netsilingmiut society and possibly the formation of such closed Bilateral-Dravidianate societies as the Beaver Dene (cf. Ives 1990).

In terms of the paralysis of the system, an arbitrary mechanism, a sort of 'sociological *clinamen*' (Levi-Strauss 1969: 475) if you will, may be introduced to mask, and thus support, the fundamental structure. In this connection, the pan-Inuit myth of the destitute orphan (e.g., Qivijuk) who overcomes all odds to become a great leader of his race represents as much a transfiguration of this problem into mythology as does the Indo-European myth of the princess who would marry a commoner should he only perform some extraordinary feat for her father, the king. Alternatively, as hypergamy->endogamy is an abrogation of reciprocity, we can see how it might lead to the systemic rejection of structures underpining this feature. Yet, it is not too difficult to see how the negation of social asymmetry as instituted through these marriage practices and consequent acceptance of social equality might lead inevitably and directly to the formation of direct exchange systems. This, not so much at the level of the group, but of the individual, appears to be the Copper Inuit development.

**Consanguineal Solidarity and Organization**

Unencumbered by positive rules of marriage, unilineal descent, and the burden of maintaining reciprocity and exchange parity, each regional group eventually sought its own particular solutions to the problems of social and material reproduction across the central and eastern Arctic. Thus perceived, complex structures are exploratory structures, and the reason why they are so is that biological propinquity has replaced institutionalized reciprocity as the foundation of society. Having no preordained social obligations to others outside the consanguineal unit, institutionalized rules of reciprocity—and moiety formations in particular—became less important in the organization of society. Reciprocal exchange relationships were still needed in the acquisition of mates and in the reproduction of society from one generation to the next, but they were established in other ways, and usually with non-kin. In this context, the solidarity and inherent organizational features of the consanguineal unit were allowed to develop and flourish. Solidarity of this unit was accomplished through increasing affectional bonds within the group—*ungayuq* in the Central Inuit vernacular. Alternatively, lines of authority and decision-making in the consanguineal unit were established by gender, generation, and birth order, or *naalaqtuq*. As individuals married into the group, consanguineal-affinal boundaries became important in the organization of society.

Yet, do not the two systemic tendencies inherent within complex systems, closeness-affection on the one hand, respect-obedience on the other, have implications for the development of more elementary structures? More specifically, just as *ungayuq*-oriented societies (where egalitarian behaviours prevail) may inevitably encourage the emergence of direct exchange systems, *naalaqtuq*-directed societies (where the accumulation of wisdom, wealth, and influence is sanctioned) may announce the development of generalized exchange systems. And, in this regard, we conclude our examination of Central Inuit social organization with a brief consideration of Labrador Inuit society.

The Labrador Inuit, as described by Taylor (1974), lived traditionally in small winter encampments centred around large virilocal extended families. While surpluses were common and starvation apparently unknown in aboriginal times, settlement size was only one-third that of the Copper, Netsilik, and Iglulingmiut. Yet extended families were normally 50 per cent larger than those among the latter regional group (1974: 79). These extended family units lived in large communal houses containing several agnately related nuclear families (i.e., *nukariik* or *irniriik* cores). However, such units frequently split along male sibling lines, indicating that *naalaqtuq* directives over-shadowed *ungayuq* directives in the organization of households. Although leadership was well-developed within this unit, its *angajuqqaq* had no authority over other extended family households, and disputes were normally resolved through common council. Indeed, there was no authority figure capable of guaranteeing harmonious relations between different households (1974: 81). While marriage within the extended family household was forbidden, neither exogamy nor endogamy was enforced at the level of the local group when, for whatever reasons, several extended family groups came together (e.g., during the fall whaling season when a dozen or more men were needed to form a whaling crew). While bride-price and the levirate were common, marriages occurred primarily among otherwise autonomous extended family households living in different, though closely situated, sealing camps.

Perhaps the most notable feature of aboriginal Labrador Inuit society was the very high rate of polygyny. In fact, the demand for wives was far greater than the gender ratio allowed—the occurrence of 54 females to 46 males recorded by Taylor (1974: 60) is consistent with moderate levels of adult male death through hunting accidents and an absence of female infanticide. Indeed, Labrador appears to be one of the few regions where the goal of polygyny was achieved among the Central Inuit (1974: 70). However, the desire for extra wives led to equally high incidences of wife-stealing, revenge murder, and blood feud.

In virtually all the above respects the Labrador Inuit are identical to the Gilyak of the lower Amur in northeastern Siberia (Black 1983)—one of the quintessential and oft-cited examples of a society integrated by generalized exchange (Levi-Strauss 1969). Although lineage development appears to have been more pronounced among the Gilyak—two or three agnately related families descended from a common ancestor often occupied a single dwelling—these large extended family households also tended to be patrilocal and patrilineal in structure (Black 1983). Likewise, lineage solidarity was pronounced and the household was organized hierarchically according to age and genealogy (1983: 76). At the same time, relationships among autonomous extended family units were largely egalitarian. Leadership was well-developed within the household, but did not extend beyond the local lineage. Marriage with the matrilateral cross-cousin, both real and classificatory, was prescribed. This practice, along with bride-price and child betrothal, especially involving MBD, were highly ritualized (1983: 80–1). The levirate was also formalized in order to keep women within the group. In turn, polygyny was the goal of every man. Yet, like the Labrador Inuit, polygyny led to wife-stealing and murder—the acquisition of women and slaves enhanced both a man's and his lineage's status (1983: 77).

The Gilyak, who express their principle of reciprocity as 'the exchange of unlike items' (1983: 75), represent a simple and egalitarian form of generalized exchange. While Gilyak marriage and kinship patterns have been examined fully by Levi-Strauss (1969: 255–68), the marked similarities between this culture and the Labrador Inuit prompt the question whether the latter did not also practise some form of generalized exchange. The Labrador Inuit during the mid-nineteenth century A.D. apparently did not distinguish cross from parallel cousins—both were called *kattangutiarsuk*, or 'little sibling' (Rasmussen 1985: 153). Nonetheless, Scheffel (1984) provides evidence that cousin marriage existed along the Labrador coast around the middle of the nineteenth century, and that it developed as a direct consequence of European trade and the emergence of Inuit middlemen. While the increased frequency of cousin marriage in this context seems plausible given a well-developed proclivity towards hypergamy->endogamy, Taylor and Taylor (1985) question Scheffel's (1984: 64) premise that cousin marriage was not present in early contact Labrador. Indeed, they provide evidence to suggest that cousin marriage was a common feature of Labrador Inuit society during the late eighteenth century. More importantly, of the four cases of cousin marriage they were able to reconstruct from incomplete early Moravian records, three represent marriage with the matrilateral cross-cousin—the other case being patrilateral parallel cousin marriage, recalling Netsilingmiut marriage preferences. Although the total percentage of cousin marriages in early contact Labrador can never be accurately reconstructed (Taylor and Taylor 1985: 186), it is probably no coincidence that 75 per cent of these marriages were with MBD. This is more than just intriguing. Even in societies with the most simple forms of generalized exchange, marriage with the matrilateral cross-cousin normally constitutes less than 30 per cent of all marriages because of demographic and other factors (Levi-Strauss 1969). Thus, even if the Taylors had recorded twice as many incidences of cousin marriage, three cases of matrilateral cross-cousin marriage would be considered high, and possibly even indicative of a system of generalized exchange.

Did the Labrador Inuit once practise generalized exchange? While all indications, with the exception of their cousin terminology, point to this conclusion, it does not take a leap of imagination to see how such a system of exchange might, under the right circumstances, emerge out of complex structures, which define the very essence of Central Inuit society.

## Conclusion

The history of Central Inuit society over the last millennium has been a history of exploration; of complex structures exploring new possibilities and possibly older ones in the formation and maintenance of productive relationships; of groups seeking better ways within societal parameters to reproduce material and social relations of production from one generation to the next. While we have seen that intrasystemic contradictions inherent within generalized exchange announce more complex systems, we have also discovered that the two structural tendencies which flourish within the consanguineal unit under such regimes—closeness-affection and respect-obedience—may lead to the emergence of simpler forms of exchange. However, I offer no simple formulae for these structural transformations. Unbridled by institutionalized rules of reciprocity, complex structures are susceptible to the spectre of dialectical materialism as it may rear up at any fissure in the system. Nonetheless, future investigations of Central Inuit society hold considerable potential for the study of kinship, social structure, and the relationship between biology and reciprocity. And towards this end, I hope others will be motivated to pursue these avenues.

Much has been said, and more written, about structural coherence and variability in Central Inuit social organization in this study, the main intent of which was to offer a viable alternative to environmental determinism in the explanation of Central Inuit society. Whether it has been entirely successful, however, now seems somehow less important than acknowledging that Inuit society, past and present, across the Canadian Arctic is a product not just of environment, but of history and social structure. Acceptance of this fact can only lead to a more informed understanding of contemporary Inuit communities and the social, economic, and political challenges that they face.

# Chapter 12

## The Politics of Survival: Central Inuit Social Structure and Nunavut

We have seen that Central Inuit social organization does indeed possess an identifiable and coherent structure. But does this finding have implications for the development of the new territory and government of Nunavut in the central and eastern Canadian Arctic?

For years, Inuit land claims negotiators had sought a number of proprietary and political rights from the federal government. Among these were surface and subsurface rights to specific lands; financial entitlements for surrendering the remaining lands to the Crown; explicit hunting, fishing and trapping rights; guaranteed representation on joint Inuit/government management boards; and the creation of a new territory known as Nunavut ('our land'). Under Section 35 of the *Constitution Act* of 1982, the rights of Canada's aboriginal peoples were 'recognized and affirmed'. In the view of the government at the time, however, the right to aboriginal self-government was not explicitly defined. Thus, Inuit negotiators were forced to link their land claims with a commitment to create a new territory and government through division of the Northwest Territories (NWT). Despite the fact that negotiators emphasized the fact that Nunavut would be a 'public' government, answerable to a legislative assembly elected by all its citizens, the federal government resisted any formula linking the two (NIC 1995: 2). Nonetheless, following the conclusion of a land claims agreement-in-principle, a compromise was found, and in 1993 two pieces of legislation were enacted by Parliament: the *Nunavut Land Claims Agreement Act*, which ratified the Nunavut Agreement, and the *Nunavut Act* which creates a Nunavut Territory by 1999.

Inuit living in small communities across the Nunavut settlement area, from Coppermine in the west, to Grise Fiord in the north, to Sanikiluaq in the south, are almost entirely descended from Central Inuit groups. Thus, it would seem that the Nunavut Government could benefit by incorporating some of the principles and features of traditional Central Inuit social organization into its design, particularly with respect to customary law, justice, social services and decision-making. Its merits notwithstanding, there are substantial barriers to this proposal.

## Barriers to Survival

First and foremost, Nunavut will be a 'public' not an aboriginal self-government. While legislation ensures that Inuktitut will become an official language in Nunavut, and that Nunavut will be able to participate directly in international matters concerning Inuit interests, Inuit gain certain rights only because of their majority status—presently 80 to 85 per cent of the residents of the Nunavut territory are Inuit—not because their aboriginal rights are explicitly recognized and guaranteed in law.[1] Consequently, it would be exceedingly difficult for the Nunavut government to adopt any traditional organizational features that were uniquely Inuit and not reconcilable with British Parliamentary law.[2]

Another barrier to incorporating, in whole or in part, structural elements of Central Inuit social organization into modern government concerns the fact that they developed as a means to cope with the challenges of material production and social reproduction in small, self-sufficient, closely-related social groups. That is, Inuit social organization developed along different paths in various areas to meet both individual and collective needs. However, with the movement to larger, centralized communities, the breakdown of the extended family unit, government emphasis on the individual, the undermining of leadership and decision-making by state and church, and declining participation in the subsistence economy, many of these features may be unworkable in modern communities. As the traditional collective basis of Inuit society has been essentially rendered irrelevant the question that arises is this: Can Inuit society in Arctic Canada survive when a large part of its *raison d'être* no longer exists?

Today, Inuit face a challenge different than that confronted by previous generations. No longer are cultural and physical survival one and the same. Once again, in order to remain Inuit, they must change. And, once again, they must do it by someone else's rules. How to preserve Inuit cultural values, customs, and traditions under such circumstances is one of the greatest challenges that Nunavut faces.

## A Proposal for a Nunavut Government

One solution to this dilemma has been proposed by the Nunavut Implementation Commission (NIC). The NIC has recommended a basic design for the operation of the Nunavut government (NIC 1995). The strengths of this proposal are considerable. Specifically, the NIC has recommended that the Nunavut government should:

(a) be staffed eventually by as many Inuit as there are proportionately in the population (i.e., 80 to 85 per cent);

# The Politics of Survival

(b) consist of only two levels of government: community and territorial, with no new law-making bodies introduced at the regional level;
(c) be decentralized at the territorial level, creating as many employment opportunities in as many communities as possible;
(d) be composed of 10 government departments, which is far fewer than presently exist in the NWT;
(e) be answerable to a legislative assembly composed of two-member constituencies guaranteeing equal numbers of male and female MLAs.

An emphasis on community government recognizes the facts that each of the 27 communities of Nunavut has its own distinctive history, culture, and ways of doing things, and that the needs, goals, and priorities of each community vary substantially (NIC 1995: 24). Through community governments, local residents have the opportunity to participate in the effective and fair delivery of programs and services devolved to the community level, thereby enhancing local decision-making in setting priorities for, allocating funds to, and co-ordinating essential programs and services. Through enhanced decision-making and accountability at the local level, the unique needs of each community can be met.

A legislative assembly composed of a man and a women from each constituency is an idea that, as much by parsimony as by design—the NIC's background paper for this proposal (NIC 1995: Appendix A-8) was preoccupied with the 'political correctness' of the concept—may help to preserve the traditional complementary nature of men and women in Inuit society. Formerly, Inuit men and women occupied distinct, but complementary, roles. Both contributed equally, but differently, to perpetuating forces and relation of production. While men engaged in hunting and other tasks related to daily survival, women looked after the domestic sphere, cooking, making and repairing clothing, and raising children, among other things. However, with the introduction of wage labour, the erosion of traditional leadership roles, and the decline of the subsistence economy, the roles of men and women have become disjunctive, contributing to social and other problems. Women, in adapting more readily to wage labour positions, have found that their role has expanded. At the same time, the traditional role of men has become more attenuated and less relevant or meaningful. By representing the needs and interests of both, a legislative assembly composed equally of men and women can explore creative solutions to help restore the balance that has been lost.

## A Missing Link

Despite these positive recommendations, and the fact that the Nunavut government will have fewer departments than the existing NWT government, the NIC's vision of Nunavut is still based on a territorial model. The need to

and running', to design the legislative, administrative and other
)f government, to make Nunavut as 'public' a government as
oubtedly constrained the NIC's vision. While the NIC proposal is a
ght direction, one of its most serious drawbacks lies in the fact
that there is no explicit recognition of the central role that the renewable
resource economy continues to play in Inuit society.

Even if hunting and other forms of subsistence are not carried out to the
extent they once were, they continue to provide substantial nutritional, social,
cultural, emotional and spiritual benefits to Inuit. In 1989, 5250 Inuit actively
engaged in such activities in the Nunavut area, contributing $62.4 million in
'country food' to the Nunavut economy (RT and Associates 1993: 38). Today,
the figure is likely much higher. Inuit have always depended on animals as well
as other Inuit to maintain their culture and perpetuate their society. In an
increasingly changing world, hunting and the sharing of animals continue to
establish and reaffirm productive and co-operative social relationships that are
crucial to Inuit cultural survival. Subsistence activities, then, are as much a set
of culturally established rights, responsibilities, and obligations as a way of
physical survival (M. Freeman, personal communication, 1994). Thus, Inuit
continue to derive a significant part of their diet, as well as a sense of culture,
identity, and spiritual attachment to the land, from hunting and other forms of
subsistence, even if opportunities to maintain this connection have waned.

The NIC has proposed the creation of a Department of Sustainable Development. But its tenuous view of the relationship between the renewable
resource economy and Inuit cultural survival is apparent in its proposal to separate this department from the departments of Culture, Language, Elders and
Youth; Education; and Health and Social Services. The efficacy of the new
government of Nunavut in serving the needs of its majority may be improved
significantly if the operations and functions of these departments were intimately integrated around the renewable resource economy, in effect creating a
mega-department (e.g., the Department of Sustainable Society and Economy)
in which other proposed departments (e.g., Executive and Intergovernmental
Affairs; Finance and Administration, Public Works and Government Services,
etc.) could be designed to support.

OTHER SOLUTIONS

Other solutions that might be entertained within the NIC's proposed structure
of government include what I term the Kudlu (Pitsualuk) and Etuangat
(Aksayuk) Solutions (Stevenson 1994), after the late matriarch and late patriarch of the Pangnirtarmiut, who originate from Umanaqjuaq and Kekerten,
respectively. The contemporary imbalance in Inuit men's and women's contributions to Inuit society has the potential to be corrected through two-member
constituencies. However, why stop here? Would it not be in the best interests

of the preservation of Inuit values, customs, and traditions to have elder representatives in the legislative assembly? These individuals need not be explicitly Inuit, as eligibility for this position would be determined by age and residency (e.g., >55 years), regardless of race. Elders would probably not make the most appropriate government ministers—the rigorous demands of such positions may be best met by younger individuals. However, elected elders could form an advisory group, a shadow Senate, to advise the ministers and members of the legislative assembly on Inuit issues and affairs. This Senate would not have veto power over legislative assembly decisions. Nonetheless, as elders are still highly cherished (*ungayuk*) and listened to (*naalaqtuq*) in contemporary Inuit society, their advice would carry some weight with the assembly's younger members. This proposal, after the form of decision-making characteristic of her group, is the Kudlu Solution.

The Etuangat Solution also utilizes the principle of *naalaqtuq*, while avoiding giving Inuit special ethno-national rights. Inuit society and culture in Nunavut today is at a crossroads. With a population dominated by people who have not grown up on the land, a birth rate that far exceeds the national average, and a recent explosion in mineral exploration activities which threatens to bring thousands of Qallunaat north, there is a very real possibility that Inuit knowledgeable in Inuit ways (Inuktitut) will soon become a minority in their homeland. Therefore, a system of voting should be considered whereby, in agreement with Inuit ways and traditions, individuals with more knowledge and life-experience (*isuma*) are given more voting power in community, and perhaps even territorial, elections. The question that should be asked is this: Should those with the least understanding of, investment in, and commitment to Inuit values, customs, and traditions and their preservation, have as much decision-making power as those with the most?

Although perhaps a challenge to the *Canadian Charter of Rights and Freedoms*, such a system could be best accommodated by giving individuals, whether Inuit or non-Inuit, more voting shares as they pass various residency milestones. Everyone would acquire the same voting power upon residence in Nunavut, whether they moved to or were born into the territory. However, every so often (e.g., a generation or so), an additional voting share would be acquired. In this system, an Inuk elder would have more voting shares than an Inuk half his/her age, while the latter would have more voting power than a transient worker who plans to make Nunavut his/her home for only a few years. In effect, this voting system would restore some of the influence elders have lost in the context of colonization, while formalizing traditional Inuit styles of decision-making in those communities with this heritage. No one of eligible voting age would be exempt from voting, as original proposals contemplated. However, those with the most invested in Nunavut, as measured by length of residence, whether Inuit or non-Inuit, would have the most voting power. There are few existing legal provisions that protect Inuit rights and effectively safeguard Inuit culture. In these regards, not only are the Kudlu and Etuangat Solutions just, equitable and reflective of Inuit values and traditions, they do not give Inuit any special rights based on race.

## Conclusion

Traditional Central Inuit social organization as a distinct and coherent structure, will probably never again be possible. Inuit do not want to return to the past. But they do want to retain their cultural identity and values, if not traditions, while gaining more control over their affairs. Will the NIC proposal fulfil these needs? There may be sufficient flexibility in the NIC proposal to incorporate some of the traditional principles governing Inuit social relations, to restore the some of the collective foundation of Inuit society, and/or to move in directions along the lines proposed above. However, this will take considerably more reflection and public consultation than heretofore has been the case.[3] In the final analysis, the success of Nunavut will be determined by the extent to which its residents participate in its design, and in decision-making at the community and territorial levels.

However, what if Nunavut fails to meet the needs and aspirations of its Inuit majority? Fortunately, all will not be lost. The Inuit of Nunavut have yet to play their 'self-government' card. Issues of aboriginal self-government have taken a back seat to the practical challenges of setting up the Nunavut territory and government. Nonetheless, the establishment of a Nunavut territory does not relieve the federal government of its fiduciary obligations and responsibilities to Inuit (Nunavut Tunngavik Incorporated 1994). If Inuit do not gain a significant measure of political control through the creation of a Nunavut territory and government, self-government is a very real option.

Yet, it is in the best interests of Inuit and all Canadians that Nunavut fulfil the expectations on which it is based. Consideration of the Kudlu and Etuangat Solutions, and other more creative solutions, under existing legal and constitutional arrangements, will allow Inuit to better manage change and chart their future at a time when such control is needed most.

# Bibliography

Adams, C.
1972 'Flexibility in Canadian Eskimo social forms and behavior: a situational and transactional appraisal'. Pp. 9–16 in *Alliance in Eskimo Society*. Proceedings of the American Ethnological Society 1971, supplement. Edited by L. Guemple. University of Washington Press, Seattle.

Akulujuk, M.
1976 'Things from a long time ago and nowadays'. Pp. 73–7 in *Stories from Pangnirtung*. Hurtig (editor and publisher), Edmonton.

Amundsen, R.E.G.
1908 *The Northwest Passage*. E.P. Dutton, New York (2 vol.).

Ansel, W.D.
1983 *The Whaleboat: A Study of Design, Construction, and Use from 1850 to 1970*. New Bedford Free Public Library, New Bedford, Mass.

Asch, M.I.
1979 'The ecological-evolutionary model and the concept of mode of production: two approaches to material reproduction'. Pp. 81–99 in *Challenging Anthropology*. Edited by D. Turner and G. Smith. McGraw-Hill, Ryerson.

1988 *Kinship and the Drum Dance in a Northern Dene Community*. The Circumpolar Research Series, the Boreal Institute for Northern Studies, Academic Printing and Publishing, University of Alberta, Edmonton.

Balikci, A.
1960 'Some acculturative trends among Eastern Canadian Eskimos'. *Anthropologica* n.s. 2: 139–53.

1964 *Development of Basic Socioeconomic Units in Two Eskimo Communities*. Anthropological Series 69, National Museum of Canada Bulletin 202, Ottawa.

1967 'Female infanticide on the Arctic coast'. *Man* n.s. 2: 615–25.

1970 *The Netsilik Eskimo*. The Natural History Press, New York.

Bandi, H.G.
1995 'Siberian Eskimos as whalers and warriors'. Pp. 165–83 in *Hunting the Largest Animals: Native Whaling in the Western Arctic and Subarctic*. Studies in Whaling No. 3, The Canadian Circumpolar Institute, University of Alberta, Edmonton.

Barron, W.
1895 *Old Whaling Days*. Hull Press, Hull.

Befu, H.
1964 'Eskimo systems of kinship terms—their diversity and uniformity'. *Arctic Anthropology* 2, 1: 84–98.

Bilby, J.W.
1923 *Among Unknown Eskimo: An Account of Twelve Years of Intimate Relations with the Primitive Eskimo of Ice-bound Baffin Land, with a Description of Their Ways of Living, Hunting Customs and Beliefs*. Seeley Service, London.

Binford, L.R.
1980 'Willow smoke and dog's tails: hunter-gatherer settlement systems and archaeological site formation'. *American Antiquity* 45: 1–17.

Birket-Smith, K.
1924 *Ethnography of the Egedesminde District, with Aspects of the General Culture of West Greenland.* Meddelelser om Grønland 66, Copenhagen.

1929 *The Caribou Eskimos: Material and Social Life and Their Cultural Position.* Report of the Fifth Thule Expedition 1921–1924, 5 (1–2), Copenhagen.

1932 *The Caribou Eskimo.* Report of the Fifth Thule Expedition 1921–1924, 3, Copenhagen.

Black, L.
1973 'Nivkh (Gilyak) of the Sakhalin and the lower Amur'. *Arctic Anthropology* 10, 2: 1–118.

Boas, F.
1883-1884 Franz Boas' Letter Diary. American Philosophical Society, Philadelphia.

1884 'A journey in Cumberland Sound and on the west shore of Davis Strait in 1883 and 1884'. *American Geographical Society Bulletin* 26: 242–72.

1907 *The Eskimo of Baffin Land and Hudson Bay* (From notes collected by Capt. G. Comer, Capt. James S. Mutch, and Rev. E.J. Peck). Bulletin of the American Museum of Natural History 15.

1964 *The Central Eskimo.* University of Nebraska Press, Lincoln. (Originally published by the Bureau of American Ethnology, 6th Annual Report, 1888, Washington.)

Bradby, B.
1975 'The destruction of natural economy'. *Economy and Society* 4: 127–61.

Briggs, J.L.
1982 'Living dangerously: the contradictory foundations of value in Canadian Inuit society'. Pp. 109–31 in *Politics and History in Band Societies.* Edited by E. Leacock and R.B. Lee. Cambridge University Press, London.

Brody, H.
1987 *Living Arctic: Hunters of the Canadian North.* Douglas and McIntyre, Vancouver.

Burch, E.S. Jr
1975 *Eskimo Kinsmen: Changing Family Relationships in Northwest Alaska.* The American Ethnological Society Monograph 59, West Pub., St Paul, Minn.

1978 'Caribou Eskimo origins: an old problem reconsidered'. *Arctic Anthropology* 15, 1: 1–35.

1979 'The Thule-historic Eskimo transition on the west coast of Hudson Bay'. Pp. 189–211 in *Thule Eskimo Culture: An Anthropological Retrospective.* National Museum of Man Mercury Series, Archaeological Survey of Canada Paper 88. Edited by A.P. McCartney. Ottawa.

1980 'Traditional Eskimo societies in northwest Alaska'. Pp. 253–304 in *Alaska Native Culture and History*. Edited by Y. Kotani and W.B. Workman. National Museum of Ethnology, Senri Ethnological Studies 4, Osaka.

Burch, E.S. and T.C. Correll
1972 'Alliance and conflict: inter-regional relations in North Alaska'. Pp. 17–39 in *Alliance in Eskimo Society*. Proceedings of the American Ethnological Society 1971, supplement. Edited by L. Guemple. University of Washington Press, Seattle.

Clark, A.H. and J.T. Brown
1887 *The Whale Fishery, Vol. 2, part 15*. Commissioner of Fisheries, Government Printing Office, Washington.

Colby, B.L.
1935 'Capt. J.O. Spicer imported Eskimos to win court case here'. *The Day*, New London, Connecticut, 9 March 1935.

Cole, D.
1983 '"The Value of a Person lies in His Herzenbildung", Franz Boas' Baffin Island Letter Diary, 1883–84'. Pp. 13–52 in *Observers Observed: Essays on Ethnographic Fieldwork*, History of Anthropology 1. Edited by G.W. Stocking Jr. The University of Wisconsin Press, Madison.

1986 'Franz Boas in Baffin Land'. *The Beaver* (August-September): 4–15.

Correll, T.C.
1968 'Demes and dialects among the Central Eskimo'. Paper presented at the Canadian Linguistic Association Annual Meeting, June 13, Calgary.

Damas, D.
1963 *Iglulingmiut Kinship and Local Groupings: A Structural Approach*. National Museum of Canada Bulletin 196, Ottawa.

1964 'The patterning of the Iglulingmiut kinship system'. *Ethnology* 3: 377–88.

1968a 'Iglulingmiut kinship terminology and behavior, consanguines'. Pp. 85–105 in *Eskimo of the Canadian Arctic*. Edited by V.F. Valentine and F.G. Vallee. McClelland and Stewart, Toronto.

1968b 'The diversity of Eskimo societies'. Pp. 111–17 in *Man the Hunter*. Edited by R.B. Lee and I. Devore. Aldine Pub., Chicago.

1969a 'Characteristics of Central Eskimo band structure'. Pp. 116–34 in *Contributions to Anthropology: Band Societies*. Edited by D. Damas. Anthropological Series 84, National Museum of Canada Bulletin 228, Ottawa.

1969b 'Environment, history and Central Eskimo society'. Pp. 40–64 in *Contributions to Anthropology: Ecological Essays*. Edited by D. Damas. Anthropological Series 86, National Museum of Canada Bulletin 230, Ottawa.

1972a 'Central Eskimo systems of food sharing'. *Ethnology* 11: 220–40.

1972b 'The structure of Central Eskimo associations'. Pp. 40–55 in *Alliance in Eskimo Society*. Proceedings of the American Ethnological Society 1971, supplement. Edited by L. Guemple. University of Washington Press, Seattle.

1975a 'Demographic aspects of Central Eskimo marriage practices'. *American Ethnologist* 2: 409–18.

1975b 'Social anthropology of the Central Eskimo'. *Canadian Review of Sociology and Anthropology* 2: 252–66.

1975c 'Three kinship systems from the Central Arctic'. *Arctic Anthropology* 12: 10–30.

1984a 'Central Eskimo: Introduction'. Pp. 391–6 in *Handbook of North American Indians, Arctic, 5*. Edited by D. Damas. Smithsonian Institution, Washington.

1984b 'Copper Eskimo'. Pp. 397–414 in *Handbook of North American Indians, Arctic, 5*. Edited by D. Damas. Smithsonian Institution, Washington.

1988 'The contact-traditional horizon of the Central Arctic: reassessment of an era and reexamination of an era'. *Arctic Anthropology* 25: 101–38.

Damkjar, E.
1987 'Late Dorset "longhouse" occupations of Creswell Bay, Somerset Island and their possible relationship to the Thule migration'. Paper presented at the 20th annual meeting of the Canadian Archaeological Association, 22–26 April, Calgary.

Davis, C.H. (ed.)
1876 *Narrative of the North Polar Expedition, U.S. Ship Polaris, Captain Charles Francis Hall Commanding*. Washington.

Department of Fisheries and Oceans (DFO)
1994 'Co-management Plan for Southeast Baffin Beluga'. Report prepared for the Minister. Ottawa.

Dundee University Library
n.d. Printed Annual Returns of Whaling Voyages, 'Kinnes Manuscripts', Dundee.

Dunning, D.W.
1966 'An aspect of domestic group structure among eastern Canadian Eskimo'. *Man* n.s. 1: 216–25.

Eevic, K.
1976 'Things that We Used to Do in the Old Days That No Longer Exist'. Pp. 78–89 in *Stories from Pangnirtung*. Hurtig (editor and publisher), Edmonton.

Fainberg, L.
1967 'On the question of the Eskimo kinship system' (translated by C.C. Hughes). *Arctic Anthropology* 4: 244–56.

Fienup-Riordan, A.
1983 *The Nelson Island Eskimo: Social Structure and Ritual Distribution*. Alaska Pacific University Press, Anchorage.

Fleming, A.L.
1913– 'Census of Eskimo Names of Eskimo Living in Baffin Land in the District of
1914 Blacklead Island, Cape Haven, Frobisher Bay and Baffin Land Shore of Hudson Strait'. Department of Northern Affairs Library, Ottawa.

1932 *Perils of the Polar Pack, or the Adventures of the Reverend E.W.T. Greenshield KTON of Blacklead Island, Baffin Island.* Missionary Society of the Church of Anglican in Canada. The Church House, Toronto.

Foster-Carter, A.
1978 'The modes of production controversy'. *New Left Review* 107: 47–77.

Franklin, J.
1828 *Narrative of a Journey to the Shores of the Polar Sea in the Years 1819, 20, 21, and 22.* John Murray, London.

Freeman, M.M.R.
1967 'An ecological study of mobility and settlement patterns among the Belcher Island Eskimo'. *Arctic* 20, 3: 154–75.

1970 'Ethos, economics, and prestige: a re-examination of Netsilik Eskimo infanticide'. *Proceedings of the International Congress of Americanists* 2, pp. 247–50, Publication of the 38th annual meetings, Stuttgart and Munchen, 12–18 August 1968.

1971 'A social and ecological analysis of systematic female infanticide among the Netsilik'. *American Anthropologist* 73: 1011–18.

1979 'A critical view of Thule culture and ecological adaptation'. Pp. 278–85 in *Thule Eskimo Culture: An Anthropological Retrospective.* National Museum of Man Mercury Series, Archaeological Survey of Canada Paper 88. Edited by A.P. McCartney. Ottawa.

Friedman, J.
1974 'Marxism, structuralism and vulgar materialism'. *Man* n.s. 9: 444–69.

1984 'Tribes, states, and transformations'. Pp. 161–202 in *Marxist Analyses and Social Anthropology.* Edited by M. Bloch. Tavistock, New York.

Gagné, R.C.
1961 'Tentative Standard Orthography for Canadian Eskimos'. Department of Northern Affairs and National Resources, Ottawa.

Gardner, D.
1979 '1979 site survey of Cumberland Sound and the Clyde River area, Baffin Island'. Pp. 365–86 in *Archaeological Whale Bone: A Northern Resource.* Edited by A.P. McCartney. University of Arkansas Anthropological Papers 1, Fayetteville.

Giddings, J.L. Jr
1952 'Observations on the "Eskimo Type" of kinship and social structure.' *Anthropological Papers of the University of Alaska* 9, 1: 5–10.

Godelier, M.
1966 'Comments on the concepts of structure and contradiction'. *Aletheia* 4: 178–88.

Goldring, P.
1984 'Arctic whaling study: 1984 site inspection report'. Historic Sites and Monuments Board of Canada Report 1984: 54, pp. 465–538, Ottawa.

1986 'Inuit economic responses to Euro-American contacts: southeast Baffin Island, 1824–1940'. *Historical Paper 1986 Communications Historiques*: 146–72.

1989 'The triumvirate in Pangnirtung: career paths and cross-cultural relations, 1921–1945'. Paper presented at the Canadian Historical Association Annual Meetings, 1989.

Goody, J. (editor)
1966 *The Developmental Cycle in Domestic Groups*. Cambridge Papers in Social Anthropology 1, Cambridge University Press, Cambridge, England.

Graburn, N.H.H.
1964 *Taqamiut Eskimo Kinship Terminology*. Northern Coordination and Research Council, NCRC 64–1, Department of Northern Affairs and Natural Resources, Ottawa.

1969 *Eskimos without Igloos: Social and Economic Development in Sugluk*. Little, Brown and Company, Boston.

Greenshield, E.W.T.
1914 'An Arctic advance'. *The Church Missionary Gleaner*, 1 January 1914: 13–14.

Guemple, L.D.
1961 *Inuit Spouse Exchange*. Occasional Paper, Department of Anthropology, University of Chicago, Chicago.

1965 'Saunik: name sharing as a factor governing Eskimo kinship terms'. *Ethnology* 4: 323–35.

1966 'Kinship Reckoning Among the Belcher Island Eskimos'. Unpublished PhD dissertation, University of Chicago, Chicago.

1969 'The Eskimo ritual sponsor: a problem in the fusion of semantic domains'. *Ethnology* 8: 468–83.

1972a 'Eskimo band organization and the "D.P. Camp" hypothesis'. *Arctic Anthropology* 9: 80–112.

1972b Introduction, in *Alliance in Eskimo Society*. Pp. 1–8, Proceedings of the American Ethnological Society 1971, supplement. Edited by L. Guemple. University of Washington Press, Seattle.

1972c 'Kinship and alliance in Belcher Island Eskimo society'. Pp. 56–78 in *Alliance in Eskimo Society*. Proceedings of the American Ethnological Society 1971, supplement. Edited by L. Guemple. University of Washington Press, Seattle.

1979 *Inuit Adoption*. Canadian Ethnology Service Paper 47, National Museum of Man Mercury Series, Ottawa.

1988 'Teaching social relations to Inuit children'. Pp. 131–9 in *Hunters and Gatherers 2: Property, Power and Ideology*. Edited by T. Ingold et al. Berg, Oxford.

Hallendy, N.
1985 'Reflections, shades and shadows'. Pp. 126–67 in *Collected Papers on the Human History of the Northwest Territories*. Edited by M.J. Patterson, R.R. Janes, and C.D. Arnold. Prince of Wales Northern Heritage Centre Occasional Paper 1, Yellowknife.

Haller, A.A.
1967 'A Human Geographical Study of the Hunting Economy of Cumberland Sound, Baffin Island, N.W.T.'. Unpublished Masters thesis, McGill University, Montreal.

Haller, A.A., D.C. Foote, and P.D. Cove
1966 *Baffin Island—East Coast: An Area Economic Survey*. Edited by G. Anders. Industrial Division, Northern Administration Branch, Department of Indian Affairs and Northern Development, Ottawa.

Hanbury, D.T.
1904 *Sport and Travel in the Northland of Canada*. Macmillan, New York.

Hantzsch, B.A.
1977 *My Life Among the Eskimos: The Baffin Land Journals of Bernhard Adolph Hantzsch 1909–1911*. Edited by L.G. Neatby. Institute for Northern Studies, Mawdsley Memoir 3, University of Saskatchewan, Saskatoon.

Harper, K.
1981 'The Moravian Mission at Cumberland Sound'. *The Beaver* (Summer): 43–7.

n.d. 'Uqarmat: the life of Reverend Edmund James Peck'. Anglican Church of Canada General Synod Archives, Peck Papers, MM 52.5 H37.

Harrington, R.
1954 *The Face of the Arctic*. Hodder and Stoughton, London.

Harritt, R.K.
1995 'The development and spread of the whale hunting complex in Bering Strait: retrospective and prospects'. Pp. 33–50 in *Hunting the Largest Animals: Native Whaling in the Western Arctic and Subarctic*. Studies in Whaling No. 3, The Canadian Circumpolar Institute, University of Alberta, Edmonton.

Hegarty, R.B.
1959 *Returns of Whaling Vessels Sailing from American Ports: A Continuation of Starbuck's 'History of the American Whale Fishery 1876–1908'*. Old Dartmouth Historical Society, New Bedford, Mass.

Heinrich, A.C.
1960 'Structural features of northwestern Alaskan kinship'. *Southwestern Journal of Anthropology* 16: 110–26.

1965 'Eskimo Type Kinship and Eskimo Kinship: An Evolutionary and Provisional Model for Presenting Data Pertaining to Inupiat Kinship Systems'. Unpublished PhD dissertation, University of Washington, Seattle.

Heinrich, A.C. and R. Anderson
1971 'Some formal aspects of a kinship system'. *Current Anthropology* 12: 541–64.

Helm, J. and D. Damas
1963 'The contact-traditional all-native community of the Canadian north'. *Anthropologica* n.s. 5: 9–21.

Hennigh, L.
1983 'North Alaskan Eskimo alliance structure'. *Arctic Anthropology* 20: 23–32.

Hickey, C.
1979 'The historic Beringian trade network: its nature and origins'. Pp. 411–34 in *Thule Eskimo Culture: An Anthropological Retrospective*. National Museum of Man Mercury Series, Archaeological Survey of Canada Paper 88. Edited by A.P. McCartney. Ottawa.

1984 'An examination of processes of culture change among the nineteenth century Copper Inuit'. *Études/Inuit/Studies* 8: 13–35.

1992 'Earliest contacts between Copper Inuit and Europeans: a consideration of the possible effects'. Paper presented at the Society for Historical Archaeology annual meetings, Kingston, Jamaica, 8–12 January.

Hickey, C. and M.G. Stevenson
1990 'Structural Variation in Traditional Canadian Inuit Marriage, Adoption, and Social Customs'. Family Law Reform Commission of the N.W.T., Yellowknife.

Holland, C.A.
1970 'William Penny, 1809–92: Arctic whaling master'. *The Polar Record* 15: 25–43.

Holm, G.F.
1914 *Ethnological Sketch of the Angmagsalik Eskimo*. Meddelelser om Grønland 39, Copenhagen.

Honigmann, J.J.
1962 *Social Networks in Great Whale River*. National Museum of Canada Bulletin 178, Anthropology Series 54, Ottawa.

Honigmann, I. and J. Honigmann
1959 'Notes on Great Whale River Ethos'. *Anthropologica* 1: 106–21.

Howgate, H.W. (editor)
1879 *The Cruise of the Florence, or Extracts from the Journal of the Preliminary Arctic Expedition of 1877–78* (From the journal of G.E. Tyson). J.J. Chapman, Washington.

Hughes, C.C.
1958 'An Eskimo deviant from the "Eskimo" of type of social organization'. *American Anthropologist* 60: 1140–7.

1963 Review of James W. Van Stone's 'Point Hope, An Eskimo Village in Transition'. *American Anthropologist* 65: 452–4.

1984 'History of ethnology after 1945'. Pp. 23–6 in *Handbook of North American Indians, Arctic, 5*. Edited by D. Damas. Smithsonian Institution, Washington.

Ives, J.W.
1990 *A Theory of Northern Athapaskan Prehistory*. Westview, Calgary.

Jenness, D.
1921 'The cultural transformation of the Copper Eskimo'. *Geographical Review* 11: 541–50.

1922 *The Life of the Copper Eskimo*. Report of the Canadian Arctic Expedition 1913–18, 12(A), Ottawa.

1923 'Origin of the Copper Eskimo and their copper culture'. *Geographical Review* 13: 540–51.

1924 *Eskimo Folklore*. Report of the Canadian Arctic Expedition 1913–18, 8, Ottawa.

Johnson, G.A.
1982 'Organizational structure and scalar stress'. Pp. 389–421 in *Theory and Explanation in Archaeology: The Southampton Conference*. Edited by C. Renfrew et al. Academic Press, New York.

Kay, G.
1976 *Development and Underdevelopment: A Marxist Analysis*. Macmillan, London.

Kellerman, B. (ed.)
1984 *Leadership: Multidisciplinary Perspectives*. Prentice Hall, Englewood Cliffs, New Jersey.

Kilabuk, J.
1976 'The things that Eskimos did'. Pp. 34–40 in *Stories from Pangnirtung*. Hurtig (editor and publisher), Edmonton.

Kjellstrom. R.
1973 *Eskimo Marriage: An Account of Traditional Eskimo Courtship and Marriage*. Nordeska Museets Handlinger 80, Norway.

Kumlien, L.
1879 *Contributions to the Natural History of Arctic America, Made in Connection with the Howgate Polar Expedition, 1877–78*. Government Printing Office, Washington.

Lantis, M.
1946 'The Social Culture of the Nunivak Eskimo'. *Transactions of the American Philosophical Society* 35, 2, Philadelphia.

1972 'Factionalism and leadership: a case study of Nunivak Island'. *Arctic Anthropology* 9: 43–65.

1987 'Important roles in organizations and communities'. *Human Organization* 46: 190–9.

Leacock, E.B.
1981 *Myths of Male Dominance*. Beacon, Boston.

Leach, E.R.
1964 *Political Systems of Highland Burma*. Boston, London.

Levi-Strauss, C.
1963 *Structural Anthropology*. Basic Books, New York.

1969 *The Elementary Structures of Kinship*. Beacon, Boston.

Lewis, A.
1904 *The Life and Work of the Rev. E.J. Peck Among the Eskimos*. Hodder and Stoughton, London.

Linton, R.
1936    *The Study of Man*. D. Appleton Century, New York.

Low, A.P.
1906    *Report on the Dominion Government Expedition to Hudson Bay and the Arctic Islands on Board the D.G.S. Neptune, 1903–1904*. Government Printing Bureau, Ottawa.

Lubbock, B.
1955    *The Arctic Whalers*. Brown, Son, and Ferguson, Glasgow.

Lyon, G.F.
1824    *The Journal of Captain G.F. Lyon*. John Murray, London.

McGhee, R.J.
1970    'Speculations on climatic change and Thule culture development'. *Folk* 11–12: 172–84.

1972    *Copper Eskimo Prehistory*. National Museum of Man Publications in Archaeology 2, Ottawa.

1982    'The past ten years in Canadian Arctic prehistory'. *Canadian Journal of Archaeology* 6: 65–78.

McLaren, I.A.
1961    'Population dynamics and exploitation of seals in the eastern Canadian Arctic'. Pp. 168–83 in *The Exploitation of Natural Animal Populations*. Edited by E. D. LeCren and M. W. Hoedgate. Blackwell Science Pub., Oxford.

Manning, T.H.
1943    'Notes on the coastal district of the eastern barren grounds and Melville Peninsula from Igloolik to Cape Fullerton'. *Canadian Geographic Journal* 26: 84–105.

Markham, A.H.
1880    *The Voyages and Works of John Davis, the Navigator*. The Hakluyt Society, Burt Franklin, New York.

Mary-Rousselière, G.
1991    *Qitlarssuaq: The Story of a Polar Migration*. (Translated by A. Cooke). Wuerz, Winnipeg.

Mathiassen, T.
1927    *Archaeology of the Central Eskimos*. Report of the Fifth Thule Expedition 1921–24, 4 (1 and 2), Copenhagen. (Reprinted by AMS Press, New York, 1976).

1928    *Material Culture of the Iglulik Eskimos*. Report of the Fifth Thule Expedition, 1921–24, 6 (1), Copenhagen.

1976    *Report of the Expedition*. Report of the Fifth Thule Expedition 1921–1924, 1 (1). AMS Press, New York.

Mauss, M. and M.H. Beuchat
1904–   *Essai sur les Variations Saisonnières des Sociétés Eskimos*. L'Année Sociologique.
1905

Maxwell, M.S.
1985   *Prehistory of the Eastern Arctic*. Academic Press, New York.

Mayes, R.G.
1978   'The Creation of a Dependent People: The Inuit of Cumberland Sound, Northwest Territories'. Unpublished PhD dissertation, McGill University, Montreal.

M'Donald, A.
1841   *A Narrative of Some Passages in the History of Eenoolooapik, a Young Esquimaux Who Was Brought to Britain in 1839, in the Ship 'Neptune' of Aberdeen: An Account of the Discovery of Hogarth's Sound*. Fraser and Company, London.

Metayer, M.
1973   *Unipkat: Tradition Esquimaude de Coppermine, Territoires-du-Nord-Ouest, Canada*. Collection Nordicana Centre d'Études Nordiques, Université Laval, Quebec, 42 (3 vol.).

Morgan, L.H.
1870   *Systems of Consanguinity and Affinity of the Human Family*. Smithsonian Contributions to Knowledge 17 (2), Washington.

Morrison, D.
1983   *Thule Culture in Western Coronation Gulf*. National Museum of Man Mercury Series, Archaeological Survey of Canada Paper 116, Ottawa.

1990   *Iglulualumiut Prehistory: The Lost Inuit of Franklin Bay*. Canadian Museum of Civilization, Archaeological Survey of Canada Mercury Series Paper 142, Hull.

1991   'The Copper Inuit soapstone trade'. *Arctic* 44: 230–46.

Munn, H.T.
1932   *Prairie and Arctic By-Ways*. Hurst and Blackett, London.

Murdock, G.P.
1949   *Social Structure*. Macmillan, New York.

Mutch, J.
1886   Letter to Franz Boas, 14 January 1886. American Philosophical Society Library, Philadelphia.

1906   'Whaling in Pond's Bay'. Pp. 485–500 in *Boas Anniversary Volume: Anthropological Papers Written in Honor of Franz Boas*. G.E. Stechert and Co., New York.

Needham, R.
1974   *Remarks and Inventions: Skeptical Essays about Kinship*. Tavistock, London.

Nichols, P.
1954   'Boat-building Eskimos'. *The Beaver* (Summer): 52–5.

NIC (Nunavut Implementation Commission)
1995   'Footprints in New Snow': a comprehensive report from the Nunavut Implementation Commission to the Department of Indian and Northern Development, Government of the Northwest Territories and Nunavut Tunngavik Incorporated concerning the establishment of the Nunavut Government. Iqaluit.

NTI (Nunavut Tunngavik Incorporated)
1994   'Implementation of the inherent right to self-government in Nunavut'. Iqaluit.

O'Laughlin, B.
1975   'Marxist Approaches in Anthropology'. *Annual Review of Anthropology* 1975 (9565): 341–70.

Old Dartmouth Historical Society
1859   Scrapbook No. T-1, 1859.

1860–1 Whaling Museum Library, Log 771, *Antelope*.

Oosten, J. G.
1976   *The Theoretical Structure of the Religion of the Netsilik and Iglulik*. Rijksuniversiteit, Groningen.

Parmi, E.S.
n.d.   Introduction. Microfilm Edition of Thirteen Arctic Logbooks in the Stefansson Collection. Dartmouth College Library, Dartmouth.

Parry, W.E.
1969   *Journal of a Second Voyage of Discovery of the Northwest Passage from the Atlantic to the Pacific: Performed in the Years 1821, 22, 23. . . .* Greenwood, New York (originally published in 1824 by John Murray, London).

Peck, E.J.
1922   *The Eskimo*. Anglican Church of Canada General Synod Archives, Peck Papers, M56–1 Series 37(11).

Penny, W.
1840   Letter to Captain Beaufort, 15 October 1840. Scott Polar Research Institute, MS. 116/63/46.

1849   General Correspondence. *Periodical Accounts Relating to the Missions with the Church of the United Brethren Established Among the Heathen* 19: 19–23.

1854a  Letter to the editor. *Periodical Accounts Relating to the Missions of the Church of the United Brethren Established Among the Heathen* 21(225): 268–9.

1854b  Letter from Captain Penny. *Literary Gazette*: 1966, 23 September 1854: 829.

1856   General Correspondence. *Periodical Accounts Relating to the Missions with the Church of the United Brethren Established Among the Heathen*. 22(233), 25 September 1856: 143.

Pitsualak, M.
1976   'How we hunted whales'. Pp. 18–25 in *Stories from Pangnirtung*. Hurtig (editor and publisher), Edmonton.

RAND
1963   *A Report on the Physical Environment of Southern Baffin Island, Northwest Territories, Canada*. Department of Geography, McGill, RAND Corporation, Santa Monica.

RT and Associates
1993    Nunavut Harvest Support Program, Background Document. Yellowknife.

Rasmussen, E.H.
1985    'First cousin marriage: social consistency or social change in Hopedale, Labrador'. *Études/Inuit/Studies* 9: 153–9.

1986    'Some aspects of the reproduction of the West Greenlandic upper social stratum, 1750–1950'. *Arctic Anthropology* 23: 137–50.

Rasmussen, K.
1929    *Intellectual Culture of the Iglulik Eskimos*. Report of the Fifth Thule Expedition 1921–24, 7 (1), Copenhagen (reprinted by AMS Press, New York).

1930    *Iglulik and Caribou Eskimo Texts*. Report of the Fifth Thule Expedition 1921–24, 7 (3), Copenhagen (reprinted by AMS Press, New York).

1931    *The Netsilik Eskimos Social Life and Spiritual Culture*. Report of the Fifth Thule Expedition 1921–24, 8 (1–2), Copenhagen.

1932    *Intellectual Culture of the Copper Eskimos*. Report of the Fifth Thule Expedition 1921–24, 9, Copenhagen.

1942    *The Mackenzie Eskimo (After Knud Rasmussen's Posthumous Notes)*. Edited by H. Ostermann. Report of the Fifth Thule Expedition 1921–24, 10 (2): 5–164, Copenhagen.

1976    *Observations on the Intellectual Culture of the Caribou Eskimos*. Report of the Fifth Thule Expedition 1921–24, 7 (2), AMS Press, New York.

Reeves, R. R. and E. Mitchell
1981    'White whale hunting in Cumberland Sound'. *The Beaver* (Winter): 42–9.

Remie, C.
1985    'Towards a new perspective on Netsilik Inuit female infanticide'. *Études/Inuit/Studies* 9: 67–76.

1988    'Flying like a butterfly, or Knud Rasmussen among the Netsilingmiut'. *Études/Inuit/Studies* 12: 101–27.

Rey, P.P.
1971    *Colonialisme, Néocolonialisme, et Transition au Capitalisme*. Maspéro, Paris.

Riches, D.
1974    'The Netsilik Eskimo: a special case of selective female infanticide'. *Ethnology* 12: 351–61.

1982    *Northern Nomadic Hunter-Gatherers: A Humanistic Approach*. Academic Press, New York.

Ricoeur, P.
1975    'Phenomenology and hermeneutics'. *Nous*: 83–103. Indiana University, Bloomington.

Rink, H.
1975 *The Eskimo Tribes: Their Distribution and Characteristics, Especially in Regard to Language*. AMS Press, New York. (Originally published in 1887–91.)

Roberts, H.H., and D. Jenness
1925 *Eskimo Songs: Songs of the Copper Eskimo*. Report of the Canadian Arctic Expedition 1913–18, 14, Ottawa.

Robinson, S.
1973 'The Influence of the American Whaling Industry on the Aivilingmiut, 1860–1919'. Unpublished Masters thesis, McMaster University, Hamilton, Ontario.

Ross, W.G.
1975 *Whaling and Eskimos: Hudson Bay 1860–1915*. National Museums of Canada Publications in Ethnology 10, Ottawa.

1979a 'Commercial Whaling and Eskimos in the Eastern Canadian Arctic 1819–1920'. Pp. 242–66 in *Thule Eskimo Culture: An Anthropological Retrospective*. National Museum of Man Mercury Series, Archaeological Survey of Canada Paper 88. Edited by A.P. McCartney. Ottawa.

1979b 'The annual catch of Greenland (bowhead) whales in waters of North Canada 1719–1915: a preliminary compilation'. *Arctic* 32: 91–121.

1981 'Whaling, Inuit, and the Arctic Islands'. Pp. 33–50 in *A Century of Canada's Arctic Islands, 1880–1980*. Edited by M. Zaslow. The Royal Society of Canada, Ottawa.

1985a 'Reflections of a whaling captain (by G. Tyson)'. Pp. 183–200 in *Arctic Whalers, Icy Seas: Narratives of the Davis Strait Whale Fishery*. Compiled and edited by W.G. Ross. Irwin Pub., Toronto.

1985b 'Snug for Winter (Journal of A.J. Whitehouse)'. Pp. 255–74 in *Arctic Whalers, Icy Seas: Narratives of the Davis Strait Whale Fishery*. Compiled and edited by W.G. Ross. Irwin Pub., Toronto.

1985c 'The inhabitants of the Arctic (Extracts of an account by D. Cardno)'. Pp. 227–42 in *Arctic Whalers, Icy Seas: Narratives of the Davis Strait Whale Fishery*. Compiled and edited by W.G. Ross. Irwin Pub., Toronto.

Royce-Peterson, A.
1982 *Ethnic Identity: Strategies of Diversity*. Indiana University Press, Bloomington.

Sahlins, M.D. and E. Service
1960 *Evolution and Culture*. The University of Michigan Press, Ann Arbor.

Salter, E.M.
1984 'Skeletal Biology of Cumberland Sound, Baffin Island'. Unpublished PhD dissertation. University of Toronto, Toronto.

Scheffel, D.
1984 'From polygyny to cousin marriage? Acculturation and marriage in the 19th century Labrador Inuit society'. *Études/Inuit/Studies* 8: 61–75.

Schledermann, P.
1975    *Thule Eskimo Prehistory of Cumberland Sound, Baffin Island, Canada*. National Museum of Man Mercury Series, Archaeological Survey of Canada Paper 38, Ottawa.

1976    'Thule culture communal houses in Labrador'. *Arctic* 29: 27–37.

1979    'The "Baleen Period" of the Arctic Whale Hunting Tradition'. Pp. 134–48 in *Thule Eskimo Culture: An Anthropological Retrospective*. National Museum of Man Mercury Series, Archaeological Survey of Canada Paper 88. Edited by A.P. McCartney. Ottawa.

Schledermann, P. and K. McCullough
1980    'Western elements in the early Thule culture of the eastern high Arctic'. *Arctic* 33: 833–41.

Service, E.
1962    *Primitive Social Organization*. Random House, New York.

Spalding, A.
1979    *Learning to Speak Inuktitut: a Grammar of North Baffin Dialects*. Native Language Research Series 1, Centre for Research and Teaching of Canadian Native Languages, The University of Western Ontario, London.

Spencer, R.F.
1959    *The North Alaskan Eskimo: A Study in Ecology and Society*. Bureau of American Ethnology Bulletin 171, Smithsonian Institution, Washington.

Sperry, J.
1952    'Eskimo Kinship'. Unpublished Masters thesis, Columbia University.

Spier, L.
1925    'The distribution of kinship systems in North America'. *University of Washington Publications in Anthropology* 1, 2: 69–88.

Stanford, D.
1976    *The Walakpa Site, Alaska*. Smithsonian Contributions to Anthropology 20, Smithsonian Institution, Washington.

Starbuck, A.
1964    *History of the American Whale Fishery from its Earliest Inception to the Year 1876* (2 vol.). Argosy-Antiquarian, New York.

Steenhoven, G. van den
1959    *Legal Concepts Among the Netsilik Eskimos of Pelly Bay N.W.T*. Northern Coordination and Research Centre, NCRC-3, Department of Northern Affairs and Natural Resources, Ottawa.

1962    *Leadership and Law Among the Eskimos of the Keewatin District, Northwest Territories*. The Hague, Uitgeverij Escelsior.

Stefansson, V.
1919    *The Stefansson-Anderson Arctic Expedition: Preliminary Ethnological Report*. Anthropological Papers of the American Museum of Natural History 14 (1), New York.

1942  *The Friendly Arctic: The Story of Five Years in Polar Regions*. Macmillan, New York.

1964  *My Life With the Eskimo*. Macmillan, Toronto.

Stenton, D.
1989  'Terrestrial Adaptations of Neo-Eskimo Coastal-Marine Hunters on Southern Baffin Island, N.W.T.'. Unpublished PhD dissertation, University of Alberta, Edmonton.

Stevenson, D.
1972  'Social Organization of Clyde Inlet Eskimo'. Unpublished PhD dissertation, University of British Columbia, Vancouver.

Stevenson, M.G.
1984  'Kekerten: Preliminary Archaeology of an Arctic Whaling Station'. Manuscript on deposit, Prince of Wales Northern Heritage Centre, Yellowknife, N.W.T.

1986  'The emergence of class structure at an Arctic whaling station'. Paper presented at the Canadian Archaeological Association 19th annual meeting, 26 April, Toronto.

1989  'Sourdoughs and cheechakos: the formation of identity-signaling social groups'. *Journal of Anthropological Archaeology* 8: 270–312.

1992  'Two Solitudes? South Amundsen Gulf History and Prehistory, N.W.T.'. Manuscript submitted to *Archaeology, Architecture and History*. Canadian Parks Service, Environment Canada, Ottawa.

1993  'Central Inuit Social Structure: The View from Cumberland Sound, Baffin Island, Northeast Territories'. Unpublished PhD thesis. University of Alberta, Edmonton.

1994  'Traditional Inuit decision-making structures and the administration of Nunavut'. Paper prepared for the Royal Commission on Aboriginal Peoples. Ottawa.

Stone, B.
1990  'The acculturative role of sea woman'. *Meddelelser om Grønland* 13: 1–34.

Sutherland, P.C.
1852  *Journal of a Voyage in Baffin's Bay and Barrow Straits in the Years 1850–51* (2 vol.). Longman, Brown, Green and Longman's, London.

1856  'On the Esquimaux'. *Journal of the Ethnological Society* 4: 193–214 (London).

Taylor, J.G.
1974  *Labrador Eskimo Settlements of the Early Contact Period*. National Museum of Man Publications in Ethnology 9, Ottawa.

1979  'Inuit whaling technology in eastern Canada and Greenland'. Pp. 189–211 in *Thule Eskimo Culture: An Anthropological Retrospective*. National Museum of Man Mercury Series, Archaeological Survey of Canada Paper 88. Edited by A.P. McCartney. Ottawa.

Taylor, J.G., and H. Taylor
1985  'Cousin marriage in traditional Labrador Inuit society'. *Études/Inuit/Studies* 9: 183–6.

Taylor, W.E. Jr
1963 'Hypotheses on the origin of Canadian Thule Culture'. *American Antiquity* 28: 456–64.

1966 'An archaeological perspective on Eskimo economy'. *Antiquity* 40: 114–20.

Terray, E.
1984 'Classes and class consciousness in Abran Kingdom of Gyaman'. Pp. 85–136 in *Marxist Analyses and Social Anthropology*. Edited by M. Bloch. Tavistock, New York.

Thalbitzer, W.
1941 *The Angmassalik Eskimo: Contributions to the Ethnology of East Greenland Natives*. Meddelelser om Grønland 39, 40, 53, Copenhagen.

Thibert, A.
1970 *English-Eskimo Dictionary Eskimo-English*. Canadian Research Centre for Anthropology, Saint Paul University, Ottawa.

Trott, C.G.
1982 'The Inuk as object, some problems in the study of Inuit social organization'. *Études/Inuit/Studies* 6: 43–108.

Turner, D.H
1978 'Ideology and Elementary Structures'. *Anthropologica* n.s. 20: 223–47.

Usher, P.J.
1965 *Economic Basis and Resource Use of the Coppermine-Holman Region, N.W.T.* Northern Coordination and Research Centre, NCRC 65–2, Department of Northern Affairs and National Resources, Ottawa.

1971 *Fur Trade Posts of the Northwest Territories 1870–1970*. Northern Science Research Group, NSRG 71–4, Department of Indian Affairs and Northern Development, Ottawa.

Valentine, C.A.
1952 'Toward a Definition of Eskimo Social Organization'. Unpublished Masters thesis, University of Pennsylvania, Philadelphia.

Van de Velde, F.
1956 'Les regles du partage du phoques pris par la chasse aux aglus'. *Anthropologica* 3: 5–15.

Van Stone, J.W. and W. Oswalt
1960 'Three Eskimo Communities'. *Anthropological Papers of the University of Alaska* 60: 12–56.

Wakeham, W.
1898 *Report of the Expedition to Hudson Bay and Cumberland Gulf in the Steamship 'Diana' Under the Command of William Wakeham, Marine and Fisheries Canada in the Year 1897*. Ottawa.

Wareham, M.
1843 'Appendix to Captain Belcher's paper, Northumberland Inlet: an extract from the journal of a whaling voyage'. *Journal of the Royal Geographical Society* 12: 21–8.

Warmow, M.
1859 'Extract from Br. M. Warmow's journal of his residence in Cumberland Inlet, during the winter of 1857–58'. *Periodical Accounts Relating to the Missions of the Church of the United Brethren Established Among the Heathen* 23, 242: 87–92, March 1859.

Wenzel, G.W.
1981 *Clyde Inuit Adaptation and Ecology: The Organization of Subsistence*. National Museum of Man Mercury Series, Canadian Ethnology Service Paper 77, Ottawa.

Weyer, E.M.
1932 *The Eskimos: Their Environment and Folkways*. Yale University Press, New Haven, Connecticut.

Willmott, W.E.
1960 'The flexibility of Eskimo social organization'. *Anthropologica* n.s. 2: 48–59.

1961 *The Eskimo Community at Port Harrison, P.Q.* Northern Coordination and Research Centre, NCRC 61-1, Department of Northern Affairs and Natural Resources, Ottawa.

Wood, D.
n.d. Abstracts of Whaling Voyages. Compiled by Dennis Wood, New Bedford Free Public Library.

Yorga, B.
1979 'Migration and adaptation: a Thule culture perspective'. Pp. 286–91 in *Thule Eskimo Culture: An Anthropological Retrospective*. National Museum of Man Mercury Series, Archaeological Survey of Canada Paper 88. Edited by A.P. McCartney. Ottawa.

# GLOSSARY OF COMMON INUKTITUT TERMS USED IN TEXT

**aggutik:** Boatsteerer, helmsman (*aggutiit* pl.).

**aglu:** Breathing-hole maintained by seals through sea ice.

**ai:** In-marrying females in Ego's generation (male Ego) (e.g., BW, male cousins' wives); in-marrying females in first ascending generation (male Ego, Iglulik only); MZH (both Egos, Cumberland Sound, Morgan's terminology); WZ, wife's female cousins, and WBD and WZD (male Ego); in-marrying males in Ego's (e.g., HB and husband's male cousins), and first descending generations (e.g., HZS, HBS) (female Ego); *aiit* (pl.).

**aigiik:** Dyad consisting of a resident person of one gender and an in-marrying person of the opposite gender (Iglulik) (e.g., brother-in-law/sister-in-law). Frequently, this relationship is characterized by avoidance.

**airaapik**(*kuluk*): MZH and FZH (Cumberland Sound, modern terminology).

**aivik:** walrus.

**aiyak:** MZ and MBW (Cumberland Sound, Morgan's terminology); MZ/mother's female cousin (Iglulik).

**akka:** FB; FB/father's male cousin (Iglulik); FB and FZH (Netsilik).

**amauk:** Great-grandparent; extended to entire generation.

**anaana:** Mother; *amaama* (Netsilik).

**angajuqqaq:** Boss, leader, one who is more substantial and must be obeyed (*angajuqqat*, pl.).

**angak:** MB; MB/mother's male cousin (Iglulik, Netsilik, and Copper); MZH (Netsilik and Copper).

**angaquk:** Shaman (*angaqut*, pl.).

**angayuk:** Older brother (male Ego); older sister (female Ego).

**angutikattigiit:** Children of two brothers (Iglulik and Copper).

**angutiqat:** FBch (male and female Egos, Copper); FBS (male Ego, Iglulik), FBD (female Ego, Iglulik).

**anik:** B (female Ego); also male cousin (female Ego, Iglulik and Netsilik).

**apak:** Father (Netsilik); *aappak* (Copper).

**arnaqat:** FZch, MBch, MZch (Egos male and female, Copper only); MZS (male Ego, Iglulik, also Netsilik); MZD (female Ego, Iglulik).

**arngnakattigiit:** Children of two sisters (Iglulik).

**arvik:** Bowhead whale.

**ataata:** Father.

**atchun:** FZ (Cumberland Sound, Morgan's terminology).

**aton:** Copper Inuit dance.

**attak:** FZ; father's female cousin; also FBW and MBW (Netsilik).

**avatak:** Float made from seal skin.

**avik:** To come apart, to separate.

**avikgiit:** Those who are separate.

**ga:** Possessive, 'my'; my mother = *anaanaga*.

**iglorek:** Song cousins (Rasmussen); *igloriit* (dual).

**iglu:** House.

**ilagiit:** Kinsmen, i.e., consanguineal and affinal relatives; *ilagit* (single).

**illuaqjuk:** Male cousins (male Ego, Cumberland Sound).

**illukuluk:** Female cousins (female Ego, Cumberland Sound).

**illulik:** Great grandchild (Cumberland Sound); *illuliarut* (Iglulik, Netsilik, Copper).

**illuq:** Female cousins (male Ego), male cousins (female Ego) (Cumberland

Sound and Netsilik); FZS and MBS (male Ego, Iglulik), FZD and MBD (female Ego, Iglulik).

**illuriik:** Rough joking relationship between cousins.

**inngutaq:** Grandchild (consanguines only).

**Inuit:** People.

**Inuk:** Person.

**Inuuk:** Two people.

**irniq:** Son.

**irniriik:** Father-son dyad. Traditionally one of the most respectful and affectionate relationships in Inuit society.

**iqaluk:** Arctic char, fish.

**isumataq:** Leader, the one who thinks; *ihumataq* (Netsilik).

**ittuq:** Grandfather; extended to all males in grandparental generation including affines.

**maqtaak:** Layer between whale's skin and blubber.

**mangniriik:** Rough joking relationship between distant relatives or non-kin.

**mauliqtuq:** Breathing-hole sealing.

**naalaqtuq:** 'To listen to': respect, obedience, deference to authority. One of two major principles upon which social relations are formed in Central Inuit society.

**nanuk:** Polar bear.

**nayagiik:** Brother-sister dyad.

**nayak:** Sister (male Ego); female cousin (Iglulik and Netsilik, male Ego); younger sister (Copper, male Ego).

**nekaishutu:** Sharing on a village-wide basis with distribution usually by one individual (Cumberland Sound).

**netsiavinik:** Young or 'silver jar' seal.

**nettik:** Ringed seal.

**nirqi:** Marine mammal meat.

**ningaugiik:** Relationship between an in-marrying male and parents-in-laws, (e.g., father-in-law and son-in-law).

**ningauk:** ZH/female cousin's husband (male Ego) and all in-marrying males in descending generations (male and female Egos) (e.g., DH, ZDH); also all in-marrying males in first ascending generation (male and female Egos, Iglulik only); *ningaut* (pl.).

**niutang:** Traditional drogue or sea anchor used to slow the escape of whales and other large sea mammals.

**nukaq:** Younger brother (male Ego); younger sister (female Ego).

**nukariik:** Older brother-younger brother dyad, older sister-younger sister dyad.

**nuliaq:** Wife.

**nuna:** Land.

**nunatakatigiit:** Members of a group of people living together on the land.

**Nunavut:** 'Our land'.

**nurraq:** Sister's or female cousin's child (female Ego); also female cousin's child (Netsilik only).

**panik:** Daughter.

**panniriik:** Mother-daughter dyad. Traditionally one of the most respectful and affectionate relationships in Inuit society.

**pisik:** Copper Inuit dance.

**piutuq:** The act of one family inviting another to share a meal (Cumberland Sound).

**qaggi:** Festive or song/dance house.

**qailertetang:** Head shaman and master of ceremonies during feast of 'Sedna'.

**qairulik:** Harp seal.

**qammaq:** Fall/winter house (*qammat*, pl.).

**qamutiik:** Sled pulled by dog team.

**qangiariik:** Paternal uncle-nephew dyad.

**qaniaq:** Brother's and male cousin's child (male Ego).

**qaqivak:** Trident, traditional fish spear.

**qilalugaq:** Beluga or white whale; *qilalugaq tuugaalik* (narwhal)

**qirniqtuk:** Black, also name for narwhal.

**qudlik:** Sea mammal oil lamp, traditionally made of soapstone.

**qujannamik:** Thank you!

**sakiaq:** WB, wife's siblings' sons, wife's male cousins; husband's sisters, husband's sisters' daughters, husband's female cousins; *hakiaq* (Netsilik and Copper).

## Glossary of Common Inuktitut Terms

**sakik**: Parent-in-law; (*sakkiik*, dual) (*hakik*, Copper and Netsilik).

**(saq)**: Post-base which has the gloss of 'has the potential to become' (e.g., *irniqsaq* = adopted son, step-father = *ataatasaq*).

**sarbuk**: Open-water area among winter ice; (*sarbut*, pl.).

**sakurpang**: Traditional whaling harpoon.

**savik**: Man's knife (blade).

**Sedna**: The sea goddess; more correctly known among most Central Inuit as Nuliajuk, Arnaluk Takanaluk, or Takanakapsaluk.

**sivataqviq**: Saturday, 'time for getting *sivat* (biscuits)'.

**sivutik**: Harpooner.

**sina**: Edge of the land-fast sea ice.

**tinu**: Tides, or more appropriately, lowering of tides.

**tiriganirk**: Arctic fox.

**tuktu**: Caribou.

**tupik**: Summer tent.

**tuugak**: Tusk, ivory.

**ugjuk**: Bearded seal.

**ui**: Husband.

**ukuaq**: Brother's/male cousin's wife (female Ego); in-marrying females in descending generations (male and female Egos); in-marrying females in ascending generations (both Egos, Cumberland Sound; female Ego only, Iglulik); *ukuvak* (Netsilik).

**ulu**: Woman's knife (crescent-shaped).

**umialik**: Leader, boat-owner (Inupiat).

**umialiqtak**: Boat-owner.

**umiaq**: Skin boat (*umiat*, pl.).

**ungayuq**: Emotional closeness, fondness, affection, endearment. One of two major principles upon which social relations are formed in Central Inuit society.

**uqsuq**: Oil from blubber.

**uttuq**: Basking seal hunting.

**uyuruk**: Sister's/female cousin's child (male Ego); also wzch (Copper and Netsilik).

# GLOSSARY OF COMMON ANTHROPOLOGICAL TERMS USED IN TEXT

**affine:** A relative through marriage.

**affinal-incorporating systems:** A kinship terminology that merges one's affinal relatives with blood relatives (e.g., in-marrying aunts and uncles are often referred to by kin terms appropriate to parent's siblings), while **affinal-excluding terminologies** separate in-marrying from blood relatives.

**agamous marriage:** A marriage practice that, apart from prohibiting marriage to other nuclear family members, has no specific rules.

**agnate:** A relation through the male or paternal line.

**avuncular relation:** Mother's brother, or a relative through mother's brother.

**band (types of):** **Anomalous band**=an eclectic mixture of people and families forming a local group. **Composite band**=a collection of families related in various ways forming a local group. **Patrilocal band**=a group formed on the basis of male kin relations (father-son, brother-brother, etc.) (Service 1962).

**bifurcate collateral system:** A kinship terminology that distinguishes one's collateral relatives along maternal and paternal lines.

**bifurcate merging system:** A kinship terminology that distinguishes between maternal and paternal kin, but ignores the distinction between lineal and collateral kin.

**bilateral kin group:** A group that traces its lines of descent through both the maternal and paternal lines.

**bride-price:** Compensation paid, usually in the form of gifts, by the groom and/or his group to the bride's family upon marriage.

**bride-service:** Service rendered, usually for a specified period of time, by the groom to the bride's family upon marriage.

**clan:** A noncorporate descent group that traces its lineage and/or membership to a common ancestor or totem. Patrilineal and matrilineal clans are social groups which trace this lineage/membership through either the father's or mother's lines, respectively.

**collateral relations:** Blood relatives not in the direct line of descent.

**commensal unit:** A unit that assembles regularly to share and eat food.

**consanguine:** Blood relative.

**conjugal family:** A family composed of a man, a woman, and their dependent children.

**cousin (types of):** **Cross-cousin**=a cousin related to Ego through two people of the opposite sex (e.g., mother's brother's child, father's sister's child). **Parallel cousin**, a cousin related to Ego through two people of the same sex (e.g., mother's sister's child). **Maternal** or **matrilateral cousin**=cousin on mother's side. **Paternal** or **patrilateral cousin**=cousin on father's side. **Matrilateral cross-cousin**=mother's brother's child. **Patrilateral parallel cousin**=father's brother's child.

**cultural materialism:** A form of anthropological research that regards the manner in which a culture adapts to its environment as the most significant factor in its development.

**deme:** A population separated from others by geographical and social barriers to interbreeding (also, genetic or marriage isolate).

**descriptive kin terms:** F=father, M=mother, B=brother, Z=sister, S=son, D=daughter, H=husband, W=wife, FZ=father's sister, MBS=mother's brother's son, etc.

**diachronic:** Respective of, or across, time. Refers to the study of phenomena with temporal change as a major variable.

**dialectical materialism:** An approach to anthropological analysis in which societies are viewed as the resolution or synthesis of contradictions (thesis and antithesis) born out of previous relations of production. The dialectic when applied to human societies assumes either that there is intrinsic contradiction in the nature of all phenomena, which leads inevitably to change (the materialism of Marx and Engels), or that there is contradiction in the set of ideas by which phenomena are envisaged or conceived (the dialectic of Hegel). In the Hegelian mode, ideas not material forces are the generative elements.

**direct exchange:** A type of marriage system wherein the exchange of individuals, goods, and services is restricted to two groups practising dual exogamy, i.e., moieties (also symmetrical or restricted exchange).

**dyad:** Two people that share a particular and exclusive relationship.

**egalitarianism:** A characteristic whereby all individuals in a society have equal access to status and strategic resources.

**Ego:** One's self.

**emic:** The actor's view or perception. A form of research whose validity depends on interpretations that reflect the 'native' point of view.

**endogamy:** Marriage to others within one's own social group (however defined) or residential group.

**etic:** The observer's view or interpretation. A form of research that does not depend on distinctions that reflect 'native' perspectives or realities.

**environmental determinism:** A theory that holds that the environment and material conditions ultimately determine the shape of human adaptation, including social organization.

**exogamy:** Marriage to others outside one's own social group (however defined) or residential group.

**extended family:** A collection of nuclear families related by ties of blood, that live together; often incorporates multiple generations.

**genealogy:** A listing, often diagrammatic, showing specific kinship linkages among individuals. Genealogies can be unilineal or bilateral, and used as charts of group membership.

**generations (types of):** First ascending generation=one's parents generation (including aunts and uncles). **Ego's generation**=one's own generation (e.g, siblings, cousins, etc.). **First descending generation**=one's children, nieces and nephews, etc. (Note: actual age may not be a necessary criterion for membership in any generation.)

**generalized exchange:** A marriage arrangement involving the exchange of individuals, goods, and services in which reciprocity is delayed or indirect, and thus difficult to maintain (also asymmetrical exchange).

**groups (types of):** Local group=a group of families, usually related through blood or marriage, living together as a socioeconomic unit (also residential group or local band). **Subregional group**=two or more allied local groups living within a fixed area (also subregional division or regional subdivision). **Regional group**=two or more subregional groups living within a fixed area and sharing a common language, culture, etc. (see tribe).

**Hawaiian system:** A mode of kinship reckoning in which all relatives of the same sex and generation are referred to be the same name (e.g., sibling and same-sex cousins terms are often the same).

**historical particularism:** The thesis that each culture is unique. A form of research that rejects generalization, theorizing, and focused enquiry in favour of recording and describing all aspects of a culture.

**hypergamy:** Marriage into a group or class of the next highest economic, social, or political standing and advantage.

**Iroquoian system:** A kinship terminology that has one term for father and father's brother, one term for mother and mother's sister, but separate terms for father's sister and mother's brother, while equating parallel cousins with one's brothers and sisters.

**kindred:** A group of people closely related to one individual through both parents.

**levirate:** A marriage custom whereby a widow marries the brother of her dead husband.

**lineage:** A corporate descent group whose

members claim, and can trace, descent from a common ancestor.

**linear descent:** Descent traced through either the matri- or patri-lines (also unilinear descent).

**lineal terms:** Kin terms that are exclusive to either line of descent.

**matri-kin:** Kin traced through the mother's line.

**matrilineal descent:** A system wherein descent is traced exclusively through the female line for purposes of group membership.

**mode of production:** The way in which a society reproduces its material and social requirements and conditions.

**moiety system:** The division of society into two halves (e.g., clans) on the basis of descent. Such groups usually exchange marriage partners or practise dual exogamy, which is characteristic of direct or restricted exchange systems.

**monogamy:** Marriage whereby an individual has but one spouse.

**nepotics:** Collateral relations in the first descending generation (e.g., one's sibling's children).

**nuclear family:** A family unit consisting of a mother, a father, and dependent children.

**Omaha system:** A mode of kinship reckoning usually associated with patrilineal descent whereby mother's patrilineal kin are equated across the generations. The Crow system is the matrilineal equivalent of the Omaha system.

**patri-kin:** Kin traced through the father's line.

**patrilineal descent:** Descent traced exclusively through the male line for purposes of group membership.

**polyandry:** The custom of a woman having two or more husbands at the same time.

**polygamy:** Marriage in which an individual has more than one spouse at the same time.

**polygyny:** The custom of a man having two or more wives at the same time.

**polythetic:** The operation of many variables or processes.

**relations of production:** Social relationships that drive or structure a particular mode of production (i.e., who works, who determines how the product of labour is to be shared and distributed, who controls the surplus, etc.). Not to be confused with organization of production.

**residence (types of):** **Ambilocal**, a custom whereby there are no particular rules of residence after marriage, i.e., the married couple is free to choose where they live. **Bilocal**, residence whereby a couple lives in the same local group as their parents. **Matrilocal**, residence whereby a couple goes to live with the wife's relatives or local group. **Neolocal**, residence whereby a couple resides with neither the husband's or wife's local group. **Patrilocal**, residence whereby a couple goes to live with the husband's relatives or local group.

**sibling:** One's brother or sister.

**sororate:** The practice whereby a man takes his deceased wife's sister as a spouse.

**sororal polygyny:** The custom of a man marrying two sisters.

**synchronic:** Literally, at the same time. Refers to the study of phenomena holding time constant.

**syncretism:** The blending of old and new traits (beliefs) in acculturative contexts to form a new (religion) system.

**transhumance:** The movement of groups between higher (mountainous) and lower (lowland) ecological settings during different seasons. More often associated with pastoralists than hunter-gathers.

**tribe:** A group of local bands sharing a common language, culture, and religion.

**virilocal:** A residence pattern whereby the newly married couple resides with the groom's kin group, indicating the prevalence of male relevant ties.

**uxorilocal:** A form of residence where the married couple lives with the bride's family. Similar to matrilocality or bride-service, but less formal in content, and usually undertaken for practical reasons (e.g., a man may go to live with his wife's parents until she is of an age or capable enough to support a family).

# Notes

## Chapter 1

1 In 1862, the explorer Charles F. Hall returned to America with two Inuit from the settlement of Kingmiksoo on the southwest shore of Cumberland Sound, Joe and Hanna, or Epeokepe and Takaretu (among other spellings), where they were subsequently interviewed by Morgan.

2 M. Freeman (personal communication, 1993) suggests that, because the institution whereby a man goes to live with his wife's parents after marriage is less formal among the Central Inuit than most cultures, uxorilocal residence might be a more appropriate term than bride-service or matri-patrilocal residence. The latter terms are retained in this study, however, so long as the reader bears in mind the variable and informal nature of this practice among most Central Inuit groups.

3 Cross-cousins are cousins linked by two people opposite in sex, while parallel cousins are cousins linked by two people of the same sex. For example, my father's sister's daughter would be my cross-cousin, while my mother's sister's son would be my parallel cousin. Such distinctions are often important in determining social relationships in non-western societies.

4 Under *naalaqtuq* directives, pairings in Iglulingmiut society are referred to by the kin terms appropriate to the subordinate member, with the suffix '*giik*' or '*riik*' usually attached (Damas 1963, 1964). Thus, the father-son (*ataata-irniq*) relationship becomes *irniriik*, while the older brother-younger brother (*angayuk-nukaq*) relationship becomes *nukariik* with similar combinations obtaining for mother-daughter (*panniriik*), brother-sister (*nayagiik*), father's brother-brother's son (*qangiariik*), and so on.

5 The reader is referred to Damas' (1963, 1964, 1968a, 1975c) excellent studies of Iglulingmiut kinship for fuller treatment of the operation of the principles of *naalaqtuq* and *ungayuq* among a wider circle of kinsmen.

6 Guemple's perspectives appear to have developed after most Belcher Islanders moved into two centralized communities. Here, composite configurations resulted in an increased emphasis on non-kinship alliances in strategies of affiliation. It is clear from Freeman's (1967) earlier work in the Belcher Islands that greater importance was attached to kinship in forming socioeconomic relationships prior to the centralization of the population.

7 Hughes (1984) provides a comprehensive listing of studies devoted to these and other issues for the period 1945 to 1984.

8 Among most groups, strangers are either repelled because they are not kinsmen or accepted into the local unit, in which case a real or fictive kinship connection is usually found or created. Few Inuit groups use personal names in forms of address and/or reference. In fact, among many groups there exist prohibitions against using personal names among certain categories of kinsmen (e.g., Damas 1963, Stefansson 1919). The high incidence of divorce and trial marriage in Inuit society indicates the comparative weakness of affinal as opposed to consanguineal relations in Inuit society. If kinship was unimportant, one would expect that the husband-wife bond would be as strong or stronger than any other. This is, and was, clearly not the case (Burch 1975: 298).

## Chapter 2

1 HBCA A97/6 fo. 90–92, 'Recommendations for the Season, 1927–1928', Milling.

2 The beluga whale may have formerly frequented Cumberland Sound in numbers of several thousand or more, and may still, although the number of beluga sighted at any

one time has decreased markedly since the late 1960s. Considering the facts that over 5400 belugas were commercially harvested in Cumberland Sound between 1923 and 1941, and that as many as 800 whales were taken in one year alone (see Chapter 3), Reeves and Mitchell's (1981: 41) estimate of 5000+ may be conservative, as they acknowledged. While Inuit believe beluga are now more dispersed and harder to hunt than in previous decades, they also feel that the number of beluga entering the Sound has not decreased appreciably over the years (DFO 1994).

3  HBCA 'Ungava Annual Reports', 28 July 1939, Stewart.

4  HBCA D.FTR. 27, Annual Report, Pangnirtung Post, Outfit 264 (1933–34).

5  HBCA D.FTR. 27, enclosure, 6 February 1935, Commissioner to Manager of St Lawrence-Ungava District.

6  If Penny's (1840) statements are correct that the Cumberland Sound Inuit took an average of 8 to 12 whales a year to feed a population of about 1000 people, and assuming that the average whale (8 to 9 m in length) produced 22,727 kg (25 tons) of meat and blubber, this species could have theoretically supplied 227 kg per year, or *c.* .62 kg a day, to each individual.

7  Penny first met Eenoolooapik and his family (which included his siblings, father, and the latter's two wives) at Durban Island in the late 1830s. Eenoolooapik's family came from Kingmiksoo, where M'Donald (1841: 101) observed that 'they were the finest tribe we had hitherto seen; and Eenoo's (*sic*) near relations in particular were much superior in point of personal appearance to the rest.' Also, M'Donald (1841: 94–5) first encountered the prominent *angaquk*, Anniapik, from Anarnitung at Durban Island in 1835.

8  Aboriginal whaling in the Sound was a co-operative enterprise involving groups of up to a dozen or more men, whether *umiat* or kayaks were used. The kayak method, as described by the late Jimmy Kilabuk of Umanaqjuaq, involved pursuing the whale in kayaks and lancing it near the kidneys at every opportunity, until, through loss of blood and resistance of the floats, the animal eventually died. Alternatively, *umiaq* whaling in Cumberland Sound was probably similar to that in Labrador whereby the *umiaq*, manned by an all-male whaling crew of 12 men or more, was used to carry the whaling harpoon (*sakurpang*), sealskin float(s) (*avatak*), and drogue (*niutang*) as these items were simply too bulky to be used from a kayak (Taylor 1979). In the *umiaq* method, the whale would be harpooned from the *umiaq* and subsequently followed and lanced in the heart with a special spear (*kalugiang*) when the animal surfaced to blow.

9  For a more detailed account of a typical annual round for this subgroup of Talirpingmiut see Boas (1964: 23–4).

10  Boas (1964: 24) reported that small parties of Inuit left Nettilling Lake for Iglulik in 1750, 1800, 1820, and 1935. Also, in 1846, Sutherland (1856: 202) met two men at Kingmiksoo who had come all the way from Iglulik when they were boys.

11  Uxorilocality is probably a more appropriate term than bride-service when discussing residence among the Central Inuit (M. Freeman, personal communication, 1993). However, bride-service is retained here with the understanding that it may serve functions not directly related to exchange, e.g., a man may go to live with his wife's family until she is old enough to set up her own household.

12  In this connection, it is interesting to consider where Eenoolooapik's wife came from and what happened to his whaleboat. From M'Donald (1841) we know that Eenoolooapik was willing to exchange his whaleboat for the 'beautiful Coonook'. Yet, he may have soon lost interest in her (or, her in him). Having established the price he was willing to pay for a wife, some time prior to 1846 Eenoolooapik appears to have lost a whaleboat, but gained a wife. The possibility that his wife came from another

settlement is suggested by the fact that, while Sutherland found six *umiat* at Kingmiksoo, no whaleboats were recorded.

13 This figure is not so much arbitrary as supported by the Kingmiksoo data. The youngest married woman was 13 years of age, while the youngest mother, Eenoolooapik's wife, was 15. Conversely, the youngest married male was 19 years old. Among most Central Inuit groups, females customarily had children soon after puberty, while males were several years older before they took a wife.

## Chapter 3

1 No years were worse than 1830 when 19 ships went down, and 1835 when 135 lives were lost (Holland 1970: 26, Lubbock 1955: 278).

2 The whalers referred to bowhead whales as 'fish' and whaling as 'fishing'. These terms are adopted here, and used along with whale and whaling.

3 Penny had tried, with Inuit assistance, to find this new whaling ground on a number of previous occasions, but was unsuccessful.

4 Penny's discovery did little to dispel the general feeling of malaise that hung over the whaling industry after the heavy losses of the previous decade. In addition, having taken no whales, Penny's voyage of discovery was a financial failure.

5 Penny took 19 whales in 1846, many if not most of which probably came from Cumberland Sound (Lubbock 1955: 345).

6 *Periodical Accounts Relating to the Missions of the Church of the United Brethren* . . . 19: 19–23 (1849, London).

7 Ibid.

8 Ibid.

9 Although the crew of the *McLellan* were left at Kingmiksoo, in late winter they shifted their base to Nuvujen as the floe edge formed well up the Sound over the winter of 1851–52.

10 'Trying out' refers to the process whereby blubber is rendered into oil by boiling.

11 Old Dartmouth Historical Society Scrapbook No. T-1, 1859, p. 23.

12 Umanaqjuaq appears to have served as a fall whaling base for ships anchored at Naujateling during the 1860s (Goldring 1984: 488–9).

13 Ross (1985b: 156) speculates that the *Emma* over-wintered at Naujateling. However, it is clear from the place names provided in Whitehouse's journal that the harbour is Union (or 'Middle') Harbour (i.e., Mitilnarvik), 3 km north of Penny's Harbour and shore-based whaling station at Kekerten.

14 'Yacks' was the whalers' term for the Inuit, and may have been a slightly derogatory reference to their method of communication and apparent unintelligibility of their language.

15 While drinking appears to have been a persistent problem among the whalers, the Inuit were given rum only on special occasions such as a wedding or after a whale was caught (see Ross 1985b: 155–73).

16 Interestingly, like the 16 whales taken at Nuvujen by the crew of the *McLellan* in 1852, the whales taken off the *sina* between Nuvujen and Miliakdjuin/Kekerten in the spring of 1860 appear to have been all young animals; the longest length of whalebone taken was only 2 m (24 May 1860, Ross 1985b: 168).

17 Although Whitehouse's journal states that at least six native boats took part in spring whaling at the floe edge off Miliakdjuin, it is not clear whether Inuit-owned whaleboats or *umiat* are meant.

18 For example, in October of 1861 the *Hannibal* was towed to Arctic Harbour at Tuapait in the Kekerten Islands, hauled up above the low water mark, auctioned off, and dismantled by the crew of the *Daniel Webster* (Goldring 1984: 506). While most of the usable wood and metal on the ship was stripped quickly by local Inuit, its lead sheathing was scavenged for decades for the manufacture of lead bullets (Etuangat Aksayuk, personal communication, 1983).

19 Old Dartmouth Historical Society, Whaling Museum Library, Log. 771, *Antelope*, 22 January-22 February, 1860–61.

20 In this connection, whaling in 1859 was felt to be inferior to previous years, even though the *Daniel Webster* and *Hannibal*, at Kekerten in 1858–59, took 14 whales between them (Old Dartmouth Historical Society, 1859, No. T-1, p. 23).

21 Information in Starbuck (1964: 630–1) suggests that by 1877 the American schooner *Helen F.* had been stationed permanently in the Sound for 10 years.

22 The loss of many whaling ships to the American civil war effort also contributed to the marked decline in the number of whaling vessels visiting the Arctic after 1870.

23 Other American vessels that may have been stationed in or around Cumberland Sound include the *Isabella* and *S.B. Howes*, which were sent to Cumberland Sound in 1870 and 1875, but were wrecked there, respectively, in 1873 and 1884 (Hegarty 1959, Starbuck 1964).

24 A continuing demand for Arctic 'whalebone' on world markets, the declining availability of bowheads, and the loss of 30 Arctic whaling ships off Point Barrow, Alaska, in 1876 kept the price of baleen high for the rest of the decade and century.

25 According to Clark and Brown (1887: 18), in former years Cumberland Sound whales averaged 'about 120 barrels (@ approximately 50 gal. per barrel) each, the bull 100 barrels, the cow 140 barrels; but of late years they have been smaller and scarcer. The yield of bone is usually about 1,300 pounds to 100 barrels of oil.'

26 The Inuktitut word for Saturday in the Baffin Island dialect is *sivataqviq*, which means 'the time for getting biscuits (*sivak*)'.

27 In the 1860s Scottish ports began to send auxiliary-screw vessels to the Arctic. As steamers were faster than sailing ships, they could venture into waters where the latter could not. By the mid-1880s the efficiency of the steamer had all but exterminated most bowhead in waters west and north of Baffin Bay (Lubbock 1955).

28 The 340 belugas taken at the head of the Sound in 1892 by the *Aurora* probably represents an average or slightly above average catch for these years (Lubbock 1955: 425).

29 While Peck and Parker were running into opposition from the *angaqut*, their church 'went to the dogs'. During the late fall of 1894 the missionaries purchased a quantity of seal skins from the Inuit in order to build a small church. However, as the weather had been bad and the hunting poor throughout most of January, towards the end of the month a large pack of ravenous dogs devoured the tasty morsel in a feeding frenzy (Lewis 1904: 225). Undeterred that the first church in the eastern Canadian Arctic was eaten by dogs, Peck rebuilt the facility and used it until a more substantial wooden building was acquired.

30 Anglican Church of Canada, General Synod Archives (ACC): Peck Papers, M56–1, Series XXXVII, No. 4, 4 November 1900.

31  ACC: Peck Papers, M56–1 Series XXVVII, 1899.
32  Ibid., No. 4, 19 October 1900.
33  Ibid., 7 December 1900, 21 December 1900; No. 5, 4 February 1904.
34  Ibid., 9 December 1900. Later in the decade, the phonograph (ACC: Peck Papers, M56–1 Series XXXVII, No. 5, 24 December 1903) and movie projector (Kudlu Pitsualuk, personal communication, 1988) were used by the missionaries to 'bring home to the people's minds the truths already attended to' (ACC: Peck Papers, M56–1 Series XXXVII, No. 5, 21 March 1904).
35  Ibid., No. 4, 14 and 15 January 1901.
36  Ibid., 16 July 1901, 3 and 4 Sept. 1901.
37  Ibid., 4 April 1901.
38  Ibid., 17 May 1901; (also Peck 1922: 34).
39  Ibid., No. 5, 29 November 1903, 13 Dec. 1903.
40  Ibid., 21 February 1904, 28 March 1904.
41  ACC: Peck Papers, M56–1, Series XXXVII No. 5, 2 March 1904.
42  Ibid., 28 May 1904.
43  Ibid., 30 June 1904.
44  Ibid., Series XXXV No. 8, 12 February 1905.
45  Ibid., Series XXXVII No. 4, 26 November 1904
46  Ibid., Series XXXV No. 8, 9 December 1904.
47  Ibid., Series XXXVII, 1899, 9 December 1900.
48  Ibid., Series XXXVII No. 4, 30 January 1900.
49  For instance, the *Heimdal* and *Jantina Agatha*, both supply vessels, were lost in the Sound, in 1905 and 1909 respectively.
50  As Cumberland Sound had been without a missionary for four years, Greenshield was sent temporarily to Blacklead Island in 1909.
51  Pawla was the son of a former American station manager at Umanaqjuaq, Paul Roche, and brother-in-law of Ittirq, another 'sort of a foreman in the service of the trading station'. Pawla learned English in America and returned to Blacklead after his father's death. 'Shrewd and vigorous, 40 years of age', Pawla managed the post here, directed its whaling operations, and 'bargained for furs on behalf of the whites at Kekerten' (Hantzsch 1977: 53, 93, 94).
52  HBCA B455/a/1, 6 February 1922.
53  HBCA B455/a/3, 22 December 1922.
54  HBCA B455/a/1, 6, 13, 16 February 1922.
55  Ibid., 22 February 1922.
56  Ibid., 1 December 1921.
57  PAC RG85/610, file 2712, 5 March 1923, Greenshield to Finnie.
58  HBCA RG3/26B/8, p. 4, Stewart, 1939.
59  PAC RG18 Acc. 85–86/1048, file. TA 500–8–1–11, 19 June 1957, Barr to Officer Commanding (Off. Comm.) 'G' Division.

60 PAC RG85/1044, file 540–3 [3B], 15 April 1925, Treadgold; Qatsu Eevic, personal communication, 1984; PAC RG85/1069, file 25–1, 30 October 1924, Burwash to Finnie.
61 PAC RG85/1069, file 25–1, 30 October 1924, Burwash to Finnie.
62 This method of indenturing hunters often resulted in bitter feelings. For example, in some cases, although enough foxes were taken to pay for the boat, title was not given to the hunter. In others, the trader (Nichols) destroyed old whaleboats belonging to such prominent hunters as Veevee and Keenainak, only to make them pay for the new ones. In addition, the trader indentured Inuit by taking up to five years to deliver promised whaleboats, as he did with Attaguyuk. Finally, while the hunter remained in debt to the trader, the latter could 'expel any member of the crew from the boat' he wished (PAC RG85/069, file 252–1, pt.1, 1 February 1925; RG85/771, file 5410, 20 August 1927).
63 PAC RG85/755, file 4687; vol. 64, file 164–1 [1]; vol. 1044, file 540–3 [3], 31 July 1925.
64 PAC RG85/815, file 6954.
65 PAC RG85/1084, file 401–2, pt.1, 4 June 1938, McKeand to Gibson.
66 PAC RG85/815, file 6954 [3], MacKinnon to Turner, 14 September 1936, p.9; HBCA RG 3/26 B/23, 'Annual Report Pangnirtung Post, Outfit 270', Stewart.
67 PAC RG85/815, file 6945 [2], Medical Report 1934, p. 17.
68 PAC RG85/1084, file 401–1, pt. 1, 28 February 1938, Livingstone.
69 PAC RG18 Acc. 85–86/1048, file TA-500–8–1–11, 1 April 1953, Daoust.
70 HBCA A97/6 'Report on Visit to Pangnirtung, 1927–28', Milling, pp. 11, 47, 85–6.
71 Ibid., p. 18.
72 Ibid., p. 15.
73 PAC RG85/815, file 6954 [1], 9 February 1926, Livingstone to Finnie.
74 PAC RG85/1045, file 540–3, pt.3-c, 5 August 1937, McDowell to Off. Comm. 'G' Div.
75 PAC RG85/815, file 6954 [1], 23 October 1923, Livingstone to Craig.
76 PAC RG85/64, file 164–1 [1], 20 September 1926, General Report of J.E.F. Wight.
77 HBCA RG 3/26B/8, p. 3, 'Annual Report Pangnirtung Post, Outfit 269', 28 July 1939.
78 PAC RG85/2147, Interim box 2, 21–24, 26 September 1946, Wight.
79 PAC RG85/1044, file 540–3 [3B], Sick and Destitute Eskimo: report 21, 5 April 1930, Petty.
80 PAC RG85/1044, file 540–3 [3A], Wight to 'HQ' Division, 31 March 1925.
81 While leaders of the whale hunt would sometimes receive a whaleboat and a rifle after a successful season (Pitsualak 1976: 24), Etuangat Aksayuk (personal communication, 1988) has described the bonus system in effect at Kekerten during the second decade of this century. At the conclusion of the spring whaling season and before the Inuit left for their annual caribou hunt, each woman of the settlement would choose, in order of her husband's productivity during the previous year, articles from a stockpile of provisions laid out by the station manager.
82 PAC RG85/1044, file 540–3 [3B], Sick and Destitute Eskimo: report 12, 31 January 1930, Petty.
83 Ibid.: report 16, 30 January 1930, Petty.
84 Ibid.: report 5, 31 January 1930, Petty.

85 PAC RG85/1044, file 540–3 [3], 31 July 1925, Wight to 'HQ' Division.

86 Ibid., file 540–3 [3B], Sick and Destitute Eskimo: report 16, 30 January 1930, Petty.

87 Ibid., file 540–3 [3], 31 July 1925, Wilcox to 'HQ' Division.

88 Ibid., file 540–3 [3B], 30 June 1930, Petty to 'HQ' Division.

89 PAC RG18 Acc.85–86/048, file TA-500–8–1–11, 1 April 1953, Barr to Off. Comm.

90 Ibid., 8 July 1954, Johnson to Off. Comm.

91 Ibid., 1 January 1955, Annual Report, Johnson. (Family allowances became an important source of relief after 1948.)

92 Ibid., 18 January 1958, Barr to Off. Comm. 'G' Division. Other sources contributing to this sketch of Cumberland Sound Inuit society include PAC RG18 Acc.85–86/048, file TA-500–8–1–11, 1 April 1953, Daoust; 8 July 1954, Johnson; 10 March 1956, Johnson.

93 Adult ringed seal skins went from $1.50 in 1955 to $12.25 in 1963, while 'silver jars' rose from $4.00 to $17.50 during the same period (Haller et al. 1966: 87).

94 PAC RG18 Acc. 85–86/048, file TA-500–8–1–11, 24 March 1960, Nazar to Off. Comm 'HQ'. Division; 8 March, 3 May 1962, Alexander to Off. Comm. Eastern Arctic.

95 PAC RG18 Acc. 85–86/048, file TA-500–8–1–11, 30 January 1966, Grabowski to Off. Comm. Eastern Arctic Subdivision.

96 In 1958–59 the kill of ringed seals in Cumberland Sound far exceeded the number traded. However, by 1963, virtually all seal skins were traded (Haller et al. 1966: 90).

97 PAC RG18 Acc. 85–86/048, file TA-500–8–1–11, 30 January 1966, Grabowski to Off. Comm. Eastern Arctic Subdivision.

98 Native products contributed to only 41.9 per cent of the annual income of eight camps in the Clyde River area in 1965–66, while household incomes ($876.00) averaged only half that in Cumberland Sound (Haller et al. 1966: 197).

99 PAC RG18 Acc. 85–86/048, file TA-500–8–1–11, 22 January 1968, Grabowski to Off. Comm. Eastern Arctic Subdivision.

100 Ibid., 7 January 1969, Nowakowski to Off. Comm. Eastern Arctic Subdivision.

101 Mayes' (1978) *The Creation of a Dependent People: The Inuit of Cumberland Sound, Northwest Territories* should be consulted for a fuller treatment of the processes of culture change to which the Cumberland Sound Inuit were subject after 1967.

## Chapter 4

1 For example, in the hypothesized demise of capitalist society, the socialization of productive forces will no longer permit the structure of social relations based on private property to operate.

2 Although wood and metal quickly replaced many aboriginal raw materials, as late as the mid-1950s wood was still hard to come by. For example, during the winter of 1954–55 several Inuit travelled from Cumberland Sound to Frobisher Bay in order to obtain this material for their *qamutit* (PAC RG Acc. 85–86, file TA-500-8-1-11, 1 January 1955, Annual Report of H.A. Johnson).

3 In contrast to the bow hunter, who could not stalk caribou over snow because of the sound made by his footsteps, the rifle hunter was not as concerned with this problem as he could now procure caribou from much greater distances (Balikci 1964: 48).

4 Even though caribou hunting became less seasonal in character after the introduction of the rifle, with the shortening of the autumn caribou hunt and the need to provide clothing for wintering whalers, it too may have intensified temporarily in the late 1850s and 1860s.

5 If the positive correlation between *umiat* and single family dwellings at Kingmiksoo in 1846 is any indication, single family residences may, in fact, have been a symbol of productivity and economic independence, and thus a desired objective, in aboriginal society (see Figure 14).

6 PAC RG85/815, file 6954 [1], 23 August 1935, p. 12, MacKinnon to Turner, Northwest Territories (NWT) and Yukon Branch.

7 Cumberland Sound was not the only region where whaleboats and similar craft (e.g., Peterheads) played a positive role in maintaining traditional authority patterns and kin relationships within extended family units. The same process seems to have occurred among the Iglulingmiut of northern Hudson Bay (Damas 1963: 157–9) and the Puvirniturmiut of eastern Hudson Bay (Balikci 1964: 98).

8 By 1915, whaling crews at Kekerten were organized by Angmarlik and three or four other *aggutit*. With Angmarlik's approval, each *aggutik* selected his own crew, most of whom were related to him in some way or another (Etuangat Aksayuk, personal communication, 1988).

9 PAC RG85/815, file 6954 [3], 14 September 1936, MacKinnon to Turner, NWT and Yukon Branch.

10 In particular, it seems that more than half the Inuit in Pangnirtung trace their relationship to either Etuangat Aksayuk or the late Jimmy Kilabuk, the two patriarchs of the community. While Etuangat originated from Kekerten, Kilabuk came from Umanaqjuaq.

11 HBCA B455/a/3, Pangnirtung (Netchilik) Post Journal Diary, 1, 6, 21, 26, March 1924.

12 PAC RG85/775, file 5648, 30 April 1929, Petty to 'HQ' Division.

13 Pangnirtung's first trader, J.W. Nichols, described Kanaaka as treating his men as 'servants (who were) issued rations once or twice a week'. But Nichols used the term in the HBC sense, i.e., as an endentured employee, implying moderate status and security (Goldring 1986: 186).

14 PAC RG18 Acc. 85–86/048, file TA-500-8-11, 26 February 1959, Barr to Off. Comm. 'G' Division.

15 This man is probably the same 'Pakak' to whom Mathias Warmow took exception in 1857. In the fall of that year, a 'very conceited' man in command of an *umiak* of seven people came to trade 'whalebone' with Penny. This man referred to himself as 'Captain Pakak' and made such an unfavourable impact on Warmow (1859: 89–90) that the latter wrote, 'I thought this was indeed a wonderful captain ... still he appeared by no means stupid. But I never had so bad an impression of any Esquimaux as of this man. In this opinion, our Captain quite agreed with me, although this man was treated with much attention on board, probably because he was skilful in the whale fishery.' Interestingly, Pakaq appears to have been Boas' principal informant.

16 PAC RG 85/610, file 2712, 5 March 1923, Greenshield to Finnie.

17 PAC RG85/1044, file 540–3 [3B], 23 April 1936, McDowell to Off. Comm. 'HQ' Division.

18 Ibid. [3A], 1 November 1927, Dunn to 'HQ' Division.

19 Etuangat Aksayuk, personal communications, 1983, 1988; PAC RG85/815, file 6954 [1], 21 Jan. 1935, 23 August 1935, MacKinnon to Turner, NWT and Yukon Branch.
20 PAC RG85/1044, file 540–3 [3A], 31 October 1928, Petty to 'HQ' Division; 85/815, file 6954 [4], 21 August 1931, Stuart to Finnie.
21 PAC RG85/815, file 6954 [3], 14 September 1936, p. 9, MacKinnon to Turner, NWT and Yukon Branch.
22 PAC RG85/799, file 6615–1, 6 September 1936, MacKinnon to Turner, Off. Comm. 'HQ' Division.
23 ACC: Peck Papers, M56–1, Series XXXVII no. 5, 24 February 1904.
24 Leadership in the commercial whale fishery was not without its drawbacks as headmen, having greater contact with wintering whalers, may have been more prone to foreign diseases than the average individual. In this regard, it was probably not fortuitous that disease carried off the 'chief' of Naujateling and seven of his kinsmen over the winter of 1853–54 (see previous chapter).
25 PAC MG30 D123 'An Arctic Diary, Being Extracts from the Diaries of the Rev. Edgar Greenshield', 20 November 1909.
26 At the same time, the missionaries recognized that the more productive members of society were also the most influential. As such, successful hunters such as Niaqutsiak in 1857 (Warmow 1859) and Tooloogakjuaq in 1903 were selectively chosen by the Christian authorities to lead their people in religious instruction.
27 PAC RG85/1044, file 540–3 [3A], 31 October 1928, Petty to 'HQ' Division.
28 Ibid., 31 January 1925, Wight to 'HQ' Division.
29 PAC RG85/609, file 2704, 'Home Bay Murders 1922–25'.
30 Belief in the polluting effect of human blood began to diminish sometime after 1900 when women were no longer banished to separate dwellings to give birth or menstruate. At Kekerten this change took place sometime between Qatsu Eevic's birth in 1897–98 in an secluded snowhut and Etuangat Aksayuk's birth in 1906–07 in the comfort of his parents' *qammaq* (personal communications, 1983, 1988).
31 PAC RG85/1044, file 540–3 [3A]; 20 July 1924, Wilcox; 31 January 1925, Wight; to 'HQ' Division.
32 PAC RG85/815, file 6954 [1], Medical Report for 1934.
33 While only traditional names were recycled in the past, Christian names (e.g., Markoosie, Peterosie, etc.) are also now passed on.
34 PAC RG85/815, file 6954 [3], 1 September. 1938, p. 8, Orford to Turner, NWT and Yukon Branch.
35 Ibid., 14 September 1936, p. 10, MacKinnon to Turner, NWT and Yukon Branch.
36 PAC RG85/815, file 6954 [3], 1 September 1938, Orford to Turner, NWT and Yukon Branch.
37 PAC RG85/1044, file 540–3 [3B], Sick and Destitute Eskimo: report no. 27, 30 May 1930, Petty to 'HQ' Division.
38 PAC RG18 Acc. 85–86/048, file Ta-500-8-1-11, 1 April 1953, Daoust to Off. Comm. 'G' Division.
39 After the death of Kingudlik in 1932, Etuangat Aksayuk appears to have been torn between returning to his mother's (and step-father's) camp in Cumberland Sound, or

continuing to provide for his wife's relatives at Padloping Island. It was not until his first wife died that he finally moved permanently to Cumberland Sound, where he was hired by St Luke's Anglican Mission hospital as the medical officer's assistant and driver. Interestingly, however, Etuangat's obligations to his deceased wife's relatives did not terminate with her death; he married her sister after the latter's first husband died (personal communications, 1983, 1988).

40 It is interesting that of all the first ascending generation kinship terms found among the Central Inuit, only those of the consanguineal cross-relations, i.e., MB (*angak*) and FZ (*attak*), are universal.

41 As noted previously, the merging of affines with consanguines in the first ascending generation is consistent with systems of cousin marriage, especially those sanctioning cross-cousin marriage whereby dual exogamy was practised (Levi-Strauss 1969). In such systems, consanguineal aunt-uncle terms are extended to affines as the latter would have assumed the roles of one's parents' cross-sex siblings who lived among the other group. Yet, cousin marriage was always forbidden in Oqomiut society.

42 With the introduction of health services by the Anglican missionaries, first at Umanaqjuaq and later at Pangnirtung, life expectancy increased. Concomitantly, third ascending and descending generation terms were adopted, where before they were merged with second generation terms. Thus, *amauk* was adopted for great-grandfather and great-grandmother, while the term *illulik* (great-grandchild) became its reciprocal.

43 Kendall Whaling Museum Log No. 111, *Milwood* 1867–68, April 1868 (cited in Goldring 1986).

44 PAC RG85/64, file 164–1 [1], 3 March 1925, Burwash to Finnie.

45 PAC RG85/1044, file 540 3 [3A], 31 October 1928, Petty to 'HQ' Division.

46 That age or seniority was a particularly important device in structuring social behaviour during the twentieth century was recently demonstrated to me during a whale hunt in Pangnirtung Fiord. The eldest man in the hunt probably did not shoot the whale. He may not even have been related directly to the hunter(s) who did. Yet, he helped himself to the choicest pieces of *maaqtak*, while overseeing the butchering and distribution of the rest of the skin and blubber (Figure 1).

## Chapter 5

1 As just one example, even though the settlement of Etelageetok was occupied for only one winter in 1930, Kudlu Pitsualuk (born in 1903) was able to recall not only the exact number of occupants at this site—as substantiated by RCMP records (PAC RG85/64, file 164–1 [1], 30 June 1930, Petty to Off. Comm. 'HQ' Division)—but their names, approximate ages, and primary kinship connections as well.

2 For example, Nichols took serviceable whaleboats from Veevee and Keenainak, which he used for firewood. While Nichols promised them new boats, he also forced them to pay for these craft. In another case, Nichols took five years to deliver a promised whaleboat to Attaguyuk, 'who thought the boat itself was payment for work but who ... (in 1927, was still) paying foxes down on it...' (PAC RG85/771, file 5410, 20 August 1927, Friel to 'HQ' Division).

3 Principal informants for Nunaata were Elija Keenainak and Jamasie Mike, while translators were Meeka Mike, July Papatsie, and Simionee Akpalialuk.

4 The association between leaders and whaleboats was so well recognized that one elder felt that a boat was the only qualification one had to have to become a leader (Koraq Akulujuk, 'Pangnirtung Interviews', 1984, p. 12).

5 This assumes that households A, B, and C formed the economic centre as well as social centre of the camp.

6 For example, Pangnirtung (Netchilik) Post Journal Diaries, HBCA B455/a/1, 31 March 1922; B455/a/2, 10 February 1923; B455/a/5, 23 May 1925; B455/a/6, 5 March 1926; B455/a/7, 21 February 1927.

7 Indeed, Boas (1883–84) noted that the old sod dwellings here (Bon Accord and Anarnitung) each had individual names, implying a sense of permanency, substantial construction, and exclusive ownership.

8 Principal informants for Idlungajung were Qatsu Eevic, Etuangat Aksayuk, Pauloosie Angmarlik, Jamasie Mike, and Livee Koodlooalik, and translators included Meeka Mike, July Papatsie, Meeka Kilabuk, and Moe Keenainak.

9 PAC RG85/815, file 6954 [1], 23 August, 1935, MacKinnon to Turner, NWT and Yukon Branch.

10 For example, HBCA Pangnirtung (Netchilik) Post Journal Diaries B455/a/3, 24 December 1923, 25 January 1924; B455/a/6, 21 July 1925, 7 May 1926.

11 HBCA Pangnirtung (Netchilik) Post Journal Diaries B455/a/1 through 12.

12 For example, HBCA Pangnirtung Post Journal B455/a/8, 12 January, 20 February, 1928.

13 Ibid., B455/a/7, 23 March 1927, 29 April 1927; B455/a/8, 12 January 1928; B455/a/9, 22 May 1929.

14 At Kekerten, Angmarlik owned the largest *qammaq* and cache. And, although his house was located at the back of the settlement near a hill, no one lived between him and the shoreline, for when he set out to go hunting he could not restrain his dogs from trampling everything in their path (Etuangat Aksayuk, personal communication, 1983).

15 Informants for Avatuktoo were Mary Batte, Kannea Etuangat, and Jamasie Mike, while July Papatsie served as translator.

16 PAC RG85/815, file 6954 [1], 21 January 1935, MacKinnon to Turner, NWT and Yukon Branch.

17 HBCA RG 3/74B/10, 'Summary of Events for January, 1943'.

18 Principal informants for Tuapait were Michael Kisa and Jaco Koonooloosie, with July Papatsie serving as primary translator.

19 Principal informants for Sauniqtuajuq include Towkee Maniapik, and Evee and Peter Anaaniliak. July Papatsie and Simionee Akpalialuk were translators.

20 PAC RG85/1044, file 540–3 [3B], Petty to 'HQ' Division, Sick and Destitute Eskimo, 31 January 1930.

21 After a decade of living in Pangnirtung as the 'most valuable native in the employ of the Hudson's Bay Company', Veevee left for Nettilling in the fall of 1934 to trap, along with four other families, including those of his sons (PAC RG85/815, file 6954 [1], 1 August 1935, MacKinnon to Turner, NWT and Yukon Branch).

22 HBCA B455/a/11, 19 March 1932.

23 Principal informants for Naujeakviq were Etuangat Aksayuk, and Evee and Peter Anaaniliak. July Papatsie was the primary translator.

24 PAC RG85/876, file 8839, 10 January 1936, McDowell to Off. Comm. 'G' Division.

25 The principal informant and translator for Keemee were Jaco Eevic and July Papatsie, respectively.

26  For example, HBCA B455/a/11.
27  Principal informants for Ussualung include Jamasie Mike and Simon Shamiyuk, while primary translators were July Papatsie and Simionee Akpalialuk.
28  PAC RG85/7624958 (William Duval, General Correspondence).
29  PAC MG30 D123, 6 March 1910, 'An Arctic Diary, Being Extracts from the Diaries of Rev. Edgar Greenshield'.
30  PAC RG85/1044, file 540-3 (3B), Petty to 'HQ' Division, 31 January 1930, Sick and Destitute Eskimo.
31  Principal informants for Iqalulik were Annie Alivaktuk and Towkie Maniapik, while translators included July Papatsie and Simionee Akpalialuk.
32  Principal informants for Kingmiksoo include Annie Alivaktuk, Pauloosie Nowyook, and Charlie Akpalialuk. Principal translators were July Papatsie and Simionee Akpalialuk.
33  PAC RG85/1044, file 540-3 [3B], 20 April 1924, Patrol Report of C.E. Wilcox.
34  Conversely, the correlation between census data and informant memory for those camps discussed above is considerably greater, approaching 80 per cent or more. This discrepancy is owing not to informant memory as my informant was unusually clear on other issues, but probably to other, more fundamental factors (see next chapter).
35  At Umanaqjuaq, 'men would get together outside as to where they were to go hunting and discussed other things to be decided as a community' (Kudlu Pitsualuk, 'Pangnirtung Interviews', 1984, p. 13). Conversely, at Kekerten, everyone, including the more substantive members of the community followed Angmarlik's instructions faithfully (Etuangat Aksayuk, 'Pangnirtung Interviews', 1984, p. 23).
36  PAC RG85/1044, file 540-3 [3B], 23 April 1936, McDowell to Commanding Officer.
37  Principal informants for Opinivik include Tashugaq Nakashuk and Pauloosie Nowyook. July Papatsie and Margret Karpik were translators.
38  Principal informants for Kipisa were Tashugaq Nakashuk and Pauloosie Nowyook, while translators were July Papatsie and Margret Karpik.
39  Principal informants for Illutalik include Kudlu Pitsualuk and Pauloosie Nowyook. July Papatsie and Meeka Mike assisted as translators.
40  Kudlu Pitsualuk, 'Pangnirtung Interviews', 1984, p. 12.
41  The principal informant and translator, respectively, for Seegatok were Pauloosie Nowyook and July Papatsie.
42  HBCA B455/a/7, 1 and 12 April 1927.
43  The principal informant for Etelageetok was Kudlu Pitsualuk with July Papatsie and Meeka Mike serving as translators.
44  Kudlu Pitsualuk, 'Pangnirtung Interviews', 1984, p. 4.

## Chapter 6

1  Etuangat Aksayuk, 'Pangnirtung Interviews', 1984, p. 21.
2  HBCA B455/a/7; 1, 12 April 1927.
3  In addition to statements made by my informants, this observation finds support in the fact that Kipisa was not abandoned until 1984, i.e., 14 years after all other camps in the Sound had relocated to Pangnirtung.

4 HBCA B455/a/1, 2, 6, 7, 8, 9, 11, 14, numerous entries.

5 Ibid. /a/7, 28 January 1927, 12 February 1927; 455/a/11, 18 February 1932, 21 May 1932, 23 March 1935.

6 PAC RG85/815, file 6954 [3], 14 September 1936, MacKinnon to Turner.

7 PAC RG85/1044, file 540-3 [3B], 1 November 1927, Patrol Report of O.J. Petty.

8 Ibid., file 540-3 [3B], 16 October 1931, Extract from Patrol Report.

9 Ibid., file 540-3 [3B], 15 April 1925, Patrol Report of T.H. Tredgold.

10 Ibid., file 540-3 [3B], 23 April 1936, McDowell to Comm. Officer.

11 PAC RG85/815, file 6954 [3], 20 June 1939, McKeand to Gibson.

12 Kudlu Pitsualuk, 'Pangnirtung Interviews', 1984, p. 13; also Annie Alivaktuk, p. 9; Simon Shamiyuk, p. 10.

13 Etuangat Aksayuk, 'Pangnirtung Interviews', 1984, pp. 22-3.

14 Ibid., p. 23.

15 Qatsu Eevic, personal communication, 1984.

16 Nowyook Nicketimoosie, 'Pangnirtung Interviews', 1984, p. 10.

17 Koraq Akulujuk, 'Pangnirtung Interviews', 1984, p. 11.

18 'Pangnirtung Interviews', 1984, Kunugusiq Nuvaqiq, p. 9; Simon Shamiyuk, p. 10; Koraq Akulujuk, pp. 11-12.

19 'Pangnirtung Interviews', 1984, Martha Kakee, p. 13; Etuangat Aksayuk, p. 22; Malaya Akulujuk, p. 10.

20 PAC RG85/1044, file 540-3 [3B], 31 October 1931, Petty to Headquarters.

21 Ibid.

22 Simon Shamiyuk, 'Pangnirtung Interviews', 1984, pp. 3-4.

23 Annie Alivaktuk, personal communication, 1989.

24 PAC RG815/6954 [3], 1 September 1938, Orford to Turner, NWT and Yukon Branch.

25 Kudlu Pitsualuk, 'Pangnirtung Interviews', 1984, p. 10.

26 Clearly, the break-up of Kingmiksoo after the death of Tooloogakjuaq was owing to the fact that people exercised this option, not to the possibility that the egalitarian structure of productive relationships within this camp could not be maintained under such circumstances.

# Chapter 8

1 Most reconstructions of Central Inuit social organization, despite being based on informant recall, are assumed to describe the aboriginal or precontact period. In other words, no significant organizational changes are thought to have occurred subsequent to contact and prior to documentation. While few regional populations had as intense an association with Qallunaat as did the Cumberland Sound Inuit, we cannot assume *a priori* that other groups experienced fewer modifications as a result of contact; the complex nature of Inuit-Qallunaat interaction needs to be assessed for each region. Nonetheless, the intent of most early observers of Inuit societies was to document traditional lifeways and customs. Thus, the following descriptions are considered to be characteristic of aboriginal social organization.

2 Parry (1969: 549) reported four other major sites and several lesser camps outside of the Iglulik area which gave him reason to believe that another three or four hundred may have belonged to the Iglulingmiut 'tribe'. If so, the aboriginal population of the Iglulingmiut may have numbered 500 to 600.

3 Regrettably, while Parry (1969: 500) mentions that these features were usually occupied by several related families, he does not describe their kin relationships. Fortunately, Mathiassen indicated to Damas (personal communication to Damas 1963: 103) that such features in the 1920s were usually occupied by 'two or more families which were in some way related to each other, mostly parents and children'.

4 Indeed, Parry (1969: 519) infers that the average number of dogs per team among the Iglulingmiut amounted to six or seven, though he made reference to much larger dog teams of 15 or more on several occasions—a sure sign of prosperity.

5 However, the Iglulingmiut of this region had for some time prior to 1920 used a skin-shelled craft modelled after the whaleboat, but smaller in size (Damas 1963: 22).

6 While Manning (1943) figured the population of the Iglulik 'Eskimo' to be around 540 in 1821–23, Mathiassen's (1928: 15–21) census in 1921–22 put the number at 504.

7 I have omitted, for brevity, discussion of other modifications of Iglulingmiut culture. However, in regard to changes in house forms brought about the increased use of wood, it is noteworthy that in 1960 the internal arrangement of the household and its major source of heat, the *qudlik* or seal oil lamp, remained unchanged from the precontact period.

8 In 1846 Sutherland (1852: 229) encountered at Kingmiksoo in Cumberland Sound two men from Iglulik and a third 'who had come all the way from Pond's Bay'.

9 The Cumberland Sound Inuit apparently added another layer to this symbolic ordering whereby the drummer/singer was encircled first by adult men, then unmarried women, and finally married women, with children seated near the door (Boas 1964: 192–3).

10 The implications of this development are discussed in a latter chapter.

11 The splitting of male sibling groups has been looked upon as a structural defect. However, its positive aspects must not be overlooked. Specifically, the splitting of kin groups along male sibling lines may have served socioeconomic functions in areas with occasional variable resource distributions by dispersing people across the landscape. In this regard, split sibling groups may not be so much a structural deficiency as a structural solution. While the productivity of the Iglulik region in recorded times argues against this interpretation, it is possible that the local environment may not have been as abundant in the past.

12 As noted in Chapter 5, name sharing and joking partnerships existed in Cumberland Sound during the contact-traditional period. In addition, Boas (1964) documented the occurrence of trading and spousal exchange partnerships, though the latter appears to have occurred mostly within the context of the 'Sedna' ceremony, rather than as a formal basis upon which to establish socioeconomic alliances at other times.

13 If so, here we have a case whereby kinship behaviours have changed without an attendant alteration in terminology. The possibility that most Central Inuit groups were derived originally from a culture with different organizational principles (including prescriptive or positive marriage rules) will be considered in a following chapter.

14 In 1902 George Comer, working for Franz Boas (1907: 377–8), estimated that the Netsilingmiut numbered about 446, of which 257 were males and 189 were females—

an estimate that surely includes those individuals who emigrated from the western Netsilik district to the west coast of Hudson Bay during the late nineteenth century.

15 The Netsilingmiut did not possess any ocean-going kayaks nor a sea mammal hunting technology (e.g., throwing boards, throwing harpoons, floats, etc.). Interestingly, the Netsilik, 'people of the seal', apparently received their name, not because of the abundance of seal in the region—which there is not—but from the Back River Inuit with whom they still had contact. As Rasmussen (1931: 85) observed, they received the name from these people because, 'after a life in the interior, they have for some reason or another separated from the Caribou Eskimos and moved to the coast.'

16 Note that this anecdote refers to the winter period when two or more extended *ilagiit* often assembled in large snowhouse villages to hunt seals co-operatively at breathing holes. The basis for co-operation among these otherwise antagonistic groups will be discussed below.

17 In her otherwise excellent treatment of the acculturative processes of the Central Inuit as reflected in the evolution of the 'Sedna' myth, Stone (1990) fails entirely to grasp the significance of this distinction.

18 As just one example, of the four boys and six girls born to woman no. 6 (Mangumagluk) on page 141 of Rasmussen (1931), four girls were put down.

19 In this respect, the apparent increase in female infanticide in the western Netsilingmiut district after people began to emigrate to Hudson Bay, might reflect an attempt, under intense demographic pressure, to maintain the integrity of adult male ties and the closed structure of local groups.

## Chapter 9

1 This behaviour is in marked contrast to that observed by Parry (1969: 524–5) among the Iglulingmiut. Not only did the latter apparently not reciprocate gifts, but they often displayed considerable ingratitude to gift-givers after accepting them.

2 An increase in the number of historic Copper Inuit sites south of Amundsen Gulf in the vicinity of the Melville Hills, relative to prehistoric and contact-traditional period sites, provides empirical support for Hickey's hypothesis (Stevenson 1992).

3 Though Stefansson (1919) suggested that this 'blondeness' was the result of intermarriage with lost Greenland Norse, later ethnographers ascribed this feature to genetic accidents and pathological causes, such as snowblindness.

4 Trapping appears to have been more readily adopted in the western Arctic than in the central and eastern Arctic, where the economy was more maritime based. In one year alone as many foxes were traded at certain posts in the western Arctic as for the entire central and eastern Arctic between 1900 and 1915 (Damas 1988: 107).

5 Damas (1988: 25–6), among others, believes local caribou shortages may have been more closely related to fluctuations intrinsic to the population dynamics of caribou. Even so, the Dolphin and Union Strait caribou herd appears to have abandoned Victoria Island as a summer calving grounds in the early 1920s (Usher 1965: 37), resulting in changes to the territories and names of some local groups (Rasmussen 1932).

6 In the region south of Queen Maud and Coronation Gulfs, winter camps were established inland at the loci of the largest caribou kills. Here, nets were set under the ice of nearby lakes and rivers, while trapping was carried out from this and surrounding cache locations. Depending on the success of the caribou hunt, the Inuit of this region usually returned to the coast for sealing as early as February or as late as April.

7 For example, a police patrol of the gulf in 1932 observed seven encampments ranging in size from 2 to 21 families (Damas 1988). While the introduction of seal hooks may have obviated, to some extent, the economic need to gather in large groups to hunt seal at breathing holes, aggregations of considerable size were nevertheless reported in the Dolphin and Union Strait area as late as 1949 (Harrington 1954: 260).

8 Elsewhere (Stevenson 1992: 93–5) I have documented differences in the organization of possible historic and contact-traditional Copper Inuit sites in the vicinity of the Melville Hills that are congruent with the decline of egalitarian relationships among nuclear families and the emergence of extended family relationships as the basic plan for the construction of social units.

9 Until this time, people lived at many different points inland south of Coronation Gulf where they subsisted throughout the year mainly on caribou (Damas 1988, Usher 1965: 160).

10 By the 1960s there was a marked reduction in the intensity of land use compared to 30 years before owing to increased employment opportunities in settlements, a decline in fur prices, and a greater dependence upon fish and maritime hunting as a result of declining caribou populations.

11 Copper Inuit intolerance of social inequality is reflected throughout the literature. However, an incident whereby two guides left a trader to freeze on the ice because he forcibly made them walk while he rode, stands out for its clarity and its comedy (Stefansson 1942: 432–3). It was this rejection of social inequality and injustice that undoubtedly lay behind the murder of several whites in Copper Inuit territory during the first quarter of this century.

12 Metayer (1973) rejects the observation that, in comparison with other groups, Copper Inuit myths and stories lacked structure. However, it is clear that from the evidence he presents, e.g., reference to incipient moiety formation, great spirits, and kayak sealing, that he has included numerous Mackenzie/Alaskan Inuit myths in his analysis.

13 As Jenness' and Rasmussen's infanticide data came from the western and eastern ends of Coronation Gulf, respectively, it is possible that the greater occurrence of this practice among the eastern Copper Inuit is, in part, a product of contact with or influence from the Netsilingmiut of King William Island.

14 The Mercy Bay depot was no longer used by Copper Inuit after about 1890, either because of depletion of its resources and/or the demise of the local musk-ox population (Hickey 1984). Whatever the case, the establishment of external trading relationships with the Dene during the early twentieth century may have continued to fuel the strict reciprocity and egalitarianism of the Copper Inuit at a time when the *Investigator* cache no longer held any attraction.

15 Asked to plot exactly where they camped, hunted, fished, etc. at specific times, or what route they took to get to where they were going, most Coppermine hunters and elders interviewed by the author could only point to a general area on a map, remarking that such details were simply unimportant to them (Stevenson 1992).

## Chapter 10

1 The Thule Inuit were not the only people to have penetrated Arctic Canada during the Neoatlantic. The Norse expanded into the north Atlantic during the same period, eventually reaching northwest Greenland, Ellesmere Island, and possibly Baffin Island.

2 For example, among the patrilocal, patrilineal Kachin of highland Burma (Leach 1964), bride-price circulates in the opposite direction of wives.

3 It is precisely the development of these features that, because of the inability of the technological base to maintain status differentiation between groups, leads to the overthrow of chiefly lineages among the Kachin (Leach 1964; Friedman 1974, 1984).

4 For example, during the 1830s social tensions in Cumberland Sound (Kingnait Fiord?) apparently forced Qitdlarssuaq and his band of 30 or so followers to embark on a migration that eventually took them to northwest Greenland (Mary-Rousselière 1991).

5 The existence of exogamous patrilineal clans among the historic Yupik 'Eskimo' of the Bering Strait region also supports this interpretation. With lineage reckoned from a common clan ancestor and marriages arranged among clans within villages, this type of organization hints at a form of asymmetrical exchange, while standing in contrast to the bilateral, extended family of Alaskan and most other Inuit groups (Harritt 1995).

6 As Royce (1982: 18) points out, 'in order for interaction to occur at all in multi-ethnic settings, there must be shared understandings and common conventions. This necessarily gives rise to ethnic stereotypes which are generalizations about the different groups they describe and which indicate appropriate attitudes and actions towards those groups.' Thus perceived, Dorset longhouses may have not only contributed to within-group solidarity, but they may have also enabled appropriate interactive behaviour to take place between the two ethnic groups (cf. Stevenson 1989).

7 The fact that affinal-including aunt-uncle terms endured in the absence of cousin marriage rules up to the present may have been an explicit demonstration of the maintenance of respect-obedience directives towards parents' siblings and their spouses.

8 In support of the greater movement and mobility of people that accompanied the development of this trading network in Alaska, Hickey (1979: 428) found a sudden nine-fold increase in transportation artifacts at sites on the Kobuk River. He also found significant, though less marked, increases in composite artifacts that would have been manufactured in 'assembly line' fashion as well as objects of personal adornment, i.e., items one would expect to differentiate trading locales and intergroup aggregations from other, more isolated contexts.

9 This is not to suggest that Morrison (1990) is incorrect when he proposes that the Iglulualumiut of Franklin Bay evolved directly from a Thule base, although the facts that the earliest occupation in Franklin Bay does not extend much further back in time than the early sixteenth century (1990: 106), and 'there is good evidence of trade in recent Franklin Bay assemblages' (1990: 112), would seem to warrant a different conclusion. Rather, what evidence there is suggests that the farther east one travels along the south shore of Amundsen Gulf, the greater is the likelihood that there will be no *in situ* development from Thule to later Mackenzie Inuit (Stevenson 1992).

10 Stefansson (1964: 320) had never seen ringed seals 'anywhere in such numbers as Darnley Bay', and the Alaskan father of an informant in Paulatuk considered Darnley Bay to be the best place for seals, fish, and caribou that he had ever seen (Stevenson 1992).

11 Morrison (1991) suggests that the trade for Coronation Gulf soapstone flourished only briefly between the 1840s and 1860s. However, early historic accounts cited by Morrison (1991: 242) more proximately place the florescence of this trade at around 1800.

12 A site containing two bilobate tent rings and dating to about 350 years ago on the south shore of Amundsen Gulf between Coppermine and Paulatuk may provide evidence of this process (Stevenson 1992).

## Chapter 11

1 Perhaps, this is the process that underlies the emergence of Bilateral-Dravidianate structures.

## Chapter 12

1 The Denendeh Proposal, in contrast to Nunavut, failed, in part, because it sought guaranteed protection of Dene rights through 30 per cent Dene representation in the National Assembly and a Denendeh Senate with veto power over National Assembly decisions adversely affecting Dene rights.

2 Ironically, in contrast to its Conservative predecessor, the Liberal government in Ottawa recognized the inherent right to self-determination in 1993, and requested that aboriginal groups bring their proposals for aboriginal self-government 'to the table'.

3 While communities were asked by the NIC about their ideas for government, there is no opportunity for the public to ratify the proposal—the decision will be made by the federal government, territorial government, and the Nunavut Tunngavik Incorporated (NTI), the organization formed to implement the land claims agreement.

# Index

**Note:**
Page numbers ending in 'f' indicate a figure. Page numbers ending in 't' indicate a table.

Abadjaq, 270
Aberdeen Arctic Company, 81
aboriginal hunting skills, 107
aboriginal self-government, and Nunavut, 335, 340
aboriginal whaling, 254; settlement, 201
Abraham Bay, 44
*achun*, 65
Adams, C., 23–4
Adelaide Peninsula, 282
Adla, 209
Admiralty Inlet, 266
adoption, 13, 20, 26–7, 244, 245t; in Avatuktoo, 169; common types of, 244; of grandchildren, 163, 274; in Idlungajung, 158, 163, 165; and Iglulingmiut, 268, 277; in Illutalik, 221, 223; in Iqalulik, 201; in Kingmiksoo, 203, 209, 210; in Kipisa, 218; in Naujeakviq, 188; and Netsilingmiut, 283; in Nunaata, 154; in Opinivik, 211; and Qikirtarmiut, 244–5; in Sauniqtuajuq, 177, 178, 181, 183–4; socioeconomic role of, 244–5; in Ussualung, 195
adult males: death rate, 67, 85, 261–2, 280, 332; ties, 270, 277
Agalik, 167–8, 195
agamous marriages, 297, 320
Aggakdjuk, 123
Aggomiut, 55
aggression: and Copper Inuit, 296–7; and Netsilingmiut, 282
*aggutik* (boat-steerer), 113, 116, 159; Attaguyuk, 176; in bowhead fishery, 160; relationship with *sivutik*, 114; Shorapik, 188, 199; *umiat* whaling, 254
*aglu* (breathing-hole), seals at, 172, 178, 259
*airaapik*, 65, 136
*aivik* (walrus), 39, 44
Aivilingmiut, 230, 254; and American whalers, 268–9; contact history of, 266–70; firearms, 280
Akatoogaq, 177
*akbeek* (bowhead whale), 33
Akilinik, 322; trade network, 323
*akka*, 7, 65, 213
Akpalialuk, 150, 195–6, 209–10, 216, 230, 238, 260
Akpalialuk, Charlie, 210
Akpalialuk, Jaypeetee, 234
Aksayuk, 131, 158–60, 185, 189, 198–9, 238

Aksayuk, Etuangat, xvii, 169, 227, 232, 243, 338, 373–4n. 39
Akulujuk, 203, 224
Akviqsurapiq, 172
Alaq, 178
Alaska: and American assimilation, 24; cousin terminology, 4; kinship systems, 5–6, 24; migration of Thule from, 46; Nelson Island social organization, 25–6; precontact tensions in, 316; social organization in, 15, 24–5; totemic groups, 6; warfare in, 58
Alaskan Inupiat, xvii
Alaskan/Mackenzie Inuit groups, 316–17
Aleut marriage region, 26
alliance formation, 13, 24, 277–8
*Alopex lagopus* (Arctic fox), 39
Alukie, 195–6, 210
American Harbour, 190
American whalers, 78, 81, 268
Amitoq, 52
Amitoq, 267, 269
ammunition, and camp bosses, 122
Amundsen, Roald, 280
Amundsen Gulf, 289, 303, 315–16, 318, 321
anarchy, concept of, 27–30
Anarnitung, 41, 43, 44, 48, 53, 118, 155; dwellings at, 252, 255, 258; feuding with Niutang, 56; headman at contact, 250; leadership at, 64; population estimates, 59–60, 63, 251
*angajuqqaq* (leader), 113; authority of, 114, 117; of Iqalulik, 198; of Kipisa, 218; origin of term, 119–20; at Tuapait, 173; at Umanaqjuaq, 205; at Ussualung, 196
*angajuqqaq* (leaders): Angutitaluk, 168; Jim Kilabuk, 151, 155f, 232; Kowna, 131, 137–8, 163; Pawla, 92, 121, 126, 179, 205, 220, 234; Pitsualuk, 220–4. *See also* Aksayuk; Angmarlik; Attaguyuk; Keenainak; Nowyook
*angak*, 158, 213
*angaqut* (shaman), 24–5, 64, 119; opposition to Anglican mission, 88–90; Uluksak, 292
Anglican Church, 88–9, 173, 270–1, 292; missionaries, 142; mission at Bernard Harbour, 294
Angmarlik, xvii, 48, 89, 92, 95, 125, 129, 131, 155f, 232, 238; and adoption, 244; at Idlungajung, 155–66, 227, 229, 235–6; and Christianity, 126–7; influence of, 121, 130,

205; kin relationships, 227, 229, 231; leadership of, 234; and spouses for kin, 134–5, 243; as trader, 118; and whaling crews, 116
Angmarlik Cultural Centre, xviii
Angmasalik, 3–4, 315
Angnalik, 206
Angnaqok, 177, 187, 197–9
*angutikattigiit* (children of two brothers), 9; and seal-sharing, 301
Angutitaluk, 168
animals: hunted, 39–45; souls of, 128–9
Anniapik (shaman), 64, 125, 250; daughter of, 132
anomalous bands, 14
*Antelope* (whaling ship), 81
anti-sealing campaign, xvii-xviii
Apiluk, 211
Aqungniq, 270
Arctic Bay, 270
Arctic char. *See* char (*iqaluk*)
Arctic cod, 42
Arctic fox, 39, 45, 91; at Iqalulik, 196; at Nunaata, 145–6; cyclical fluctuations in, 95; farm in Pangnirtung Fiord, 96; and health of Inuit in good seasons, 98; Imigen Island, 174; prices, 271; trapping, 97, 109, 110, 113, 270, 292
Arctic Gold Exploration Syndicate, 92–3, 190
*arngnakattigiit* (children of two sisters), 9
artifacts, whaling, 253
Arvertormiut, 282
*arvik* (bowhead whale), 33, 39, 40
Arviligjuarmiut, of Pelly Bay, 282–3
Asch, M.I., 131, 276, 327
Ashivak (wife of Angmarlik), 89–90, 127, 131, 158
Ashuluk, 150
*ataatasaq*, 154, 211
*atchunga*, 65
Athapaskan socioeconomy, 327
*aton* (dance), 299
Attaguyuk, 121, 129, 151, 174, 176–9, 181, 183–5, 187, 231
*attak*, 65, 136
Aukadliving, 54, 145
Aulaqeak, 167–8, 191, 194–6, 220
Aupalluktung, 54
authority, 139; of the *angajuqqaq*, 117; and decision-making, 233–4, 268; and leadership, 122, 276; patterns, 115; of whaleboat owner, 113; of white missionaries, 130
autumn houses, precontact, 252
autumn settlements, 75
Avatuktoo, 45, 59–60, 167–72, 245; marital residence pattern, 243
*avik*, 176
*avikgiit*, 176
Avingaq, 150
Avvaqmiut, 317

Back, George, 279–80
Back River, 282, 321–2
Baffin Bay whale fishery, 73, 269
Baffin Island, xvii, 33, 36, 65, 73, 191, 193; intertribal relations, 115, 304; Thule migration to, 314
Baffinlanders: mythology of, 36; totemic groups, 311
*Balaena mysticetus* (bowhead whale), 33, 39
baleen (whalebone), 40, 48; market crash, 92; in middens, 46–50; prices of, 82–3, 87, 91
Balikci, A., 8, 110, 281, 282, 283, 286
Banks Island, 290, 291, 298
Barron, W., 55–6
barter system, HBC, 99–100
basking seal hunting, 16, 272, 291
Bathurst Inlet, 292, 303
bearded seal, 39, 42
Beaufort whale fishery, 292
Beaver Dene, 331
Befu, H., 6, 310, 316
Belcher Islanders, 13–14, 28
beluga whale, 39, 43, 88; at Kingua Fiord, 156; at Milurialik, 109, 145, 262; at Ussualung, 190; Cumberland Sound, 365–6n. 2; fishery, 96; Iqalulik, 197; returns, 97t
Beringian trade network, 316–17, 330
Bering Sea region, 309, 317
Bering Strait, 311, 313, 327
Berkeley Point, 290
Bernard Harbour, 292, 294, 300
Beverly Lake, 322
Bilateral-Dravidianate, 327–8, 331
Bilby, J.W., 125
bilocal arrangement, 166
Binford, Lewis, 263
birds, as food source, 45
Birket-Smith, K., 3, 319
Birnirk culture, 309
Blacklead Island. *See* Umanaqjuaq (Blacklead) Island
Bloody Falls: massacre of Inuit, 289; murder of Catholic priests, 292
blubber: as food, 39; storage of *see* food storage
blubber-skin trade, 82, 85
Boas, Franz, xvii, 3, 32–3, 37–8, 44, 278; economic cycle 1880s, 85; feuding to 1820s, 55–6; hostility and aggression of Kingnaimiut, 256; land-use patterns, 50–8; on leadership, 63–4, 117, 119, 255, 274; on marriage, 65, 85; opposition from Nepekin, 121; on Oqomiut society, 231; population estimates, 58–9, 63, 74, 83–4; religious ideology, 125; salvage ethnography, 134; society, 132; whale-hunting methods, 47, 113–14, 254
boats, 214; at Kingmiksoo, 210; of Nowyook's son, 224. *See also* kayaks; *umiat*; whaleboats
boat-steerer (*aggutik*). *See aggutik* (boat steerer)
boat travel, Cumberland Sound, 37

Bon Accord Harbour, 40, 42, 53, 64, 155–6. *See also* Idlungajung
Boothia Gulf, 279
Boothia Isthmus, 282
Boothia Peninsula, 282
*Boreogadus saida* (Arctic cod), 42
bowhead whales, 33, 39, 40, 43, 109; at Akviqsurapiq, 172; at Bon Accord, 156; at Kingmiksoo, 201; Baffin Bay fishery, 73; decrease in, 81; Kingnait Fiord, 256; in Little Ice Age, 46–7, 70; population estimates, 42; prehistoric period, 252; and Thule economy, 46, 309
bow hunter, 108, 371n. 3
Bradby, B., 139
breathing-hole sealing, 16, 20, 52, 53–4, 172, 260, 277
bride-price, 65, 71, 132, 312, 332, 333
bride-purchase, 329
bride-service, 3, 65, 67, 71, 135, 243, 329, 365n. 2; in Avatuktoo, 168, 169; in Idlungajung, 156, 158, 165–6; Iglulingmiut, 273; in Kingmiksoo, 206; in Nunaata, 149–50; in Sauniqtuajuq, 183–4; Thule Inuit, 312
Briggs, Jean, 218, 220
British Navy, 289
British whale fishery, 73–5, 268
brothers, disagreements between, 276
Broughton Island (Kekertakjuaq), 174
Brown Inlet, 33–4, 39, 189, 201, 223
Burch, E.S. Jr, 24–5, 27, 56, 58, 319, 320–3
burial grounds, 313
burial populations, 50, 58
Burnside river, 322

Cambridge Bay, 290
camp bosses, 119, 121–2, 191, 211
Canadian Arctic Expedition, 292
*Canadian Charter of Rights and Freedoms*, 339
canvas tents, 101, 102
Cape Bathurst, 317
Cape Dyer, 73
Cape Hamilton, 289–90
Cape Haven. *See* Singnija
Cape Mercy, 44, 91
Cape Turnagain, 289
capital equipment, sharing practices, 129, 214
capitalist mode of production, 139
Cardno, 125, 134, 231
caretakership, of *inngutat*, 218
caribou, xvii, 39, 266; bones, 129; distribution of, 303; as food, 44, 112–13; near Nunaata, 145; precontact, 52–4; scarcity of, 295; in Thule Inuit culture, 321
caribou calving grounds, 293
Caribou Eskimos, cousin terminology, 5
caribou hunting, 36, 44, 85, 102, 109, 197, 272, 277, 280; Copper Inuit, 296

Caribou Inuit: and Euroamerican materials, 290; kin terminology, 314; precontact history, 316, 318–25
caribou skin clothing, 44, 101, 252, 295
cash, as medium for exchange, 139
Catholic Church, 270–1, 292; mission at Coppermine, 294
census: 1846 of Cumberland Sound, 59, 251, 253; 1846 of Kingmiksoo, 66, 68f, 69f, 262; 1927 of Kingmiksoo, 203, 235–6; 1949, 271; by Boas 1883, 84t; of Idlungajung, 235
Central Eskimo. *See* Central Inuit
Central Inuit: historical factors, 22–3; kin terminologies, 314; social organization, 23–4, 27–30, 325, 335; studies of, xvii, 16–23; topographic knowledge of, 303. *See also* Copper Inuit; Iglulingmiut; Netsilingmiut
char (*iqaluk*), 39, 44–5, 113, 145; at Avatuktoo, 167; at Iqaluit river, 211; land-locked, 196
charity, concept of, 129
Charlie (son of Akpalialuk), 210
Chesterfield Inlet, 269, 271, 280
Chidlak Bay, 33–4, 201
Chidlak Hills, 33–4; child betrothal, xvii, 3, 15, 65, 135; Belcher Islanders, 13; Central Inuit, 20–1; Copper Inuit, 300; Gilyak, 333; Iglulingmiut, 277; Netsilingmiut, 287
children: birth-order ranking of, 313; schooling of, 104, 271
Chipewyan groups, 322
cholera, at Naujateling Harbour, 76
Christianity, 72 and Copper Inuit, 294; and leadership, 90, 125–30; and Niaqutsiaq, 127
Churchill River, 322
Clachan phase, Thule culture, 315, 319
class, emergence of, 130–8
class structure, 131–3
Clearwater Fiord, 33, 43
climate: Cumberland Sound, 36–7; and migration, 322. *See also* Little Ice Age
climatic cooling: and economic shift, 47; and Thule Inuit, 321
climatic warming model, and Thule migration, 309
Clisby, Captain, 91
closed groups: and ecological necessity, 287–8; Netsilingmiut, 282–4, 286–7
clothing: caribou skin, 44, 113; store-bought, 101, 103; use of metal for, 108
Clyde Inlet, 304
Clyde River Inuit, kinship, 25
cold, western fear of, xvi
collective rationing system, of whalers, 99–100
collective soul, of seals, 129
Collinson (explorer), 290
commercial hunting, 109
commercial whaling, xviii, 32–3, 42, 43, 72, 118, 155, 205, 269; *aggutik* and *sivutik*, 113–14; as annual round, 109, 110; beluga,

90–1; and capitalist economy, 139; demise of, 142, 160, 292; inequality within, 123; and leadership, 235
common jar seals: catch, 98; price for, 102
communal houses, 257–8, 315, 332
communal living, 49, 112; in Little Ice Age, 46–7
communal longhouses, 315
community food sharing. See nekaishutu (community food sharing)
community government, Nunavut, 337
competition, among Iglulingmiut, 273–4
complex exchange system, Bering Strait, 329
complex marriage structure, 24, 328–34
composite bands, 14–16
composite bows, 108, 290
compositization, of settlements, 115–16
consanguineal solidarity, 331–4
*Constitution Act* (1982), 335
contact-traditional period: Copper Inuit, 291–5; definition of, 142
contact-traditional settlements, 260; Iglulingmiut, 272; locations of, 144–5, 146f
Coonook, 64
co-operation, in economic activity, 8, 12
co-operative hunting, 277; of bowhead, 49; caribou, 112; open water season, 149; of seal, 178; and whaleboats, 159, 270
co-operative socioeconomy, 252, 294
copper, as trade good, 317–18
Copper Inuit, xvii, 322; contact history, 289–95; cousin terminology, 330; distributions of, 267f; and egalitarianism *see* egalitarianism; homicide rate, 291; household organization, 17; impact of *Investigator* (ship), 290–2, 302; kinship, 278; kinship terminology, 275t, 301, 304, 314, 325; migration of, 313–14; murder of Catholic priests, 292; and Paleoeskimo culture, 325–6; precontact history, 316, 318–25; and soapstone trade, 318; social organization, xxi, 20, 301–3; socioeconomy, 16; theories of origins, 319–21; traditional society, 295–301; winter sealing, 293
Coppermine, 295, 303, 335
Coppermine River, 289, 321–2
Coral Harbour, 195, 230
core group structure: kin relationships, 227, 236; social relationships, 226–7, 229; Talirpingmiut and Kinguamiut, 251–3
Coronation Gulf, 290, 293, 302–3, 315–18, 320–1
Correll, T.C., 56, 58
country, use of term, 238
country food, 43, 101; caribou, 44, 112–13; *maqtaak*, xviii-xix, 43, 76; marine mammals, 39; in Nunavut economy, 338; seal meat, 103; sources, 39
cousin marriage, 17, 65–6, 311, 329; Copper Inuit, 297, 304; Labrador Inuit, 333–4; Netsilingmiut, 281, 283, 284–5, 286, 288

cousin system, 22
cousin terminologies, 4–6, 311–12, 314–15, 330; Copper Inuit, 18; Cumberland Sound Inuit, 3f, 67f; Iglulingmiut, 9f. *See also* cousin marriage; three-cousin system
Crawford Noble Company of Dundee, 81
credit: rejection of, 133–4, 139; and traders, 293–4; and traditional mode of production, 133–4
Crow-Omaha systems, 327
Cumberland Fiords, 33, 36
Cumberland Gulf Trading Company, 92–3, 155
Cumberland Sound, xvii-xviii, xxi, 40, 269; baleen trade, 48; climate, 36–7; communal dwellings in, 315; early whaling in, 74–7; Eenoolooapik's map, 41f, 61f; ice conditions, 38f; maps, 35f, 41f, 51f, 62f. population estimates, 83, 99t, *see also* specific villages; shipwrecks in, 81; site plans prehistoric villages, 57f; whale fishery decline, 81; whaling 1857–70, 77–81
Cumberland Sound Inuit: cousin terminology, 3f; kinship terminology, 32, 135–6, 279; winter camp, 298
Cumberland Straits. *See* Cumberland Sound
cycling, of life forces, 25–6

Damas, David, xvii, 4, 249, 272, 278; Central Inuit studies, 16–23; Copper Inuit, 297, 304; decision-making, 276; environment and society, 20; Iglulik region camps, 271; and Iglulingmiut, 269, 271, 273, 275–6; *Iglulingmiut Kinship and Local Groupings: a Structural Approach*, xx; kinship terminology, 8, 9; on leadership, 274; on Netsilingmiut, 281, 284–5; social groups, 12, 277
Damkjar, E., 315
dancers, 299–300
*Daniel Webster* (whaler), 368nn. 17, 20
Davis, C.H., 40
Davis Strait, 36–7; baleen trade, 48; marriage patterns, 85; seal migration, 42; winter settlement on, 54
Dease, Peter Warren, 280, 289
death: from overpopulation, 76; and rebirth, 36; and religious ideology, 128–9. *See also* adult males; murder; starvation
debt, and Qallunaat, 133–4
decision-making: and authority, 117, 233; group, 205; Nunavut, 339; and women, 268, 296
deities, female, 7
*Delphinapterus leucas* (beluga whale), 39
Dene guides, massacre of Inuit, 289
Denendeh Proposal, Nunavut, 382n. 1
Depot Island, 269
descent, 64–70
DEW (Distant Early Warning) line, 101, 271, 295

dialectical materialism, 302, 330, 334; and the structural paradigm, 329
diametric dualism, 299–300
disease, Franklin's expedition, 280. *See also* foreign diseases
dispute settlement, 15
distemper, in the Sound, 102
dog population, 103–4; Copper Inuit, 292; decimated by distemper, 102; epidemic, 172, 201; free roaming, 245; Iglulik region, 270; Netsilingmiut, 281; and settlements, 160, 162
Dolphin and Union Strait, 292, 302
dories, 210. *See also* boats
Dorset Inuit, 310; art, 325; culture, 45, 315, 325
drinking, at winter quarters, 79, 367n. 15
drogues. *See niutang* (drogues)
Drum Islands, 53
dual clan organization, 311
dual exogamy, 7, 240, 311–12, 314
dualism, of Copper Inuit, 300, 301, 325
ducks, totemic group, 311
Durban Island, 48, 59, 64, 73, 91, 92, 193
Duval, William, 91, 93, 167, 190–1, 193, 195
dwellings, 24; among Qikirtarmiut, 230; of Angmarlik, 158; at Anarnitung, 252t; at Avatuktoo, 168; at Iqalulik, 198–9; at Kingmiksoo, 205, 208–9, 210, 252t, 255; at Naujeakviq, 188; at Sauniqtuajuq, 184; at Tasaiju, 258t; at Ussualung, 194, 196; of Copper Inuit, 298, 320, 325; early Thule, 315; Idlungajung, 162; of Illutalik, 222–3; in Kipisa, 214, 216, 218; of Opinivik, 211, 213–14; Paleoeskimo, 325; platform communal, 252, 257; shared by spousal exchange partners, 299. *See also* communal houses; *qammaq*

eating patterns, 272
eco-evolutionary perspective, 278
ecological base: Central Inuit, 25; of Keemee, 239
ecology, and organization of production, 25
economic conditions, and environment, 25
economic cycle, Cumberland Sound Inuit 1880s–1920, 85–6
economic diversification, 82–8
economy. *See* socioeconomy
Eenoolooapik, 40, 59, 64, 76, 250; map of Cumberland Sound, 41f, 59–63, 61f, 62f, 70, 251, 255–7; and marriage, 132, 366–7n. 12; and William Penny, 74, 366n. 7
Eevic, 131, 167, 187; dogs and move of, 160; dwelling of, 158–9; in Idlungajung, 150, 156, 158, 163; whaleboat owner, 159
egalitarianism, 248, 249, 332–3; aboriginal whaling, 254; among Umanaqjuarmiut, 237; of Copper Inuit, 291, 293, 296–7, 299–301, 325, 380n. 11; and housing, 252; and kayak hunting, 262–3
Ekaliq, 177–9, 181
Ekalujuaq, 199

elders: attitude towards, 296; care of, 181, 203, 211, 244; informants, 234, 236–7
*Elementary Structures of Kinship* (Levi-Strauss), 327
élite: among the Iglulingmiut, 274; in Greenland, 331
Ellesmere Island, 309
Emakee, 188
emigration, and population changes, 91
*Emma* (whaler), 79–81, 118
employment: 1890s, 87–8; for hunters/guides, xviii; seasonal, 295; whale fishery, 77
endogamous marriage, 183, 203, 209, 312, 331
entertainment, at winter quarters, 79
entitlements, Nunavut, 335
entrepreneurial ideology, of Copper Inuit, 291
environment: compositization of social organization, 16; and Copper Inuit society, 20–1, 302–3; of Cumberland Sound Inuit, 33–45, 259–61; and society, 20–1
environmental determinism, xx, 334
epidemics, 14; and social change, 123–5
equality and inequality: among women, 296; commercial whaling, 123; in Idlungajung, 159; and Qikirtarmiut, xvii
*Erignatus barbatus* (bearded seal), 39
Eskimo: kinship system, 5, 32; social organization, 4, 7, 8, 15, 296
Eskimo Point, Hudson Bay, 8
Etalik, 223
Etelageetok, 220, 223, 224
Etuangat Solution, 338–40
Euroamerican contact, 269; long-term effect of, 294–5; and social organization, 72
Euroamerican materials, from *Investigator*, 290, 302
Euroamerican whaling master, 119
Europeans. *See* Qallunaat
Exaluaqdjuin, 53
Exaluqdjuaq Fiord, 53–4
exchange parity, among Copper Inuit, 290–1, 323
exchange systems, 294, 327–30, 332–4; Bering Strait, 313, 327
executioner, Pakaq as, 120
exogamy. *See* local group exogamy
exploitive areas, 21
extended family, 114; Copper Inuit, 296; decline of, 111; of Eevic, 163; households, 69, 252, 272; Labrador Inuit, 332; marriage within, 258; productive activity, 279; socioeconomic independence of, 271; solidarity of, 273; structure, 113, 138

Fainberg, L., 6, 314
fall fishery, 75
fall hunting, 256–7
fall whaling, 77, 252
family allowances, 271, 295
family-oriented kinship system, Inuit, 24

famine, in Arctic environments, 303. *See also* starvation
farming, and Inuit beliefs, 96
fast ice. *See* land-fast ice
father-son ties, 230, 232
female deities, 7
female infanticide, 20–1; among Copper Inuit, 300–1; as man's prerogative, 283, 286–7; Netsilingmiut, 280, 281, 282, 285–8; rate of, 285–6
female leaders, in Oqomiut society, 169
female relationships, 231. *See also* women
feuding: and Labrador Inuit, 332; and Netsilingmiut, 282; precontact, 54–8, 316; and warfare distinction, 56
Fienup-Riordan, A., xvi, 25–6, 29
Fifth Thule expedition, 269
firearms. *See* rifles
Fisher's Exact Test, 240, 252
fishing, 53. *See also* char (*iqaluk*)
fishing rights, Nunavut, 335
fish nets, 108, 293–4
floe-edge hunting, 54; Nuvujen, 77
followers, and leaders, 117–25
food caches: at Avatuktoo, 245; at Idlungajung, 159, 245; at Iqalulik, 198; at Keemee, 189; at Kingmiksoo, 209; at Kipisa, 218; at Nunaata, 245; at Sauniqtuajuq, 178; at Ussualung, 191; Opinivik, 211; of Umanaqjuarmiut, 246. *See also nekaishutu* (community food sharing)
food production, and rifles, 112
food. *See* country food
food sharing. *See* game sharing; *nekaishutu*
food storage, 159, 178, 189, 191, 198, 209, 211, 218, 245–6, 267, 321
forces of production, 106, 107–15
foreign diseases, 72, 84; Copper Inuit, 293; flu at Idlungajung, 163; Inuit deaths from, 76, 91, 373n. 24; and social change, 123–5
Foster-Carter, A., 139
fox. *See* Arctic fox
Foxe Basin, 8, 46, 266
Fox Islands, cousin terminology, 4
Franklin, John, 280, 289
Freeman, M.M.R., 28, 286, 287, 304
freighter canoes, 103
Friedman, J., 106
Frobisher Bay, 36, 46, 55, 81, 92, 193, 305; intermarriage with Inuit from Umanaqjuaq, 206
Fullerton Harbour, 268–9
fur trade economy, 39, 91, 92; and Copper Inuit, 294; and HBC in Pangnirtung, 93–100. *See also* Hudson's Bay Company

game butchering, 246
game depletion, and starvation, 15
game sharing: Iglulingmiut, 272; Kingmiksoo, 203; Nunaata, 147

*gamutiik* (sled), 108
Gardner, D., 60, 257; map of Kingmiksoo, 252
gender solidarity, 9–10, 12, 25, 276
general trading, 91–2, 119; and capitalist economy, 139; change from commercial whaling, 118, 292; technology in, 123. *See also* trading
Giddings, J.L. Jr, 5
gift-giving, among Copper Inuit, 290
Gilyak, of northeast Siberia, 312, 332–3
Gjoa Haven, 281
Godiliak (trading manager), 92, 137
government services, 103–4, 295; establishment of, 142
Graburn, N.H.H., 7, 8, 29, 249, 304
grandparental adoption, 26, 181, 244, 245t
Great Bear Lake, 292
Greenland, 3, 309, 315; endogamy in, 331; Thule migration to, 314
Greenshield, E.W.T., 89, 92, 193, 202
Grise Fiord, 335
group composition, 72
group exogamy. *See* exogamy
group formation, 227, 235, 238, 249; Iglulik region, 271
group leadership, in Nunaata, 147, 152
group membership: Copper Inuit, 298; stability of, 236–7, 294; young couples, 300
group solidarity: and death of a father, 271, 276; Seegatok, 223; and whaleboats, 115
Guemple, L.D., 12, 13–16, 23, 24; on adoption, 26–7, 154, 163, 184; Belcher Islanders, 144, 304; epidemics and social change, 124; on social relationships, 27–8, 112

Hall, Charles F., 42, 269, 305, 365n. 1
Haller, A.A., 40, 42, 43, 238, 262
Hall Point, 271
Hanbury, D.T., 323
hand harpoon, 108
Haneragmiut, 290
*Hannibal* (American whaler), 172, 368nn. 18, 20
Hantzsch, B.A., 91, 115, 122–3, 128, 132, 138, 202, 227
Hare Dene, 292
harpooner (*sivutik*), 113, 254
harpoon heads, 46, 309
harpoon. *See* hand harpoon; sealing harpoon; whaling harpoon
harp seals, 39, 42–3, 156, 262
Hawaiian system, 18, 19, 275, 330
headship, and leadership, 120
Hearne, Samuel, 289
Heinrich, A.C., 5–6
Hennigh, L., 24
Hickey, C., 290–1, 302, 317, 323
hierarchy, xxi, 237, 249; in aboriginal whaling, 254; at Kekerten, 234; and Copper Inuit society, 294; Iglulingmiut society, 11, 273–4
Hogarth's Sound. *See* Cumberland Sound

Holm, G.F., 3, 4
Home Bay Murders, 127
homicide rate, of Copper Inuit, 291
Hood river, 322
hostility: between Anarnitung and Niutang, 56; Copper Inuit and Hare Dene, 292; Kinguaniut and Kingnaimiut, 56; and Netsilingmiut society, 281, 288
household location: and kin relatedness, 144, 152, 177–8; and kin of Soudlu, 172; and Nukinga, 168
households: extended family, 272; multi-family, 49, 68–70, 298; organization of, 17; single-family, 69
Hudson Bay: Thule migration to, 314, 320; whaling in, 81
Hudson Bay Inuit, xvii
Hudson's Bay Company (HBC), 40, 93, 142, 145; and Aivilingmiut, 269; at Pangnirtung Fiord, 191; Bernard Harbour, 292; Churchill River, 322–3; and competition for labour, 93–4; Fullerton Harbour, 268; Imigen Island, 174; as sole supplier to Inuit, 195, 370n. 62
Hudson Strait, 46
Hughes, C.C., 5, 8
human blood, and religious ideology, 128–9, 373n. 30
hunters, Inuit as, 94
hunting: floe-edge, 54; individualization of, 8, 110–11, 113, 294; in kayaks, 245; open-water season, 245; and social structure, 262–3
hunting economy, and mechanization, 103
hunting equipment, inventory, 109t
hunting partnerships, and kinship, 167
hunting rights, Nunavut territory, 335
hypergamy, 242t, 243, 330–1; at Iqalulik, 198; at Nunaata, 151; Iglulingmiut, 274; precontact, 131–2, 313

ice, types of, 37
ice conditions, 33, 37; Cumberland Sound, 38f. *See also* pack-ice; sea ice
ideologies, clash of, 88–90
Idjorituaqtuin, 52
Idlungajung, 53, 121, 131, 155–69, 194, 238; Angmarlik at, 227; community food caches, 245; Eevic at, 150; flu at, 163; marital residence patterns, 243; population estimates, 235–6, 237f
*iglorek* (song cousins), 274
Iglulik, xvii, 46, 223, 258, 269–71; distributions of, 267f; kinship terminology, 136; voluntary alliances, 278–9
Iglulik Island, 267, 270, 277
Iglulik Point, 267, 269
Iglulingmiut, 58; bride-service, 135; child betrothal, 21; contact history, 266–70; cousin terminology, 9f; of Foxe Basin, 8; household organization, 17; kinship terminology, 9–12,
275t, 311; leadership, 18; mythology of, 36; population, 270; *qammat* of, 267; and Qikirtarmiut compared, 278–9; religious ideology, 26; social organization, xxi, 8–9, 20; socioeconomy, 16, 270–2, 324; solidarity in, 274–6; spousal exchange, 21; three-cousin system, 9, 18, 22, 274–5, 277–8, 314, 330; topographic knowledge of, 303; totemic groups, 311; *ukuaq*, 135; winter camp, 298. *See also* three-cousin system
*Iglulingmiut Kinship and Local Groupings: a Structural Approach* (Damas), xx
*ihumataq*, 283
Ikpiakjuk (Igloolik), 270–1
Ikpit Bay, 210
*ilagiit*, 63, 119, 160, 173, 174, 181; among Netsilingmiut, 283, 287; of Nowyook, 220
Ilivilermiut, 282
*illiyuariik*, 273
*illuajuk*, 138
*illukuluk*, 138
*illuq*, 279, 314–15
*illuriik* (roughhouse joking), 277
Illutalik, 197, 211, 220–3, 236, 238
Imigen, 53; dwellings at, 258; population estimates, 59, 63
Imigen Island, 174, 254
Imiyoomee, 42
implements, of Cumberland Sound Inuit, 108
incest taboo, 66
individualism, of Copper Inuit, 301
individualization: of hunting, 8, 110–11, 113, 281, 294; and rifles, 110, 274, 280
individual mobility, 235–8, 269; among Umanaqjuarmiut, 237, 260; early contact period, 251; and group membership, 298–9
individuals, of influence, 130–1
infant mortality, 84, 170
infant survival, 98, 170, 271
inheritance, 64. *See also* whaleboats
*inngutaq* (grandchild), 150, 158, 163, 211, 218
Inosiq, 179, 184, 209
Intermediate interval sites (McGhee), 320
internal integration, 278
international markets, 139
interregional group relations, 115–17
intersystemic contradiction, 106
intertribal relations, social organization, 115
intrasystemic contradiction, 106
Inuit: allegiances and kinship obligations, 93–4; cultural vitality of communities, xvi; of east Hudson Bay, 8; interaction with whalers, 74–7; kinship terminology, 6; social customs, 7
Inuit-Qallunaat interaction, 137; contact 1859–60, 79–81; and HBC, 93, 99–100; and individual influence, 130–2
Inuktitut: bibles in, 88; dual possessive form in, 311; official language in Nunavut, 336; research interviews in, 144

Inupiat, 58; marriage region, 26
Inupik-speaking groups, cousin terminology, 5
inventory of hunting equipment, 109t
*Investigator* (ship): abandoned at Mercy Bay, 290–2, 294; impact on Copper Inuit, 290–2, 302, 324
Iqalugaqdjuin Fiord, 45
Iqaluit, 45, 206
Iqaluit river, 211
*iqaluk* (char). *See* char (*iqaluk*)
Iqalulik, 150–1, 162, 181, 185, 196–201, 238; movement between Kipisa and, 214, 216; relocation of Maniapik to, 178, 196–9
*irniq*, 154
*irniqsaq*, 154
*irniriik* relationship, 12, 69, 112, 170, 176, 233, 279; among Umanaqjuarmiut, 230–1; in Inglungajung, 158, 160, 162; in Iqalulik, 199; in Kingmiksoo, 210; Labrador Inuit, 332; in Nunaata, 147, 167; in Opinivik, 211; in Sauniqtuajuq, 179; in Ussualung, 196
Iroquois, 4
Irvine Inlet, 36
*Isabel* (whaling ship), 79–80
Ishulutaq, 146, 220, 243
Isoa, 52–3
Issortuqjuaq (Clearwater) Fiord, 53, 145, 262
*issumautang* (chief), 63, 119
*isumakattiginituk* (they disagree), 276
*isuma* (life experience), xx, 339
*isumataq* (chief), 17, 63, 120, 272, 274; Tooloogakjuaq as, 205
Itibjiriaq, 269
Itiqilik, 285
Ittirq, 205
Ittusarjuaq, 2–6, 123
Ituksarjuak, descendants of, 271, 274, 331
Ives, J.W., 263, 276, 327
ivory bow drill, whaling scene on, 254f

Jenness, D., 3, 4, 290, 292, 293, 298–300, 302–3, 319, 323

Kachin, of Burma, 132, 312, 329, 380n. 2
Kaigosuit Islands, 185
Kaka, 197, 238; at Nuvujen, 224, 231
Kakatunaq, 197
*kalugiang* (whaling lances), 254
Kanaaka (chief shaman), 88, 121, 238; and Christianity, 126; leadership of, 125, 234; trading manager, 92, 121
Kaneetookjuak (HBC temporary camp), 145
Kangertlukdjuaq Fiord, 54
Kangertlung Fiord, 52
Kanghiryuachiakmiut, 290
Kanghiryuarmiut, 290
Kangiloo Fiord, 33, 42, 43, 53, 262
Karpik, 151, 198
Kasigejut, 223
*kattangutiarsuk* (little sibling), 333

Kaxoudluin Island, 38, 42
kayak hunting: adult male death rate, 262; dependence of Talirpingmiut/Umanaqjuarmiut on, 262
kayaks, 101, 108, 209; abandonment of, 280; and adult male death rate, 261–2; at Kipisa, 214; at Opinivik, 211; use of by Qikirtarmiut, 262
kayak whaling, 113, 253–4, 366n. 8
Keemee, 39, 189–90, 238–9, 260
Keenainak, 48, 121, 129, 131, 149, 231, 232, 238
Keenainak, Elija, 259
Keenainak: impact of loss of, 229; large size of descendants, 258–9; in Nunaata, 146–50, 152, 198
Kekertakjuaq, 174
Kekertelung, 190; dwellings at, 258
Kekerten, xvii, 42, 45, 53, 131, 173, 188, 199, 338; American operation at, 86–7; conversion to Christianity at, 89–90; William Duval at, 193; illness in, 76; Inuit whaling 1880s, 85; leadership at, 125, 205; local group composition, 145; population estimates, 58–9, 83, 87, 91, 92; seasonal hunting, 109; trade, 82; whaling station at, 77–80, 116
Kekerten Historic Park, Pangnirtung, xviii
Kekerten Island, 32, 37, 44
Kellerman, B., 120
Kent Peninsula, 289
Kidlapik (Luke), 90
Kikistan Islands, 43, 54, 172
Kilabuk, Jim, 155f, 232, 253, 366n. 8
Kilabuk (daughter of Veevee), 151
Kilauting, 53
Kingmiksoo, 38, 40, 42, 48, 52, 75, 160, 198, 201–10, 213, 238, 260; census 1846, 66–7; census 1927, 203; contact communal households, 252; early contact period, 250–2; leadership at, 64, 121, 236; map of Eenoolooapik, 60–1, 61f; marital residence pattern, 243; meat caches at, 209; population 1839, 251; population estimates, 59, 60, 206, 235–6, 237f; as principal settlement, 59; whaling base, 78
Kingmiksormiut, 55, 250; and *umiat*, 253
Kingnaimiut, 52, 116; big groups and big men, 255–7; conflict in society, 255–7; early 1880s, 54; feuding of, 55–7; population estimates, 58–9, 62–3, 83, 255; precontact boundaries, 53, 255; social organization, 70, 257
Kingnait, 189, 190
Kingnait Fiord, 36–7, 38, 50, 54, 55, 76, 162, 172; whaling station at, 78, 80
kin group endogamy, 154, 168, 184, 187, 194, 213, 216, 241, 258–60, 271, 287, 304; Ituksarjuak, kin of, 272, 274
kin group exogamy, 65, 68, 71, 134, 249, 271–4, 276

Index        391

kin groups, in Kingmiksoo, 202–3
Kingu, 163
Kingua, 145
Kingua Fiord, 36, 40, 191, 279; population estimates, 60; *sarbut* (open water), 259; settlement size, 251; trapping settlement, 95; whaling, 80
Kinguamiut, 52, 56–7, 116; dwellings, 258; *naalaqtuq*-structured, 255; population estimates, 63, 83; settlement at Imigen Island, 174; social organization, 70, 257; and Talirpingmiut, 250–5
Kingudlik, 122, 131, 236; and children's marriages, 134–5, 243; trading manager, 92
King William Island, 280, 282, 310
Kinnes Company of Dundee, 91–2, 155
kinship: among central core groups, 230; and economic status, 166–7; Eskimo type, 2–3; and household location, 152, 166, 188, 194; Iglulik area, 17; nuclear family, 294, 296–7; obligations and HBC, 93–4; personal name use, 297; and social organization, 2–4, 8–12, 27–30, 116, 177, 226; versus alliance, 28; and whaleboat crews, 116
kinship systems: Alaska and Canadian Arctic, 310–15; and family relations, 24, 273; and hierarchy of advantage, 169–70; Malemiut, 5; and mode of production, 134
kinship terminology, 2–3, 135, 136–8, 275t, 277, 279, 281, 285, 301; Copper Inuit, 275t, 301, 304, 314, 325; Cumberland Sound Inuit, 136f; differences, 6; Pangnirtung, 135; Qikartarmiut and Umanaqjuarmiut differences, 246; and socioeconomic organization, 21, 330; Talirpingmiut, 330
kinship ties: among Qikartarmiut, 227; of central core groups, 226; parent-child cores, 231t; Qikartarmiut and Umanaqjuarmiut differences, 229t, 230t; structure and strength of, 227, 228t
Kipisa, 39, 104, 168, 198, 206, 213, 214–20, 236, 238; marital residence pattern, 243; marriage in, 259; movement between Iqalulik and, 214; Nowyook's kin group at, 227, 232; settled by Umanaqjuarmiut, 214; social organization, 247
Kipisamiut, 218
Kisa, 172–4
Kitingujang, 54
Kivitoo, 91–2, 118, 127, 131, 137, 163
Kjellstrom, R., 26, 67
Klengenberg, C., 292
Koangoon, 158, 185, 189
Kokopaq, 211
Koodlooaktok, 198, 205–6; and group formation, 230, 235; kin group, 202–3; move to Kipisa, 214, 216
Koodlooalik, 146
Kookootok, 160, 244
Koonooloosie, 172–4

Kopalee, 160, 185, 187, 244
Kopee, 177, 179, 194
Koseaq, 176–7, 179
Koukjuaq River, 52–3
Kowna, 131, 137–8, 163, 243
Kudjak Island, 189
Kudlu, of Illutalik, 220–2, 224
Kudlu Solution, 338–40
Kugardjuk, 281
Kumlien, L., 39, 43, 55, 107, 133; population estimates, 83
Kupugmiut, 317

labour, rivalry for Inuit, 93–4
Labrador: communal dwellings in, 315; harp seal breeding ground, 42
Labrador Inuit, xvii; social organization, 332–3
Lake Harbour, 90
Lancaster Sound, whaling grounds, 73, 268
land claims, 335
land-fast ice, 37, 46; at Kingmiksoo, 201; pre-contact settlements on, 53, 309. *See also sina*
land-use patterns, precontact, 50–8
language, and origins of Copper Inuit, 319, 322
Lantis, M., 5, 120
law and order, in Pangnirtung, 100–2
Lazalusie, 150
lead, for bullets, 172
leaders: and followers, 117–25; loss of, 248
leadership, 63–4, 72, 233–5; at Nunaata, 149, 167; at Sauniqtuajuq, 184; at Tuapait, 173; and camp prosperity, 121–2, 156, 218; Central Inuit, 17; and Christianity, 90, 125–30; in egalitarian society, 249; and headship, 120; and Iglulingmiut, 272, 274; and Kinguamiut, 250; in Kipisa, 214, 216, 218; and Labrador Inuit, 332; Netsilik patterns, 7–8; of Nowlalik's kin group, 183–4; in Oqomiut society, 231; and Pangnirarmiut, xviii; sacred and secular, 125; and Talirpingmiut, 250; and Umanaqjuarmiut, 248
legends. *See* mythology
legislative assembly, Nunavut, 337
levirate tendencies, 276, 332, 333
Levi-Strauss, C., 24, 313, 327, 329, 330, 333
Leybourne Islands, 44
Linton, R., 4
Little Ice Age, 46, 316, 319, 321
local group composition, 142–5, 160, 193; Inglulingmiut, 272; Netsilingmiut, 283
local group endogamy, 17, 154, 173, 181, 194, 206, 216, 241, 242t, 258
local group exogamy, 65, 68, 71, 134, 240, 249, 257, 271–2, 274, 276, 279, 304; in Alaska, 311
local group organization, 225, 260
local group size, 58–63, 235–8, 251; Kingnaimiut, 255; Kinguamiut and Talirpingmiut, 251; Tasaiju, 258

locality: importance of, 193; and social relationships, 27–30. *See also* matrilocality; patrilocality
loitering, in Pangnirtung, 100–2
longhouse, 315
Lord Mayor Bay, 279, 282
Low, A.P., 90
Lyon, G.F., 267, 268

M'Clure, Robert, 289–90
M'Donald, A., 33, 44, 53, 250; on leadership, 64, 70, 117; population estimates, 59–60, 201, 251
McGhee, R.J., 309, 316, 318, 319–20, 325
McKeand River, 33
Mackenzie Delta, 65
Mackenzie Inuit, xvii, 316–18
Mackenzie River trading posts, 318
*McLellan* (American whaling ship), 75, 201
Magic Lantern show, 88
Malemiut, kinship system, 5
Malukaitok, 137, 178–9
Malukaitok Fiord, 118
Mamukto, 223
*mangnariik* relationship, 176
Maniapik, 129, 150, 151, 174, 176–9, 181, 183, 185, 191, 193–5, 214, 233; relocation to Iqalulik, 178, 196–9
Manirtuq, 269
man's knife, 108
maps: and Copper Inuit hunters, 303; of Cumberland Sound, 59–60; of Nunaata, 148f. *See also* Eenoolooapik
*maqtaak* (country food), xviii-xx, 43, 76
Mara river, 322
marine mammals, 39, 85, 109, 112–13
maritime hunting tradition, 95; Thule Inuit, 309
marriage, 58, 64–70; agamous, 297; arranged, 134, 231; between relatives, 134, 271, 272, 331; marrying out, 240–3; marrying up, 243; and mode of production, 134; regions, 26; and residence patterns, 135; rules, xvii, 3, 15, 314, 328; wife-givers and wife-takers circles, 312. *See also* cousin marriage
marriage exchange, and seal parties, 25
marriage patterns, 85, 240–4; Belcher Islanders, 13; between Umanaqjurmiut and Qikirtarmiut, 241t; Central Inuit, 17, 18; Gilyak, 333; and Netsilingmiut kinship terminology, 285; Thule Inuit, 312–13, 315
Mary-Rousselière, G., 130
Mary (Veevee's adopted daughter), 187
material inequity, and Qikirtarmiut, xvii
Mathiassen, T., 3, 269
matriarchs, emergence of, 137–8, 169
matri-line, 65, 136, 312
matrilineal clan organization, 6–7, 314
matrilocality, 65, 67, 135, 231; at Idlungajung, 156, 166; at Iqalulik, 199; at Kingmiksoo,
203, 209; at Kipisa, 214, 220; at Nunaata, 150; at Sauniqtuajuq, 178, 181, 183–4; at Ussualung, 195; Iglulingmiut society, 273; of Umanaqjuarmiut, 243–4
*mauliqtuq* (breathing-hole sealing), 16, 20, 52, 108, 272; Copper Inuit, 295; and the rifle, 112; Thule Inuit, 320
Mauss, M., 3–4
Mayes, R.G., 140
means of production, 105, 107; control of, 132
meat caches. *See* food storage
mechanization, and hunting economy, 103, 109, 110
mediators, Inuit as, 122
Melville Peninsula, 266
men. *See* adult males
Mequt, 243
Mercy Bay, 290, 380n. 14
metal pots, 317–18
Metiq, 121
Michiman (trading manager), 92
midden deposits, 45, 47–50
middlemen: Inuit, 333; proto-Caribou as, 323; role of, 118–19, 122
Midlikjuaq, 36, 40, 54
Midlurielung, 53
mid-summer whale drive, 97
migration: and climatic change, 309–10, 319–20, 322; Copper Inuit, 313–14; Inuit into Cumberland Sound, 74; and population changes, 91; Qitlarssuaq, 130; as result of social factors, 313, 316; Thule Inuit, 22, 46, 309–10, 313–17, 320–3
Mike (from Nunaata), 150, 165, 195–6
Miliakdjuin, 36, 40, 54; whaling station, 79, 80
Millut Bay, 43
Milurialik, 43, 96, 109, 190, 262
*Milwood* (New Bedford barque), 137
Mingoakto, 137
missionaries, 130; and Iglulingmiut, 270; Moravian, 76–7; and Netsilingmiut, 281
mission station, at Lake Harbour, 90
mode of production, 105–6, 131, 134
Mogg, 'Billy', 292
*Monodon monoceros* (narwhal), 39
Moravian Church, 76; records, 333
Moravian Mission Board, 77
Morgan, L.H., 2, 65, 135–8, 246, 284, 330
Morrison, D., 315–16, 318, 319
Mosesie, 179, 181, 184
multiple affinal ties, 241, 242t
Munn, H.T., 127, 131, 190–1, 195
murder: of Catholic priests by Uluksak, 292, 294; and Netsilingmiut, 282; and punishment, 296; of RCMP corporal and HBC trader, 294. *See also* revenge murder
murder-suicide, Kaka, 197, 224, 231
Murdock, G.P., 4, 5, 7, 8, 16–17, 24
musk-ox depletion, 290
Mutch, James, 91, 193

Index    393

mythology: of Copper Inuit, 291, 299–300, 323–4; of the destitute orphan, 284, 331; Iglulingmiut, 273; Netsilingmiut, 284, 323–4; themes of, 66

*naalaqtuq* (respect-obedience), xx, 9–12, 25, 169–70, 230, 237–8; and adoption, 244; in Copper Inuit social organization, 296, 304–5; Etuangat Solution, 339; in Idlungajung, 163; and Iglulingmiut society, 273–4, 276–8; in Nunaata, 147, 149, 152, 154; social structure, 247–9, 332; and Umanaqjuarmiut society, 232. *See also ungayuq*
Nakashuk, 211, 213
names, personal, 297
namesake relationships, 15, 277
name soul, 128–9
naming, 13; and descent, 233; in Iglulingmiut society, 273
*nangminariit* (partnerships), 288
*nanuq* (polar bear), 39, 43, 54, 91, 270
Napaskiak, Alaska, 8
Napaskiamiut, 8
narwhal, 39, 43
native products, 110
Naujateling, 52, 55, 75; early contact period, 251; population estimates, 58, 63, 83; Tesuwin as trader in, 118; whaling base, 78–9, 80
Naujeakviq, 151, 169, 185–7; patrilocality in, 243
*nayagiik*, 279
Neakunggoon, 223
Needham, R., 28
Neeoudlook, 59–60
negotiation: and Inuit social organization, 13–16; kinship as, 24
*nekaishutu* (community food sharing), 112, 133, 245; at Avatuktoo, 169, 245; at Idlungajung, 159, 245; at Iqalulik, 199; at Kingmiksoo, 203, 209; at Kipisa, 214, 218; at Nunaata, 147–8, 245; at Sauniqtuajuq, 178; at Ussualung, 191
Nelson Island: ritual distribution, 26; social structure, 25, 29
Neoatlantic period, 309
neolocal living arrangements, 195, 209
Nepekin, 118, 121
*Neptune* (ship), 59
*nerqri* (marine mammal meat), 39
nets, seal and fish, 108, 113, 293–4
*netsiavinik* (silver jars), 39, 82, 201; at Illutalik, 220; HBC, 98
Netsilik, xvii; distributions of, 267f; female infanticide in, 285; leadership patterns, 7–8; precontact history, 316, 318–25
Netsilingmiut, 322; closed groups, 282–4, 324; contact history, 279–80; contradiction and integration in, 288; cousin terminology, 330; and endogamy, 331; and Euroamerican materials, 290; household organization, 17; kinship terminology, 275t, 314; of Pelly Bay, 7–8; religious ideology, 26; relocation of, 268; rifle and culture change, 280–1; social organization, xxi, 20, 324; socioeconomy, 16, 288f
*nettik* (ringed seal). *See* ringed seals
Nettilling Fiord, 36, 43, 44, 52–3, 178, 196, 201; population estimates, 59; trapping settlement, 95
Nettilling Lake, 36, 44, 53; archaeological investigations at, 129; caribou hunt, 112; Talirpingmiut settlement, 52
Nettilling Uplands, 33–4, 36, 56
Niaqutsiaq, 131, 137; marriage of, 243; murder of, 127; trading manager, 92
Nichols (HBC trader), 94, 370n. 62, 374n. 2
Nimigen Island, 201
*ningaugiik*, 159, 279
*ningauk* (male Ego), 10–11, 65, 154
*ningaut*, 166, 312
*niqaiturasuaktut*. *See* seal-sharing partnerships
Nirdlirn Fiord, 54
Niutang, 50, 54; burial population, 54–5, 58–9, 70; feuding with Anarnitung, 56
*niutang* (drogues), 80, 253–4
Noble Company, 82, 91, 92; on Umanaqjuaq (Blacklead) Island, 87
Noodlook, 40, 53, 155
Norse materials, 310
Northumberland Inlet. *See* Cumberland Sound
Northwest Passage, 289
Northwest Territories (NWT), 335
*Nova Zembla* (whaling ship), 90
Nowdluk, 209
Nowlalik, 179, 181, 183, 185, 190, 194
Nowyook, 216, 218, 223; kin group at Kipisa, 220, 227, 230, 232
nuclear family, 69; Copper Inuit, 16–17, 296–7; houses, 112; Murdock's opinion of, 24; Oqomiut, 251; as socioeconomic unit, 111, 114, 227, 252, 281; winter hunting with, 178
Nugumiut, 55, 58, 115, 206; population estimates, 83, 91
*nukariik* relationship, 69, 112, 167, 237, 279; in Avatuktoo, 172; in Etelageetok, 224; in Illutalik, 221, 223; in Iqalulik, 199; in Kingmiksoo, 202–3, 205–6; Labrador Inuit, 332; in Nunaata, 147, 150; in Opinivik, 213; in Sauniqtuajuq, 176–7, 179; in Ussualung, 195
Nukeeruaq, 176–9, 185–6; as RCMP constable, 187; relocated to Pangnirtung, 188
Nukinga, 168
Nuliajuk, 284
Nunaata, 33, 50, 121, 145–55, 245; map, 148f; marital residence patterns, 243; marriage in, 241, 259; move to Kingua Fiord, 191; social organization, 247, 259

Nunaata Island, 145; biologically distinct population of, 258
Nunaatarmiut, label of, 193
*nunatakatigiit*, 63, 193, 283
Nunavut: barriers to survival, 336; territorial model, 337–8
*Nunavut Act*, 335
Nunavut Agreement, 335
Nunavut Government, 335–40; a proposal for, 336–7
Nunavut Implementation Commission (NIC), 336–8
*Nunavut Land Claims Agreement Act*, 335
Nunavut Territory, 335
Nuneeaguh, 137–8
Nunivagmiut, kinship system, 5
Nunivak, cousin terminology, 4
Nuvujadlung, precontact fall settlement, 52
Nuvujen, 223; early contact period, 251; population estimates, 58, 63; precontact winter settlement, 52; and spring whaling, 39, 77, 201, 224; whaling station at, 77–9
Nuvujen Inlet, 38–9
Nuvuk Point, 38, 40, 44

*Odobenus rosmarus* (walrus), 39
Okaitok, 125; and Christianity, 126; representative of Tyson, 118, 121
Okittok, 90
old copper culture, 292
Oleetivik, 33
Omaha system, 327
Ooneasagak, 126, 131, 156, 158, 160, 189
Oosten, J.G., 26
Opinivik, 45, 121, 210–14, 236; population of, 213; settled by Umanaqjuarniut, 211
Oqomiut, 50–1, 55, 125; big groups and big men, 255–7; Cumberland Sound, 26; and William Duval, 191; Idlungajung camp, 236; and matrilocality, 135; population levels, 101; topographic knowledge of, 303; totemic groups, 311; whaleboats, 270; women, 138, 169
oral histories, collection of, xviii
organization of production, 113; changes in, 110–15
Oshutapik, 185, 187, 188–9, 220; at Nuvujen, 224; move to Iqalulik, 197–8; move to Naujeakviq, 188
Oswalt, W., 8
outboard motors, 103
overpopulation, and death, 76, 91
over-wintering: American Harbour, 190; in Cumberland Sound 1851–80, 78t; in Cumberland Sound 1857–58, 78; and employment in whale fishery, 77; Kingmiksoo, 75; Penny at Kekerten, 78

pack-ice, 37
Padli, 55, 122, 236
Padlimiut, 54, 57
Padloping Island, 121, 131, 150, 169, 193–4
Padluq, 185, 187, 188
*Pagophilus groenlandicus* (harp seal), 39
Pakaq, 120–1, 125, 372n. 15
Paleoeskimo, and Copper Inuit culture, 325–6
Pangnirtarmiut, 338; and name soul, 128–9; research among, xvi, 143–5; *ukuaq*, 135
Pangnirtung: centralization of population around, 142; climate, 36–7; families move to, 178; and HBC 1921–62, 93–102; Inuit in, 100–2; *irniriik* trading parties, 232; kinship terminologies, 65, 135; local group composition, 145; political reality in, xviii; Qatsu, 48; telephone directory, 232–3; wage labour in, 110
Pangnirtung Fiord, 36–7, 42, 44, 54
Pangnirtung Post Journals, 149, 159, 185, 232, 238; and Angmarlik, 160; and family size, 98
*paniksaq*, 154
pan-Inuit social structure, 27
*panniriik* ties, 279; Avatuktoo, 172; Nowyook at Kipisa, 230; Nunaata, 147; Ussualung, 194
parent-child cores, 236–7, 258; Qikirtarmiut, 226–7, 230–1
parent-daughter ties, 230
Parker, Captain J., 74–5, 77
Parker, J.C., 88, 91
Parks Canada, 234
Parry, W.E., 267, 268, 269, 276
patri-line, 65
patrilineal clan organization, 233, 312; Gilyak, 333; St Lawrence Islanders, 5
patrilocality, 14–16, 67, 69, 71, 135; at Iqalulik, 199; at Kingmiksoo, 203, 209–10; at Kipisa, 220; at Ussualung, 195; and dwellings, 258; of Gilyak, 333; in Idlungajung, 156, 166; Iglulingmiut society, 273; in Naujeakviq, 188; Netsilingmiut, 283, 286–7; in Nunaata, 149, 150, 154; of Qikirstarmiut, 243–4; in Sauniqtuajuq, 178, 181, 183–4; Thule Inuit, 312
Pawla, 92, 121, 126, 179, 205, 220, 234, 369n. 51
peace-keeping, 14–15
Peck, E.J., 88–9, 127, 202
Peeka, 131
Pelly Bay, 8, 113, 280, 281, 282, 285
Penny, William, 38, 40, 53, 64, 156, 250; Cumberland Sound whale fishery, 73–7, 81; population estimates, 59–60; whaling stations, 77–8
Penny Highlands, 36
Penny Ice Cap, 36
Penny's Harbour, 81
Petaosie, 197–8, 220; at Nuvujen, 224, 231
Peterhead whaleboat, 270
*Phoca hispida* (ringed seal), 33, 39
*pimain* (chief), 63, 119

*pimaji* (chief), 119
Pingirqalik, 267, 269
pinnace, 214
*pisik* (dance), 299
Pitsualuk: in Etelageetok, 224; of Illutalik, 220–3, 338
*piutuq*: at Idlungajung, 159; at Nunaata, 147–8; at Ussualung, 191
place, use of term, 238
plan: of Avatuktoo, 171f; of Idlungajung, 157f, 161f, 165f; of Illutalik, 222f; of Iqalulik, 199, 200f; of Kingmiksoo, 204f, 208f; of Kipisa, 215f, 217f, 219f; of Naujeakviq, 186f; of Nunaata, 151f; of Opinivik, 212f; of Sauniqtuajuq, 175f, 180f, 182f; of Tuapait, 173f; of Ussualung, 192f
poets, 299–300
Point Hope, Alaska, 8
polar bear. *See nanuq* (polar bear)
polyandry: Central Inuit, 17, 67; Netsilingmiut, 282
polygamy, Cumberland Sound, 67, 69
polygyny: Central Inuit, 17, 65, 67, 71, 128; Gilyak, 333; Iglulingmiut, 268; Labrador Inuit, 332; Netsilingmiut, 282; Thule Inuit, 313
Pond Inlet, 131; and Christianity, 127; family size, 98; and Inuit from Umanaqjuaq, 193; Tununirmiut of, 266, 268–9; whaling station, 73, 193
Popham Bay, 33
population: Arctic depopulation causes, 321; biologically distinct, 258; dispersion of, 142; dispersion theories, 310, 313, 321–2; of Idlungajung, 237f; of Kingmiksoo, 237f
population estimates: A. M'Donald, 59–60; Franz Boas, 58–9, 84; Cumberland Sound settlements, 99t, 102t; B.A. Hantzsch, 91; L. Kumlien, 83; William Penny, 59–60; P.C. Sutherland, 59–60; Mathias Warmow, 76–7
Port Harrison, 7
Port Harrison Inuit, cousin terminology, 311, 330
pre-Christian concepts, 128–9
precontact: 1820–40, 73–4; burial populations, 50; Cumberland Sound, 32–3; Cumberland Sound and Iglulik area, 46; Kingmiksoo villages, 201
prehistory: Canadian Arctic, 308; Cumberland Sound Inuit, 45–50
prices: baleen (whalebone), 83t, 87; fur, 295; whale oil, 82, 83t
Prince of Wales Northern Heritage Centre, Yellowknife, xvii-xviii
Prince of Wales Strait, 290
productive activity: at Kingnait Fiord, 256; extended family, 279, 281; Netsilingmiut, 281; and relationships, 253–5, 297; Talirpingmiut and Kinguamiut, 253; and Talirpingmiut and Umanaqjuarmiut, 253

productivity, and prestige, 117
ptarmigans, totemic group, 311
Pudjun, 190
Pujetung, 38, 43
Pujetung island group, 53
Punuk culture, 309; conflict in, 313

*qaggi*, 273, 311–12
Qaggilortung Fiord. *See* Kangiloo Fiord
*qailertetang* (exchange of spouses), 89
*qairulik* (harp seal). *See* harp seals
Qallunaat, xxi; and band structure, 14–15; and concept of flexibility, 23–4, 138; contact and social change theory, 15–16; dependence on, 295; institutions, 122; and Inuit middlemen, 118
Qallunaat-Inuit mediators, 122
*qammat* (winter dwellings), 47; of Akpalialuk, 210; Amundsen Gulf, 318; foundations in Ussualung, 190; of Iglulingmiut, 267; of Keenainak, 149, 152; on Kekerten Island, 48; Thule period, 46; Veevee, 188. *See also* dwellings
*qamutiit* (sleds), 80
*qangiariik* relations, 152
*qaniaksaq*, 177
Qaqasiq, 179
Qaqortingneq, 285
Qarmang, 53
Qarmaqdjuin, 54
Qarmat, 271
Qasigidjen Fiord, 52
Qatsu (daughter of Angmarlik), 48, 131, 156, 158, 166
Qeqertarmiut, 282
Qikirtan. *See* Kekerten
Qikirtarmiut, xvii, 48; and adoption, 244–5; at Idlungajung, 155; at Naujeakviq, 188; at Tuapait, 172; central cores, 236; contact-traditional settlements, 144, 145–90; exogamous tendency of, 257; group formation in, 225; hypergamous tendencies of, 243; and Iglulingmiut compared, 278–9; and leadership, 233–5; and Umanaqjuarmiut, xxi, 116–17, 137, 188, 198, 225–49; whale hunting technique, 253; winter hunting, 259. *See also* Kingnaimiut; Kinguamiut
*qilalugaq* (beluga whale). *See* beluga whale
*qilalugaq tuugaalik* (narwhal), 39
*qirniqtuq* (narwhal), 39
Qitdlarssuaq, 55; migration, 130, 381n. 4
Qivijuk, 331
Qivitormiut, of Kivitoo, 127
*qudlit* (seal oil lamps), 108
Queen Maud Gulf, 279, 289, 303, 315, 320–1
Quertilik, 285

Rae, John, 289–90
Rae River, 290, 298
Ranger River, 43

*Rangifer tarandus arcticus* (caribou), 39
Rasmussen, Knud, 3, 273–4, 281, 282–7, 292, 296–7, 299–300, 302, 319
RCMP, 173, 187; and Aivilingmiut, 269; censuses, 203, 227; in Pangnirtung, 100–2, 142; and Qikirtarmiut and Umanaqjuarmiut differences, 233; Tree River, 294 reciprocal egalitarianism, 299–300
reciprocal rights: Copper Inuit, 290, 293–4; in families, 252
regional group exogamy, 240
relations of production, 105–6, 115–25, 139, 226
religious endogamy, 271
religious ideology: Copper Inuit, 296, 299–300; human blood and death, 128–9; Iglulingmiut, 273; and production, 107; structure of, 126–7; and white missionaries, 130
Remie, C., 285
renewable resource economy, and cultural survival in Nunavut, 338
Repulse Bay, 150, 266, 269–70, 280–1
residence, 64–70, 135; marital, 243–4
residential groups, 24, 112, 225; Central Inuit, 17; and mode of production, 134
residential solidarity: among Qikirtarmiut, 230; at Avatuktoo, 172; at Idlungajung, 160; at Iqalulik, 197, 220; at Kingmiksoo, 206, 208, 210; at Kipisa, 214, 216; at Nunaata, 146, 152; at Opinivik, 211, 213; at Ussualung, 191, 193
residential stability, 235–8; among Umanaqjuarmiut, 260; early contact period, 251
revenge murder: Gilyak, 333; Labrador Inuit, 332. *See also* murder
Rey, P.P., 139
Richardson River, 289
Riches, D., 286–7
rifles, 81, 108, 115, 151–2, 268; and Copper Inuit, 292, 294; from HBC, 146; and hunting practices, 8, 44, 109, 112, 117, 280, 371n. 3; Iglulik area, 269; and individualization, 110–11, 274; and organization of production, 110
ringed seals, 33, 39, 43, 103, 266; at Avatuktoo, 167; at Idlungajung, 155; at Illutalik, 220; at Kingmiksoo, 201, 260; at Kipisa, 214; at Nunaata, 145; at Opinivik, 211; at Tuapait, 172; at Ussualung, 190; Iqalulik, 196; in Little Ice Age, 46; market value, 82; Netsilingmiut district, 287; population in Cumberland Sound, 40, 262; precontact population, 309; returns, 97t; Sauniqtuajuq, 174
ritual: and ceremony, 299–300; cycling of, 25–6; as ideology in action, 25. *See also* Sedna (sea goddess)
ritual dance/song, Copper Inuit, 293–4, 298, 302, 325
ritual sponsorship, 13, 15

Robert Peel Inlet, 210
Robinson, S., 268–9
Roche, Paul, 126, 205
Roes Welcome Sound, whaling in, 81, 268–9
Ross, John, 279
Ross, W.G., 268–9
rowboats, 113; in open water, 108
rowers, *umiat* whaling, 254
Rowley, Susan, 304–5
Russian trade goods, 318

Sabellum Company of Peterhead, 91–2, 121, 137
Sadlirmiut, 268
Sagdluaqdjung, warfare at, 55
Sahlins, M.D., 16
St Lawrence Island: kin terminology, 314; patrilineal clans, 5
*sakiaq*, 158
Sakiaqdjung, 53
*sakiaqsaq*, 176
*sakkiik*, 135, 150, 166, 199
*sakurpang* (heavy-shafted harpoon), 254
salmon fishing, 53
Salter, E.M., 50, 70, 257
*Salvelinus alpinus* (char), 39
Sanikiluaq, 335
*sarbut* (open water), 33, 37, 53–4, 155–6; Drum Islands, 174; Kingua Fiord, 259; sealing at, 112, 122, 196
Saumia, 50, 134, 145
Saumingmiut, 52, 54, 57; contact-traditional settlements, 145; population estimates of, 59–60, 83
Sauniq Island, 174
Sauniqtuajuq, 121, 151, 162, 174–85, 197, 238; kin groups in, 185; patrilocality in, 243; population of, 178
Saunirtung, 53
Saunirtuqdjuaq, 53
*savik* (man's knife), 108
Scheffel, D., 333
Schledermann, P., 45–50, 56, 59–60, 70, 257; excavations of Niutang, 256; map of Anarnitung, 252; site maps, 257
sea anchor. *See niutang* (drogues)
sea goddess. *See* Sedna
sea ice conditions, 37, 290. *See also* ice conditions
seal denning sites, Sunigut Islands, 190
sealing, 16, 39–40, 101, 108, 109, 113, 293, 302; access to grounds, 122–3; at Seegatok, 223; demand for fur, 98; economic value of, 83, 103, 140; and fox trapping, 97; from whaleboats, 114; importance of to family welfare, 98; open water, 53; Thule period, 47, 49; and whaling 1870–94, 82–8; for young seal, 52, 54, 85. *See also* common jar seals; *mauliqtuq* (breathing-hole sealing); *netsiavinik* (silver jars)

sealing harpoon, 108
seal oil, 98; lamps, 108
seal party, and marriage exchange, 25
seals, souls of, 128-9
seal-sharing partnerships, 21, 278, 281, 287-8; Copper Inuit, 300-1
seal skin float (*avatak*), 108
seal skins, 103; market demise, 140; trade 1880-90s, 87
seal skin tents, 101-2
sea mammals, and Iglulingmiut, 16
sedentism, 272
Sedna (sea goddess), 128, 273, 284, 311; ceremony, 6, 26, 66, 311; Feast of Sedna, 89; revelation from, 89
Seegatok, 104, 218, 223-4
Senate, Nunavut, 339
servants: adult men as, 132-3; and Iglulingmiut, 268
Service, E., 14-16, 23
settlement patterns: Copper Inuit, 293; Kingnaimiut, 54
settlements: precontact in Cumberland Peninsula, 50-8; size of, 114-15, 332
Shamiyuk, Simon, 195
sharing practices, 21-2, 110, 112, 178; along kin lines, 129; at Kipisa, 218; at Opinivik, 211; of capital equipment, 129, 218; Copper Inuit, 296; Kipisa, 214; Nunavut, 338; and traditional relations of production, 133-4; and whaling system, 100. See also game sharing; *nekaishutu*
Shark Fiord, 145
Shimilik, 53
Shimilik Bay, 34
shipwrecks, wood and metal from, 81, 87, 279
Shorapik, 188, 199
shrimp/scallop fishery, 140
Siberia, cousin terminology, 4-5
sibling adoption, 244, 245t
sibling cores, 237; Iglulingmiut, 277; Qikirtarmiut, 230, 239; Umanaqjuarmiut, 226-7
Sikirnik (Ittusarjuaq's wife), 138
silver jars. See *netsiavinik* (silver jars)
Simpson, Thomas, 280, 289
*sina* (edge of land-fast ice), 37-8, 39, 43, 54, 155; Iqalulik, 196; Keemee, 189; Kingmiksoo, 75; marine mammals, 172; Sauniqtuajuq, 174; sealing at, 112, 259; use of kayak at, 108, 262
Singnija (Cape Haven), 91, 181, 206
*sivutik* (harpooner), 113; relationship with *aggutik*, 114; *umiat* whaling, 254
Sivutiksaq (harpooner's apprentice), 190-1, 196; affinal relations, 191-4
skiffs, 210
skin-covered kayaks, 101
skin tents, 53
slate manufacture, 318
slaves, Gilyak, 333

snowhouses, 47, 258, 267; double, 298
snowhouse villages, 295, 320
snowmobiles, 103, 110; Nunaata, 155
soapstone: from Coronation Gulf, 318; as trade good, 317-18
social composition: of Avatuktoo, 170f; of Idlungajung, 157f, 161f, 163, 164f; of Illutalik, 220, 221f, 222f; of Iqalulik, 197f, 200f; of Kingmiksoo, 204f, 207f, 210; of Kipisa, 214, 215f, 217f, 219f; of Naujeakviq, 186f; of Nunaata, 148f, 152, 153f, 154; of Opinivik, 212f, 213, 213f; of Sauniqtuajuq, 175f, 180f; Tuapait, 173f; of Ussualung, 192f
social distance: dwellings at Naujeakviq, 188; and spatial distance, 144, 154, 162
social formation, 107f; during contact-traditional period, 225
social organization, 4, 13-16, 70-1, 257; in 1883, 85; Copper Inuit, 296, 304-5; Cumberland Sound Inuit, 130-8; differences between Qikirtarmiut and Umanaqjuarmiut, 226, 247-9; and economic organization, 110; and environment, xx, 259-63; Eskimo type, 4; flexibility of, 23-4; Iglulingmiut, 272-6; intertribal relations, 115; and kinship, 2-4; legends and, 66; of Netsilingmiut, 282-7; precontact Cumberland Sound, 32-3; structure of, 142, 308; traditional, 58-70
social position: based on ability, 132; and economic productivity, 238
social relationships: based on locality, 27-30; egalitarian, xxi; hierarchical, xxi; and modes of production, 106; structure, 131
social solidarity, at Nunaata, 147
social stratification, 130-8
social structure, at Thule migration, 327
socioeconomic organization, 131, 249; Cumberland Sound Inuit 1955, 101; of living arrangements, 252
socioeconomy: and alliances, 58; Central Inuit, 21; and class and marriage, 131; Copper Inuit, 293-4, 295-6, 304; environment and Copper Inuit, 302-3; of Iglulingmiut, 271-2, 276-8; Netsilingmiut society, 282-3; Nunavut, 337-8; Thule Inuit, 309
sod houses, 252, 258; at Anarnitung, 251; at Kingmiksoo, 201, 251-2; at Tuapait, 172; excavation of, 45; Tasaiju, 257
solidarity: in Iglulingmiut, 274-6; kin group, 278, 283
song contests, 274
song/dance partnerships, Copper Inuit, 18
songwriters, 299-300
*Sophia* (whaling ship), 80, 118
sororate tendencies, 276; among the Oqomiut, 181
Soudlu, 168, 169, 172
Southampton Island, 230, 268
South Baffin caribou herd, 44

South Baffin Island, 119
Spence Bay, 281, 282
Spencer, R.F., 6
Sperry, J., 4–5, 314
Spier, L., 4, 9
spousal exchange, 7, 15, 17, 128; Belcher Islanders, 13; Central Inuit, 20–1; Copper Inuit, 18, 293, 294, 298; Iglulingmiut, 21, 268, 277; Thule Inuit, 311
spring whale hunt, 75, 77, 80
Stanford, D., 309
starvation, 74–5; Franklin's expedition, 280; and game depletion, 15; in Kingmiksoo, 209; Thule Inuit, 319–20
steel traps, 109–10, 270
Steenhoven, G. van den, 7
Stefansson, V., 3, 291–2, 298, 318, 323
Stenton, Doug, 44, 129
Stevenson, D., 304
storytellers, Copper Inuit as, 299–300
strangers, 365n. 8
string figures, and Copper Inuit ideology, 300
structural change, Cumberland Sound Inuit, 107
structural tendencies: Iglulingmiut society, 276–8; Inuit social organization, 225–47, 261; Qikirtarmiut and Umanaqjuarmiut, 247t, 308
subsistence economy, Nunavut, 337–8
subsistence patterns, 73; changes in, 108–10; Copper Inuit, 291, 293
Sugluk region, 7, 304
suicide: of Kopalee, 187; of Sukulak, 168
Sukulak, *qammaq* of, 168
summer camps: for hunters, 156; Naujeakviq, 189; precontact, 52, 54; Tuapait, 174
summer fishing site, 172
summer hunting ranges, 238
summer society, 3
Sunigut Islands, 54, 190
'Sun and the Moon', 66
surnames, adoption of, 232–3
survival ability, xvi
Sutherland, P.C., 33; census 1846 of Kingmiksoo, 59, 66–7, 69, 135, 251, 253, 262; population estimates, 59–60, 201, 251

taboos: incest, 66; and personal names, 297; and women, 126
Takamiut, kinship terminology, 7
Talirpingmiut, 44, 52; arranged marriages among, 134; cousin terminology, 330; and Kinguamiut, 250–5; population estimates, 60–1, 63, 83; precontact boundaries, 53, 57–8; settlement size, 251; social organization, 70, 116, 257; *ungayuq*-structured, 255
Talirpingmiut/Umanaqjuarmiut: adult male death rate, 261–2; and productive activity, 253

Tarrajuk, 285
Tarrionitung Peninsula, 53
Tasaiju, 50; burial population, 54, 58–9, 70, 257–8
Tautuajuk, 194
Taylor, H., 333–4
Taylor, J.G., 333–4
Taylor, W.E. Jr, 319, 320–1, 332
technology: changes in, 108–10; in commercial whaling, 114, 123; Qallunaat, 110
telescopes, 117
tents, 101, 102, 103, 139–40
Tenudiakbeek. *See* Cumberland Sound
Tern Islands, 269
Terray, E., 132
territoriality, 238–40, 281, 313
Tesseralik, 158, 189
Tesuwin (Inuit whaler), 80, 118, 120, 130
*Thalarctos maritimus maritimus* (polar bear), 39
Thalbitzer, W., 119, 314–15
Thelon River, 321–2
Thelon Woods, 321–2
three-cousin system, 5–6, 9; Iglulingmiut, 9, 18, 22, 274–5, 277–8, 314, 330
Thule Inuit, 22; culture, 45–50; disappearance of, 319; economy, 319–20; first expansion, 310–15; migration out of Alaska, 309–18; second expansion, 315–16; third expansion, 316–18; winter houses, 60
tidal rips, 33, 53, 145, 155, 259
tides (*tinu*), Cumberland Sound, 33, 37
Tiniqdjuarbing (place of great tides), 33
*tiriganirk*. *See* Arctic fox
Togaqjuaq (big tusk), 44, 54, 64; population estimates, 59–60
Tooleemaijuk, 211
Tooloogakjuaq (Peter), 90, 121, 129, 160, 202–9, 211; at HBC, 94; boat-owner, 209; and Christianity, 125–6, 202, 205; and group formation, 230, 235; impact of loss of, 229; Kingmiksoo leader, 202–3; leadership of, 203, 205
topographic knowledge, of Inuit, 303
Tornait, 76
totemic groups, 6, 311
tourism development, in Pangnirtung, xvii–xviii
Towkee, 195–6
Towkie, 181, 198, 210
trade: after 1915, 92; annual fairs, 317; baleen, 48; Beringian network, 316–17; in blubber skins, 97; with British, 49; copper, 317–18; and Copper Inuit, 291; European goods, 318; exotic goods, 317–18; and geographical mobility, 14–15; goods, 290, 317–18; Russian goods, 318; soapstone, 317–18
traders: capitalist mode of production, 139; Keenainak, 149
trader-trapper relationships, 8

trading companies, 118–19; Sabellum Company, 92
trading posts, 115, 295; Bathurst Inlet, 292; Churchill River, 322
traditional leadership roles, 117, 129–30
traditional mode of production, 133–4, 139
traditional relations of production, 139
trappers, Inuit as, 94–5
trapping: Copper Inuit, 292, 295; as individualistic activity, 281. *See also* Arctic fox
trapping rights, Nunavut, 335
trap-rifles, 108
Tree River, 294, 303, 318
Trott, C.G., 29
*Truelove* (whaling ship), 74, 75
Tuapait, 172–4, 238
Tuapait Harbour, 172
Tujarapik, 177
*tuktu* (caribou). *See* caribou
Tulakan, 53, 54
Tunnit, 315
Tununirmiut, contact history of, 266–70
turbot fishery, 140
Turner, D.H., 29
Tyson, Captain G., 76, 118

*ugjuk* (bearded seal), 39, 42
Ugjuktung, 54
Uglit, 269
Uglit Islands, 267
Ukjulingmiut, 282
*ukuaq* (female Ego), 10, 135–6
*ula* (woman's knife), 108
Uluksak, 292, 294, 303
Umana Islands, 38, 54
Umanaqjuaq, 38, 45, 52, 116, 199, 338; group formation, 225; intermarriage with Inuit from Frobisher Bay, 206; and Inuit from Pond Inlet, 193; Inuit whaling 1880s, 85; leadership at, 205, 234; local group composition, 145; and male adoption, 244; population estimates, 87, 91, 92; as proletariat at, 122; trade, 82, 121
Umanaqjuaq (Blacklead) Island, 44; Anglican mission at, 88–90; whaling station at, 78–9
Umanaqjuarmiut, xviii; at Kingmiksoo, 202; contact-traditional camps, 145; conversion to Christianity, 88–90; and leadership, 233–5; matrilocality among, 243–4; and Qikirtarmiut, xxi, 117, 137, 188, 198, 225–49; settlements, 190–224; whale hunting technique, 253. *See also* Talirpingmiut
*umialiqtak* (boat owner), 113
*umialit* (boat owners/rich men), 24–5
*umiat*, 80, 108
*umiat* whaling, 113, 253–4, 366n. 8; among Kinguamiut/Qikirtarmiut, 262–3; positions, 254
Unalit, 120; kinship structure, 5

Unaq, 220–1
Ungava Bay, Thule migration to, 314
*ungayuq* (closeness-affection), xvii, 9, 11–12, 25, 152, 154, 168–9, 194–5, 231, 237–8; in Copper Inuit social structure, 304–5; and Iglulingmiut, 274; and Netsilingmiut society, 281–2; social structure, 247–9, 331. *See also naalaqtuq*
Union Harbour, Kekerten Islands, 79
*Union* (whaling ship), 79–80
*uqsuq* (blubber), 39
Ussualung, 40, 53, 160, 162, 190–6; break-up of, 195; and Inuit labour, 93; leadership, 260–1; marital residence pattern, 243; move to Kingua Fiord, 191; trading post at, 93
Utkuhikjalingmiut, 282
*uttuq* (basking seals), 16, 272, 291
*uugaq* (Arctic cod), 42
*uyuruk*, 154

Valentine, C.A., 4, 135, 231
value systems, xvi; institutionalized, 24
Van Stone, J.W., 8
Veevee, 151, 154, 169, 185–8, 191, 193–5, 231
Victoria Island, 289–90, 292–3, 303, 320
virilocality, 269–70, 271, 300
voluntary alliances, 27, 277–8; Copper Inuit, 17–18, 20, 294, 297–8; Iglulik region, 278–9
voting system, Nunavut, 339–40

wage labour economy, 110; Copper Inuit, 295; Cumberland Sound, 103–4; Iglulik region, 271; in Nunavut, 337
Walker Bay, 290
walrus, 39; hunting of, 16, 44, 54; and Iglulingmiut, 266, 268, 270, 272, 277
walrus tusk harpoons, 254
Wareham, Captain M., 41–2, 74
warfare: and feuding distinction, 56; precontact, 55–6, 313, 316
Warmow, Mathias, 108; population estimates, 76–7
watercraft ownership, 67
wealth, 123; from *Investigator*, 290; in Kingmiksoo, 2–9; personal, 294; and trading, 323
weaponry: for aboriginal whaling, 254; and hunting economy, 108; Thule Inuit, 313
Wenzel, G.W., 25
whaleboat crews, 159; heads of, 114t; Iqalulik, 197–8, 199; Nunaata, 149; *umiak*, 113
whaleboat owners: Akpalialuk, 210; Aksayuk, 199; Angmarlik, 159; Attaguyuk, 174; William Duval, 196; Keenainak, 146–7; Kisa, 173; Koodlooaktok, 203; Nakashuk, 211; Nowyook, 218; Nukeeruaq, 187; Pawla, 221; Pitsualuk, 221; Shorapik, 188; Tooloogakjuaq, 203; Veevee, 191
whaleboats, 109, 268; at Idlungajung, 162; at Iqalulik, 197, 269; at Kingmiksoo, 209; in

commercial whaling, 113–14; decline in numbers, 92; and extended family structure, 113, 372n. 7; and Iglulingmiut, 270; inheritance of, 64, 152, 188, 196, 198, 221; and leadership, 113–14, 117; motorized 114, 159, 167, 173, 188, 210, 218; in open-water season, 245; of Oqomiut, 270; and organization of production, 110; owned by Inuit, 81. *See also* whaleboat owners; social impact of, 270–1; white whale drive, 114
whalebone (baleen). *See* baleen (whalebone)
whale fishery, 76, 292; in 1923, 96; annual routine of Inuit, 96; decline of, 81–2, 87, 92; employment in, 77; processing facility, 96; Tooloogakjuaq in, 202
whale hunt, 16; American, 268–9; dangers in, 80; economy of, 47, 49, 58; from kayaks, 253; from *umiat*, 253, 254; Inuit tradition of sharing, 100; mid-summer, 97, 160; Pangnirtung Fiord, xviii-xx; scene on ivory bow drill, 254f; and sealing 1870–94, 82–8; social order within, xx; spring, 75, 77, 80
whale oil prices, 82–3
whalers: American, 78, 81; capitalist mode of production, 139; over-wintering site, 190
whales, calving grounds, 201. *See also* specific species
whaling harpoon, 108
whaling ships: American, 37, 82; loss of, 73; selection of crews for, 116; as trading opportunities, 108
whaling station, 77–8, 115; tourist attraction, xvii

whitecoats, HBC, 98
Whitehouse, A.C., 79–81
white missionaries, 130
wife-exchange. *See* spousal exchange
wife-purchase, 329
wife-stealing: Gilyak, 333; and Labrador Inuit, 332; and Netsilingmiut, 282–3, 288
Williams, C.A., 82
Williams Co., C.A., on Umanaqjuaq (Blacklead) Island, 87
Williams and Haven Company, 82
Willmott, W.E., 7
winter hunting grounds, 178, 238, 239f, 259; Imigen Island, 174
winter settlements, 17, 269, 298; Copper Inuit, 293; precontact, 52–3; on sea-ice, 256; site plans, 56–7
winter society, 3
woman's knife, 108
women: and Christianity, 89–90, 126; and decision-making, 268, 296; equal status of, 296; and game sharing, 246; making of caribou skin clothing, 252; in Nunavut, 337; as objects of exchange, 327; in positions of leadership, 137; religious importance of, 26; surnames of, 233; and taboos, 126
wood and metal, for implements, 108, 117, 268, 371n. 2. *See also* metal; shipwrecks

York Factory, 322
Yupik, marriage region, 26
Yupik-speaking groups, cousin terminology, 5